LabVIEW®
for
Everyone

Second Edition

ISBN 0-13-065096-X

90000

9 790130 650961

 **NATIONAL INSTRUMENTS** **VIRTUAL INSTRUMENTATION SERIES**

LabVIEW®
for
Everyone

Second Edition

▲ Jeffrey Travis

Prentice Hall PTR, Upper Saddle River, NJ 07458
www.phptr.com

Library of Congress Cataloging-in-Publication Data
Travis, Jeffrey.
   LabVIEW for everyone/Jeffrey D. Travis—2nd ed.
     p.cm.—(National Instruments virtual instrumentation series)
   Original ed. of: LabVIEW for everyone/Lisa K. Wells. 1st ed. 1997.
   Includes index.
   ISBN 0-13-065096-X (pbk.)
     1. Scientific apparatus and instruments—Computer simulation. 2. LabVIEW. I. Wells, Lisa K.
LabVIEW for everyone. II. Title. III. Series.

Q.183.A1 T73 2002
006—dc21

                                                                                  2001052051

Editorial/Production Supervision: *Argosy*
Acquisitions Editor: *Bernard Goodwin*
Editorial Assistant: *Michelle Vincenti*
Marketing Manager: *Dan DePasquale*
Manufacturing Manager: *Alexis R. Heydt-Long*
Cover Design: *Bruce Kensalaar/Talar Agasyan-Boorujy*
Cover Design Direction: *Jerry Votta*
Interior Series Design: *Gail Cocker-Bogusz*

 © 2002 Prentice Hall PTR
Prentice-Hall, Inc.
Upper Saddle River, NJ 07458

Prentice Hall books are widely used by corporations and government agencies for training, marketing, and resale.

The publisher offers discounts on this book when ordered in bulk quantities.
For more information, contact
Corporate Sales Department
Prentice Hall PTR
One Lake Street
Upper Saddle River, NJ 07458
Phone: 800-382-3419; FAX: 201-236-714
E-mail (Internet): corpsales@prenhall.com

Printed in the United States of America
10 9 8 7 6 5 4   Fourth Printing

Reprinted with corrections October, 2002.

ISBN 0-13-065096-X

Pearson Education LTD.
Pearson Education Australia PTY, Limited
Pearson Education Singapore, Pte. Ltd.
Pearson Education North Asia Ltd.
Pearson Education Canada, Ltd.
Pearson Educación de Mexico, S.A. de C.V.
Pearson Education—Japan
Pearson Education Malaysia, Pte. Ltd.

To Stephanie, for her love and support, and to Maeve, Aidan, and Rachel, who are glad Daddy is done writing the book

Contents

▼**2**

Virtual Instrumentation: Hooking Your
Computer Up to the Real World   19

▼**3**

The LabVIEW Environment:
Building Your Own Workbench   37

4

LabVIEW Foundations   75

▼**5**

Yet More Foundations   115

▼6

Controlling Program Execution with Structures   147

▼7

LabVIEW's Composite Data:
Arrays and Clusters   181

▼8

LabVIEW's Exciting Visual Displays:
Charts and Graphs   215

▼9

Exploring Strings and File I/O   263

Advanced Topics

▼10

Getting Data into and out of Your Computer: Data Acquisition and Instrument Control   285

▼11

DAQ and Instrument Control in LabVIEW   335

▼14

Connectivity in LabVIEW   447

▼15

Advanced File I/O, Printing, and Reports   487

▼16

The Art of LabVIEW Programming   515

Preface

LabVIEW, or *Laboratory Virtual Instrument Engineering Workbench*, is a graphical programming language that has been widely adopted throughout industry, academia, and research labs as the standard for data acquisition and instrument control software. LabVIEW is a powerful and flexible instrumentation and analysis software system that is multiplatform (predating Java, which makes the same claim)—you can run LabVIEW on Windows, MacOS, Linux, Solaris, and HP-UX. Personal computers are much more flexible than standard instruments, and creating your own LabVIEW program, or *virtual instrument* (VI), is simple. LabVIEW's intuitive user interface makes writing and using programs exciting and fun!

LabVIEW departs from the sequential nature of traditional programming languages and features an easy-to-use graphical programming environment, including all of the tools necessary for data acquisition (DAQ), data analysis, and presentation of results. With its graphical programming language, called "G," you program using a graphical block diagram that compiles into machine code. Ideal for a countless number of science and engineering applications, LabVIEW helps you solve many types of problems in only a fraction of the time and hassle it would take to write "conventional" code.

Beyond the Lab

LabVIEW has found its way into such a broad spectrum of virtual instrumentation applications that it is hard to know where to begin. As its name implies, it began in the laboratory and still remains very popular in many kinds of laboratories—from major research and development laboratories around the world (such as Lawrence Livermore, Argonne, Batelle, Sandia, Jet Propulsion Laboratory, White Sands, and Oak Ridge in the United States and CERN in Europe), to R&D laboratories in many industries, and to teaching laboratories in universities all over the world, especially in the disciplines of electrical and mechanical engineering and physics.

The spread of LabVIEW beyond the laboratory has gone in many directions—up (aboard the space shuttle), down (aboard U.S. Navy submarines), and around the world (from oil wells in the North Sea to factories in New Zealand). And with the latest Internet capabilities, LabVIEW applications are being deployed not only physically in many places but virtually across cyberspace. More and more people are creating web-based control or monitoring of their LabVIEW applications to allow remote access and instant information about what's happening in their lab. Virtual instrumentation systems are known for their low cost, both in hardware and development time, and their great flexibility. Is it any wonder that they are so popular?

The Expanding World of Virtual Instrumentation

Perhaps the best way to describe the expansion (or perhaps explosion) of LabVIEW applications is to generalize it. There are niches in many industries where measurements of some kind are required—most often of temperature, whether it be in an oven, a refrigerator, a greenhouse, a clean room, or a vat of soup. Beyond temperature, users measure pressure, force, displacement, strain, pH, and so on, ad infinitum. Personal computers are used virtually everywhere. LabVIEW is the catalyst that links the PC with measuring things, not only because it makes it easy, but also because it brings along the ability to analyze what you have measured and display it and communicate it halfway around the world if you so choose.

After measuring and analyzing something, the next logical step often is to change (control) something based upon the results. For example, measure temperature and then turn on either a furnace or a chiller. Again, LabVIEW makes this easy to do; monitoring and control have become LabVIEW strengths. Sometimes it is direct monitoring and control, or it may be through communicating with a programmable logic controller (PLC) in what is commonly called supervisory control and data acquisition (SCADA).

The Results

You will find descriptions of LabVIEW applications interspersed throughout this book. They are written by users in a very brief style to give you the essence of the application and are chosen from industry segments in which LabVIEW tends to be very popular—automated electronics testing, semiconductor manufacturing, medical instrumentation, automotive testing, and industrial automation applications. Of the multitude of successful LabVIEW applications, these are particularly interesting current examples of virtual instrumentation at its finest!

A few of LabVIEW's many uses include:

- Simulating heart activity
- Controlling an ice cream-making process
- Detecting hydrogen gas leaks on the space shuttle
- Monitoring feeding patterns of baby ostriches
- Modeling power systems to analyze power quality
- Measuring physical effects of exercise in lab rats
- Controlling motion of servo and stepper motors
- Testing circuit boards in computers and other electronic devices
- Simulating motion in a virtual reality system
- Allowing remote navigation and feedback over the Web of a helium-filled blimp

Objectives of This Book

LabVIEW for Everyone will help you get LabVIEW up and running quickly and easily and will start you down the road to becoming an expert programmer. The book offers additional examples and activities to demonstrate techniques, identifies other sources of information about LabVIEW, and features descriptions of cool LabVIEW applications. You are invited to open, inspect, use, and modify any of the programs on the accompanying CD-ROM. The CD-ROM includes the 30-day evaluation version of LabVIEW, which allows you to do just about everything the commercial version does.

This book expects you to have basic knowledge of your computer's operating system. If you don't have much computer experience, you may want to spend a little time with your operating system manual and familiarize yourself with your computer. For example, you should know how to access menus, open and save files, make backup disks, and use a mouse.

After reading this book and working through the exercises, you should be able to do the following, and much more, with the greatest of ease:

- Write LabVIEW programs, called virtual instruments, or VIs
- Employ various debugging techniques
- Manipulate both built-in LabVIEW functions and library VIs
- Create and save your own VIs so that you can use them as subVIs, or subroutines
- Design custom graphical user interfaces (GUIs)
- Save your data in a file and display it on a graph or chart
- Build applications that use General Purpose Interface Bus (GPIB) or serial instruments
- Create applications that use plug-in DAQ boards
- Use built-in analysis functions to process your data
- Optimize the speed and performance of your LabVIEW programs
- Employ advanced techniques such as globals, locals, and attribute nodes
- Publish your data over the Internet or on the Web, using LabVIEW's features such as HTML printing or DataSocket
- Use LabVIEW to create your instrumentation applications

LabVIEW for Everyone helps you get started quickly with LabVIEW to develop your instrumentation and analysis applications. The book is divided into two main sections: *Fundamentals* and *Advanced Topics*.

The *Fundamentals* section contains nine chapters and teaches you the fundamentals of G programming in LabVIEW. The *Advanced Topics* section contains six chapters that further develop your skills and introduce helpful techniques and optimizing strategies. We suggest that you work through the beginning section to master the basics; then, if you're short on time, skip around to what you really want to learn in the advanced section.

In both sections, chapters have a special structure to facilitate learning.

- *Overview, goals,* and *key terms* describe the main ideas covered in that chapter.
- The main sections are a discussion of the featured topics.
- *Activities* reinforce the information presented in the discussion.
- *Wrap It Up!* summarizes important concepts and skills taught in the chapter.
- Additional activities in many chapters give you more practice with the new material.

Fundamentals

Chapter 1 describes LabVIEW and introduces you to some of Lab-VIEW's features and uses.

In Chapter 2, you will get an overview of virtual instrumentation: how data acquisition, instrument control, and data analysis are performed with LabVIEW. You will also learn about LabVIEW's history and a couple of real-world applications.

In Chapter 3, you will get acquainted with the LabVIEW environment, including the essential parts of a virtual instrument (or VI), the Help window, menus, tools, palettes, and subVIs.

In Chapters 4 and 5, you will become familiar with the basics of G programming in LabVIEW—using controls and indicators (such as numerics, Booleans, and strings); wiring, creating, editing, debugging, and saving VIs; creating subVIs; and documenting your work. You will also begin to understand why G is considered a dataflow programming language.

Chapter 6 describes the basic G programming structures in LabVIEW: while loops, for loops, shift registers, case structures, sequence structures, and formula nodes. It also teaches you how to introduce timing into your programs.

In Chapter 7, you will learn how to use two important data structures—arrays and clusters—in your programs. You will also explore Lab-VIEW's built-in functions for manipulating arrays and clusters.

Chapter 8 details the variety of charts and graphs available in LabVIEW and teaches you how to use them for animated and informative data presentation. It also introduces the waveform data type.

Chapter 9 discusses string data types, string functions, and tables. It also talks a little about how to save data in and read data from a file, using LabVIEW's easy File I/O VIs.

Advanced Topics

Chapter 10 teaches you more about data acquisition, GPIB, and serial communication. You will learn a bit of theory and some hardware considerations, and you will find a valuable guide to many common acronyms used in instrumentation. Chapter 10 also discusses software setup for data acquisition hardware.

Chapter 11 discusses a few basics on how to use LabVIEW to acquire data using plug-in DAQ boards and presents a brief overview on communicating with other instruments using GPIB and serial protocols.

Chapter 12 covers some invaluable advanced features such as local and global variables, property nodes, ActiveX, DLLs, data type conversions, and much more.

Chapter 13 shows you how to configure VI behavior and appearance using VI Setup options and how to access front panel controls using the keyboard.

Chapter 14 covers connectivity in LabVIEW and deals with exciting topics such as publishing to the Web from LabVIEW, data sharing with DataSocket, and communication with protocols such as TCP/IP. It also introduces the VI Server.

Chapter 15 describes some good programming techniques that you can use to make your programs run faster, use less memory, port more easily to other platforms, and behave more efficiently overall.

In Chapter 16, you will learn good LabVIEW style and some new tips, such as how to add a customized look to your applications by importing pictures and using the Control Editor.

You will find a glossary, index, and an appendix at the end of the book.

Appendix A lists some resources that can help you with LabVIEW. It also describes add-on toolkits available to enhance LabVIEW's functionality.

The following table describes the conventions used in this book:

bold   Bold text denotes VI names, function names, menus, menu items, and palettes. In addition, bold text denotes VI input and output parameters. For example, "Choose **TCP Read** from the **TCP Functions** palette."

italic   Italic text denotes emphasis, a cross reference, or an introduction to a key term or concept. For example, "A *control reference* is an object that points to a LabVIEW control or indicator and can manage its properties."

`Courier`   `Courier` type denotes text or characters that you enter using the keyboard. It also denotes files and paths. Sections of code, programming examples, syntax examples, and messages and responses that the computer automatically prints to the screen also appear in this font. For example, "In the text box, enter `c:\data\datafile.txt` as the filename."

      Note. This icon marks information to which you should pay special attention.

      Watch Out! This icon flags a common pitfall or special information that you should be aware of in order to keep out of trouble.

      Hint. This icon calls your attention to useful tips and hints on how to do something efficiently.

      On CD. This icon notes that the example or activity you're about to read is available on the book's CD.

      This information only pertains to the Windows platform.

This information only pertains to the MacOS platform.

 This information only pertains to the Unix-based platform.

A Note about Paths

Different platforms have different conventions for specifying path names. For example, Windows paths take the form X: \LABVIEW\MINE.LLB\BINGO. VI. The same path on a MacOS 9.x or earlier system would be denoted Hard Drive Name:LabVIEW:Mine.llb:Bingo.vi. On Linux machines, it would be /usr/labview/mine.llb/bingo.vi. Rather than specifying a full path, this book will list the default path from the LabVIEW directory or folder when telling you where to find an example VI. To simplify notation, we will use the Windows standard to describe paths; if you use MacOS or UNIX machines, please substitute colons or forward slashes where necessary.

What's New in This Second Edition

LabVIEW for Everyone was the first book published aimed at the beginner LabVIEW user; since then, Prentice-Hall and other publishers have produced over a dozen LabVIEW books on specific topics. This new second edition of *LabVIEW for Everyone* has been completely updated for LabVIEW 6i. Among the changes from the first edition (which was published at the time of LabVIEW 4.0) are:

- Completely revamped and updated screenshots, examples, and activities updated for LabVIEW 6i's new interface "look and feel"
- Updated text covering the new LabVIEW 6i features
- Updated section on data acquisition and instrument control, featuring the NI Measurement and Automation Explorer (MAX)

- A new chapter on LabVIEW connectivity that has extended coverage of more Internet topics, such as DataSocket, the VI Server, and publishing to the Web
- Discussion of new features such as the waveform data type, 3D graphs, report generation, and more

LabVIEW Installation Instructions

If you have the full version of LabVIEW and need instructions on how to install it, please see the release notes that came with your software. If you don't have LabVIEW yet, you can try out the evaluation version on the CD. The evaluation version expires 30 days after you install it.

In addition, you will need to access the Everyone directory from the CD-ROM in the back of this book. It contains the activities in this book and their solutions. You may want to copy the Everyone directory onto your PC so you can save your activity work there as well.

Minimum Specifications for LabVIEW and the Evaluation Software

The basic engine of the evaluation software is the same as the full version of LabVIEW, so the system requirements are also about the same.

Windows XP/2000/NT/Me/9X

- For Windows NT, use Windows NT 4.0 Service Pack 3 or later
- 32 MB RAM minimum, 64 MB recommended
- 65 MB disk space for minimal LabVIEW installation, 200 MB for full installation*
- Pentium processor or equivalent recommended

Mac OS

- Mac OS 7.6.1 or later
- 32 MB RAM minimum, 64 MB recommended

*Driver software requires additional hard disk space for installation.

- 100 MB disk space for minimal LabVIEW installation, 225 MB for full installation*
- PowerPC processor

Linux

- Linux kernel 2.0.x or later
- Any Linux distribution with GNU C Library Version 2.0.5 or later (known as glibc2 or libc.so.6), including:
  - RedHat Linux 5.0 or later
  - SuSE Linux 6.0 or later
  - SuSE Linux 5.3 with shlibs6-98.9.25-0 RPM installed
  - Caldera OpenLinux 1.3 or later
  - Debian Linux 2.0 or later
- 32 MB RAM minimum, 64 MB recommended
- 32 MB swap space storage
- 65 MB disk space for minimal LabVIEW installation, 150 MB for full installation*
- XWindows System server
- Pentium processor or equivalent recommended

Sun

- Solaris 2.5.1 or later
- 32 MB RAM minimum, 64 MB recommended
- 32 MB swap space storage
- 65 MB disk space for minimal LabVIEW installation, 150 MB for full installation*
- XWindows System server
- SPARC processor (Sun SPARCstations)

*Driver software requires additional hard disk space for installation.

HP-UX

- HP-UX 10.20 or later
- 32 MB RAM minimum, 64 MB recommended
- 32 MB swap space storage
- 65 MB disk space for minimal LabVIEW installation, 150 MB for full installation*
- XWindows System server
- PA-RISC processor (Hewlett-Packard 9000 Series 700 workstations)

All

LabVIEW uses a directory to store temporary files. Some of the temporary files are large, so we recommend that you have several megabytes of disk space available for this temporary directory. The default for the temporary directory is /tmp on Windows and Unix and inside the trash can on MacOS.

Purchasing LabVIEW

If you would like information on how to purchase LabVIEW, contact National Instruments.

National Instruments
6504 Bridge Point Parkway
Austin, Texas 78730
Telephone: (512) 794-0100
Fax: (512) 794-8411
E-mail: info@ni.com
Web: http: //www.ni.com

---

*Driver software requires additional hard disk space for installation.

Acknowledgments

As with the first edition and my other book (*Internet Applications in Lab-VIEW*), this work was made possible by the collaboration of numerous individuals:

To Bernard Goodwin, my editor at Prentice Hall, who somehow convinced me to write a third manuscript for Prentice Hall, thanks for all your support and for making this book a reality.

To Lisa Wells, friend and co-author of the first edition, who was wise enough to turn down the grueling year-long writing project this time, thanks for your support and your material from the first edition. This book would not be the success it has been without your original efforts.

To Ravi Marawar at National Instruments, who not only provided source material, figures, and DAQ hardware, but also invaluably assisted with legal and practical logistics for making this second edition, I wish to express my gratitude.

To my colleagues at Rayodyne—Bill Schwartz, Michael Tucker, George Castle, Sherry Arnold, Holley Tondre, Steven Mattison, Graeme Cloughley, Kevin Schmeisser, and the rest of the Rat Pack—you all are the finest group of people to work with. Thanks for your encouragement and support (but not for the beanie cap).

My gratitude and appreciation also goes out to my good friends whom I count myself fortunate to have: Paul and Una Davis, Travis and Leslie Hines, David Taylor, Tristan D'Artagnan, Justin and Emily Hargrave, Jim and Michelle Hasbrouck, Vivian Badillo, the men of The Loft, and other friends from Hope Chapel.

And finally, my heartfelt gratitude and love goes toward my family who always keeps my life in perspective: Stephanie, my wife, and my delightful children Maeve, Aidan, and Rachel. Last but not least, my deepest thanks go to the Author of authors, Christ Jesus, in whom there is freedom and life.

Fundamentals

OVERVIEW

Welcome to the world of LabVIEW! This chapter gives you a basic explanation of LabVIEW and its capabilities and shows how it can make your life easier.

GOALS

- Develop an idea of what LabVIEW really is
- Learn what graphical programming language and dataflow programming mean
- Peruse the introductory examples
- Get a feel for the LabVIEW environment

KEY TERMS

- LabVIEW
- Virtual instrument (VI)
- Dataflow
- Graphical language
- Front panel
- Block diagram
- Icon
- Connector
- Toolbar
- Palette
- Hierarchy

What in the World Is LabVIEW?

1.1 What Exactly Is LabVIEW, and What Can It Do for Me?

You'd probably like to know what exactly LabVIEW is before you go much further. What can you do with it and what can it do for you? LabVIEW, short for *Laboratory Virtual Instrument Engineering Workbench*, is a programming environment in which you create programs with graphics; in this regard it differs from traditional programming languages like C, C++, or Java, in which you program with text. However, LabVIEW is much more than a language. It is a program development and execution system designed for people, such as scientists and engineers, who need to program as part of their jobs. LabVIEW works on PCs running Windows, MacOS, Linux, Solaris, and HP-UX.

Providing you with a very powerful graphical programming language, LabVIEW can increase your productivity by orders of magnitude. Programs that take weeks or months to write using conventional programming languages can be completed in hours using LabVIEW, because it is specifically designed to take measurements, analyze data, and present results to the user. And because LabVIEW has such a versatile graphical user interface and is so easy to program with, it is also ideal for simulations, presentation of ideas, general programming, or even teaching basic programming concepts.

LabVIEW offers more flexibility than standard laboratory instruments because it is software-based. You, not the instrument manufacturer, define instrument functionality. Your computer, plug-in hardware, and LabVIEW comprise a completely configurable virtual instrument to accomplish your tasks. Using LabVIEW, you can create exactly the type of virtual instrument you need, when you need it, at a fraction of the cost of traditional instruments. When your needs change, you can modify your virtual instrument in moments.

Figure 1.1
The Space Industries Sheet Float Zone Furnace is used for high-temperature superconductor materials processing research in a microgravity environment aboard the NASA KC-135 parabolic aircraft. LabVIEW controls the industrialized Macintosh-based system.

LabVIEW tries to make your life as hassle-free as possible. It has extensive libraries of functions and subroutines to help you with most programming tasks, without the fuss of pointers, memory allocation, and other arcane programming problems found in conventional programming languages. LabVIEW also contains application-specific libraries of code for data acquisition (DAQ), General Purpose Interface Bus (GPIB), and serial instrument control, data analysis, data presentation, data storage, and communication over the Internet. The Analysis library contains a multitude of useful functions, including signal generation, signal processing, filters, windows, statistics, regression, linear algebra, and array arithmetic.

Because of LabVIEW's graphical nature, it is inherently a data presentation package. Output appears in any form you desire. Charts, graphs, and user-defined graphics comprise just a fraction of available output options. This book will show you how to present data in all of these forms.

LabVIEW's programs are portable across platforms, so you can write a program on a Macintosh and then load and run it on a Windows machine without changing a thing in most applications. You will find LabVIEW applications improving operations in any number of industries, from every kind of engineering and process control to biology, farming, psychology, chemistry, physics, teaching, and many others.

1.1.1 Dataflow and the Graphical Programming Language

The LabVIEW program development environment is different from commercial C or Java development systems in one important respect. Whereas other programming systems use text-based languages to create lines of code, LabVIEW uses a graphical programming language to create programs in a pictorial form called a block diagram, eliminating a lot of the syntactical details. With this method, you can concentrate on the flow of data within your application; the simpler syntax doesn't obscure what the program is doing. Figures 1.2 and 1.3 show a simple LabVIEW user interface and the code behind it.

LabVIEW uses terminology, icons, and ideas familiar to scientists and engineers. It relies on graphical symbols rather than textual language to describe programming actions. The principle of *dataflow*, in which functions execute only after receiving the necessary data, governs execution in a straightforward manner. You can learn LabVIEW even if you have little or

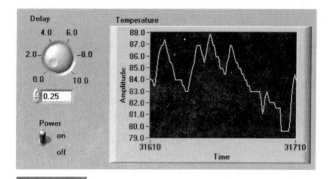

Figure 1.2
User interface.

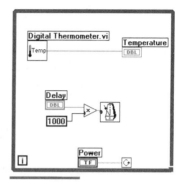

Figure 1.3
Graphical code.

no programming experience, but you will find knowledge of programming fundamentals very helpful.

1.1.2 How Does LabVIEW Work?

LabVIEW programs are called *virtual instruments* (*VIs*) because their appearance and operation imitate actual instruments. However, behind the scenes they are analogous to main programs, functions, and subroutines from popular programming languages like C or Basic. Hereafter, we will refer to a LabVIEW program as a "VI" (pronounced "vee eye," *not* the Roman numeral six as we've heard some people say). Also, be aware that a LabVIEW program is always called a VI, whether its appearance or function relates to an actual instrument or not.

A VI has three main parts:

• The *front panel* is the interactive user interface of a VI, so named because it simulates the front panel of a physical instrument. The front panel can contain knobs, push buttons, graphs, and many other controls (which are user inputs) and indicators (which are program outputs). A user will input data using a mouse and keyboard and then view the results produced by the program on the screen.

• The *block diagram* is the VI's source code, constructed in LabVIEW's graphical programming language, G. The block diagram is the actual executable program. The components of a block diagram are lower-level VIs, built-in functions, constants, and program execution control

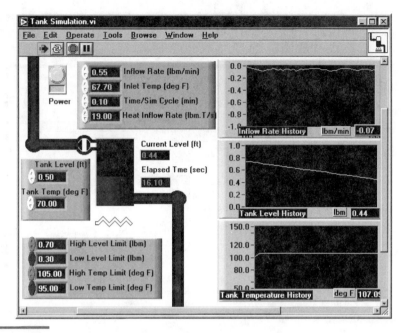

Figure 1.4

structures. You draw wires to connect the appropriate objects together to indicate the flow of data between them. Front panel objects have corresponding terminals on the block diagram so that data can pass from the user to the program and back to the user.

• In order to use a VI as a subroutine in the block diagram of another VI, it must have an *icon* and a *connector*. A VI that is used within another VI is called a *subVI* and is analogous to a subroutine. The icon is a VI's pictorial representation and is used as an object in the block diagram of another VI. A VI's connector is the mechanism used to wire data into the VI from other block diagrams when the VI is used as a subVI. Much like parameters of a subroutine, the connector defines the inputs and outputs of the VI.

Virtual instruments are *hierarchical* and *modular*. You can use them as top-level programs or subprograms. With this architecture, LabVIEW promotes the concept of *modular programming*. First, you divide an application into a series of simple subtasks. Next, you build a VI to accomplish each subtask and then combine those VIs on a top-level block diagram to complete the larger task.

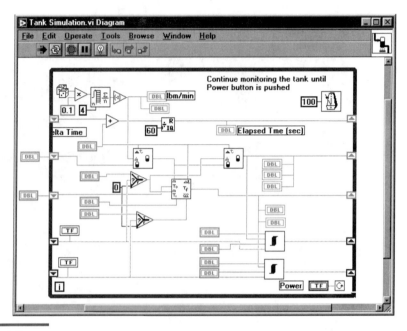

Figure 1.5

Icon   Connector

Figure 1.6

Modular programming is a plus because you can execute each subVI by itself, which facilitates debugging. Furthermore, many low-level subVIs often perform tasks common to several applications and can be used independently by each individual application.

Just so you can keep things straight, we've listed a few common LabVIEW terms with their conventional programming equivalents in Table 1.1.

1.2 Demonstration Examples

Okay, you have enough reading for now. To get an idea of how LabVIEW works, you can open and run a few existing LabVIEW programs.

Table 1.1 *LabVIEW Terms and Their Conventional Equivalents.*

LabVIEW	Conventional Language
VI	program
function	function or method
subVI	subroutine, subprogram, object
front panel	user interface
block diagram	program code
"G" or LabVIEW	C, C++, Java, Pascal, BASIC, etc.

Whether you are using the full or evaluation version of LabVIEW, just launch it. Make sure you can access the Everyone directory from the CD or your hard drive, as described in the Preface; it contains the activities for this book. After launching LabVIEW, a dialog box will appear. To open an example, select **Open VI** and choose the one you want.

Throughout this book, use the left mouse button (if you have more than one) unless we specifically tell you to use the right one. On MacOS computers, <command>-click when right-mouse functionality is necessary. In most LabVIEW situations, the <control> key on Windows will correspond to <command> on Macs, <meta> on Suns, and <alt> on Linus and HP machines.

1.2.1 Activity 1-1: Temperature System Demo

Open and run the VI called **Temperature System Demo.vi** by following these steps:

1. Launch LabVIEW. At the startup dialog, click on "**Search Examples**". Alternatively, if LabVIEW is already running, go to the **Help** menu and select **Examples...**

2. Click on "Demonstrations", then "Analysis", then "Temperature System Demo".

3. You will see the VI shown in the following figure.

> **Note:** The above steps are the process for quickly loading the example VIs that come with LabVIEW. You can also access all the LabVIEW example VIs directly in the `examples` directory which is inside your Lab-VIEW installation directory. For example, on Windows, LabVIEW is usually installed at `C:\Program Files\National Instruments\Lab-VIEW`. So the examples directory is at `C:\Program Files\National Instruments\LabVIEW\examples`. The Temperature System Demo example in particular would be located at `C:\Program Files\National Instruments\LabVIEW\examples\apps\tempsys.llb\Temperature System Demo.vi`.
>
> Generally, however, it's easier to find examples by using the Search Examples feature as just described.

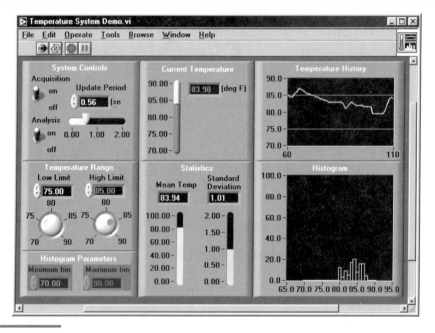

Figure 1.7
Temperature System Demo test front panel window.

Run button

Run button
(active)

4. Run the VI by clicking on the Run button. The button changes appearance to indicate that the VI is running. The *Toolbar*, which is the row of icons on the top bar of the screen, also changes, since editing functionality won't be necessary while the VI is running.

Abort button

Notice also that the Abort button becomes active in the Toolbar. You can press it to abort program execution.

Temperature System Demo.vi simulates a temperature monitoring application. The VI makes temperature measurements and displays them in the thermometer indicator and on the chart. Although the readings are simulated in this example, you can easily modify the program to measure real values. The **Update Period** slide controls how fast the VI acquires the new temperature readings. LabVIEW also plots high and low temperature limits on the chart; you can change these limits using the **Temperature Range** knobs. If the current temperature reading is out of the set range, LEDs light up next to the thermometer.

This VI continues to run until you click the **Acquisition** switch to *off*. You can also turn the data analysis on and off. The **Statistics** section shows you a running calculation of the mean and standard deviation, and the **Histogram** plots the frequency with which each temperature value occurs.

Tweaking Values

Operating tool

5. Use the cursor, which takes on the personality of the Operating tool while the VI is running, to change the values of the high and low limits. Highlight the old high or low value, either by clicking twice on the value you want to change, or by clicking and dragging across the value with the Operating tool. Then type in the new value and click on the enter button, located next to the run button on the Toolbar.

Enter button

6. Change the **Update Period** slide control by placing the Operating tool on the slider, and then clicking and dragging it to a new location.

You can also operate slide controls using the Operating tool by clicking on a point on the slide to snap the slider to that location, by clicking on a scroll button to move the slider slowly toward the arrow, or by clicking in the slide's digital display and entering a number.

Note

Even though the display changes, LabVIEW does not accept the new values in digital displays until you press the enter button, or click the mouse in an open area of the window.

7. Try adjusting the other controls in a similar manner.

8. Stop the VI by clicking on the **Acquisition** switch.

Examine the Block Diagram

The block diagram shown in Figure 1.8 represents a complete LabVIEW application. You don't need to understand all of these block diagram elements right now—we'll deal with them later. Just get a feel for the nature of a block diagram. If you already do understand this diagram, you'll probably fly through the first part of this book!

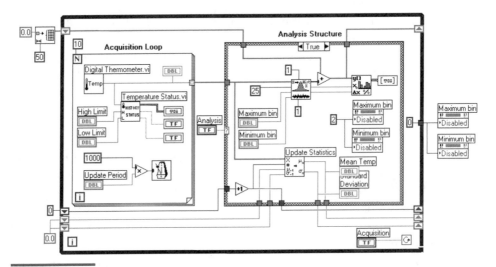

Figure 1.8

9. Open the block diagram of **Temperature System Demo.vi** by choosing Show Diagram from the Windows menu.

10. Examine the different objects in the diagram window. Don't panic at the detail shown here! These structures are explained step by step later in this book.

11. Open the contextual Help window by choosing **Show Context Help** from the **Help** menu. Position the cursor over different objects in the block diagram and watch the Help window change to show descriptions of the objects. If the object is a function or subVI, the Help window will describe the inputs and outputs as well.

Hierarchy

LabVIEW's power lies in the hierarchical nature of its VIs. After you create a VI, you can use it as a subVI in the block diagram of a higher-level VI, and you can have as many layers of hierarchy as you need. To demonstrate this versatile ability, look at a subVI of **Temperature System Demo.vi**.

12. Open the **Temperature Status** subVI by double-clicking on its icon.

The front panel shown in Figure 1.9 springs to life.

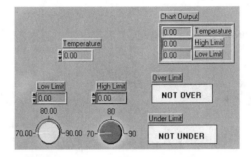

Figure 1.9

Icon and Connector

The icon and connector provide the graphical representation and parameter definitions needed if you want to use a VI as a subroutine or function in the block diagrams of other VIs. They reside in the upper-right corner of the VI's front panel window. The icon graphically represents the VI in the block diagram of other VIs, while the connector terminals are where you must wire the inputs and outputs. These terminals are analogous to parameters of a subroutine or function. You need one terminal for each front panel control and indicator through which you want to pass data to the VI. The icon sits on top of the connector pattern until you choose to view the connector.

Temperature Status Icon   Connector

Figure 1.10

By using subVIs, you can make your block diagrams modular and more manageable. This modularity makes VIs easy to maintain, understand, and debug. In addition, you can often create one subVI to accomplish a function required by many different VIs.

Now run the top-level VI with both its window and the **Temperature Status** subVI window visible. Notice how the subVI values change as the main program calls it over and over.

13. Select **Close** from the **File** menu of the **Temperature Status** subVI. Do not save any changes.

14. Select **Close** from the **File** menu of **Temperature System Demo.vi**, and do not save any changes.

*Selecting **Close** from the File menu of a VI diagram closes the block diagram window only. Selecting **Close** on a front panel window closes both the panel and the diagram.*

1.2.2 Activity 1-2: Frequency Response Example

This example measures the frequency response of an unknown "black box." A function generator supplies a sinusoidal input to the black box. (Hint: It contains a bandpass filter, which lets only certain signal components through it.) A digital multimeter measures the output voltage of the black box. Although this VI uses subVIs to simulate a function generator and a digital multimeter, real instruments could easily be hooked up to a real black box to provide real-world data. You would then use subVIs to control data acquisition, GPIB transfers, or serial port communication to bring in or send out real data instead of simulating it.

You will open, run, and observe the VI in this activity.

1. Launch LabVIEW. At the startup dialog, click on "**Search Examples**". Alternatively, if LabVIEW is already running, go to the **Help** menu and select **Examples...**

2. Click on "Demonstrations", then "Instrument I/O", then "Frequency Response" (Note: You can also find this example in the LabVIEW in-

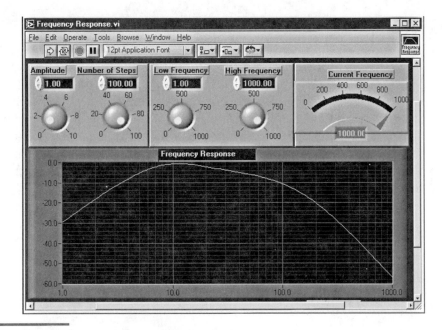

Figure 1.11

stall directory, under `examples/apps/freqresp.llb`). The front panel shown in the next illustration should appear.

Run button

3. Run the VI by clicking on the Run button. You can specify the amplitude of the input sine wave and the number of steps the VI uses to find the frequency response by changing the Amplitude control and the **Number of Steps** control, and then run the VI again. You can also specify the frequency sweep by inputting the upper and lower limits with the **Low Frequency** and **High Frequency** knobs. Play with these controls and observe the effect they have on the output of the "black box."

4. Open and examine the block diagram by choosing **Show Diagram** from the **Window** menu.

5. Close the VI by selecting **Close** from the **File** menu. These exercises should give you a basic feel for LabVIEW's programming environment. With LabVIEW, you'll find writing powerful applications (and debugging them) to be a snap! Read on to learn how!

1.3 Wrap It Up!

LabVIEW is a powerful and flexible instrumentation and analysis software system. It uses a graphical programming language, to create programs called *virtual instruments,* or VIs. The user interacts with the program through the *front panel.* Each front panel has an accompanying *block diagram,* which is the VI's source code. LabVIEW has many built-in functions to facilitate the programming process; components are wired together to show the flow of data within the block diagram. Stay tuned—the next chapters will teach you how to effectively use LabVIEW's many features.

You will find the solutions to every activity in the upcoming chapters in the EVERYONE *directory on the CD that accompanies the book. We'll trust you not to cheat!*

1.4 Additional Activities

Activity 1-3: More Neat Examples

In this activity, you will look at some example programs that ship with Lab-VIEW.

1. From the **Help** menu, choose **Examples**.
2. This will bring up the LabVIEW Help system, opened at the Examples section. Clicking on a link to a specific example will bring up the specific VI in LabVIEW.

Run button

3. Run the example by clicking on the Run button.
4. After you run an example, choose **Show Diagram** from the **Window** menu to see what the program looks like.
5. Now look through and run other VIs in the examples to try to get an idea of the LabVIEW environment and what you can do with it. Although all of the examples are extremely educational, you should investigate a few particularly interesting directories: Demonstrations, Measurement Examples, and I/O Interfaces. Feel free to browse through any VIs that strike your fancy; you can learn a lot just by

watching how they work. Also feel free to modify and use these examples for your own applications (just be sure to save them to a different location so you don't overwrite the built-in examples).

6. When you're done, select **Close** from the **File** menu to close each VI. Do not save any changes you may have made.

OVERVIEW

Virtual instrumentation is the foundation for the modern laboratory. A virtual instrument consists of a computer, software, and a plug-in board simulating the function of traditional hardware instrumentation; it's also what we call a LabVIEW program. Because their functionality is software-defined by the user, virtual instruments are extremely flexible, powerful, and cost-effective. This chapter explains how to communicate with the outside world (e.g., take measurements, "talk" to an instrument, send data to another computer) using LabVIEW. We're only giving you a very brief overview here; you can learn more about acquiring data, controlling instruments, and networking your computer with Lab-VIEW in the second half of this book. You'll also learn a little about how LabVIEW has changed over the years.

GOALS

- Understand the nature of data acquisition and instrument control
- Be able to describe the components of a typical DAQ or GPIB system
- Learn about your computer's serial, network, and USB ports
- Appreciate the usefulness of analysis functions
- Learn a little about PXI and VXI
- See how LabVIEW can exchange data with other computers and applications
- Know about some of the toolkits that enhance LabVIEW's capabilities

KEY TERMS

- Data acquisition (DAQ)
- General Purpose Interface Bus (GPIB)
- Institute of Electrical and Electronic Engineers (IEEE) 488 standard
- Serial port
- PXI

- VXI
- Internet capabilities
- Networking
- Dynamic link library (DLL)
- Code interface node (CIN)
- ActiveX
- Toolkit

Virtual Instrumentation: Hooking Your Computer Up to the Real World

2

2.1 The Evolution of LabVIEW

In 1983, National Instruments began to search for a way to minimize the time needed to program instrumentation systems. Through this effort, the LabVIEW virtual instrument concept evolved—intuitive front panel user interfaces combined with an innovative block diagram programming methodology to produce an efficient, software-based graphical instrumentation system.

LabVIEW version 1 was released in 1986 on the Macintosh only. Although the Mac was not widely used for measurement and instrumentation applications, its graphical nature best accommodated the LabVIEW technology until the more common operating systems could support it.

By 1990, National Instruments had completely rewritten LabVIEW, combining new software technology with years of customer feedback. More importantly, LabVIEW 2 featured a compiler that made execution speeds of VIs comparable with programs created in the C programming language. The United States Patent Office issued several patents recognizing the innovative LabVIEW technology.

As new graphical operating systems like the MacOS appeared, National Instruments ported the now mature LabVIEW technology to the other platforms: PCs and workstations. In 1992, they introduced LabVIEW for Windows and LabVIEW for Sun based on the new portable architecture.

LabVIEW 3 arrived in 1993 for Macintosh, Windows, and Sun operating systems. LabVIEW 3 programs written on one platform could run on another. This multiplatform compatibility gave users the opportunity to choose the development platform while ensuring that they can run their VIs on other platforms (consider that this was a couple of years before Java was introduced). In 1994, the list of LabVIEW-supported platforms grew to include Windows NT, Power Macs, and HP workstations. 1995 brought about an adaptation to Windows 95.

LabVIEW 4, released in 1996, featured a more customizable development environment so that users could create their own workspace to match their industry, experience level, and development habits. In addition, LabVIEW 4 added high-powered editing and debugging tools for advanced instrumentation systems, as well as OLE-based connectivity and distributed execution tools.

LabVIEW 5 and 5.1 (in 1999) continued to improve on the development tool by introducing a built-in Web server, a dynamic programming and control framework (VI server), integration with ActiveX, and easy sharing of data over the Internet with a protocol called DataSocket. The *undo* feature taken for granted in most programs finally was implemented.

In 2000, LabVIEW 6 (sometimes called 6i) got a face lift: a new suite of 3-D controls was introduced, appropriately at a time when the computing industry was discovering that style did matter (spearheaded by the introduction of Apple's iMac and G4 cubes). LabVIEW 6 does a very impressive job of providing both an easy and intuitive programming interface (especially for nonprogrammers!), as well as supporting a slew of advanced programming techniques, such as object-oriented development, multithreading, distributed computing, and much more. Don't let the graphical nature of LabVIEW fool you: LabVIEW is a tool that can easily rival C++ or Visual Basic as a development tool—with the benefit, as thousands have discovered, that it's much more fun!

In 2001, LabVIEW 6.1 introduced event-oriented programming, remote Web control of LabVIEW, and other improvements.

One special version of LabVIEW we should mention is *LabVIEW RT*. RT stands for *real time*. LabVIEW RT is a hardware and software combination that allows you to take portions of your LabVIEW code and download them to be executed on a separate controller board with its own real-time operating system. This means that you can guarantee that certain pieces of your LabVIEW code will keep running with precision, even if Windows crashes and your computer screeches to a halt.

Although LabVIEW is a very powerful simulation tool, it is most often used to gather data from an external source, and it contains many VIs built especially for this purpose. For example, LabVIEW can command plug-in data acquisition, or DAQ, boards to acquire or generate analog and digital signals. You might use DAQ boards and LabVIEW to monitor a temperature, send signals to an external system, or determine the frequency of an unknown signal. LabVIEW also facilitates data transfer over the General Purpose Interface Bus (GPIB), or through your computer's built-in serial port. GPIB is frequently used to communicate with oscilloscopes, scanners, and multimeters and to drive instruments from remote locations. LabVIEW software can also control sophisticated VXI hardware instrumentation systems, Ethernet, or USB-based instruments. Once you have acquired or received your dataset, you can use LabVIEW's many analysis VIs to process and manipulate it.

Often you will find it useful to share data with other applications or computers in addition to an instrument. LabVIEW has built-in functions that simplify this process, supporting several networking protocols, external calls to existing code or dynamic link libraries (DLLs), and ActiveX automation. We'll spend the rest of this chapter talking about some of the tasks LabVIEW was designed to accomplish.

2.2 What Is Data Acquisition?

Data acquisition, or *DAQ*, is simply the process of measuring a real-world signal, such as a voltage, and bringing that information into the computer for processing, analysis, storage, or other data manipulation. Figure 2.1 shows the components of a DAQ system. Physical phenomena represent the real-world signals you are trying to measure, such as speed, temperature, humidity, pressure, flow, pH, start–stop, radioactivity, light intensity, and so on. Transducers, a fancy word for sensors, evaluate the physical phenomena and produce electrical signals proportionately. For example, thermocouples, a type of transducer, convert temperature into a voltage that an A/D (analog to digital) converter can measure. Other examples of transducers include strain gauges, flowmeters, and pressure transducers, which measure force, rate of flow, and pressure, respectively. In each case, the electrical signal produced by the transducer is directly related to the phenomenon it monitors.

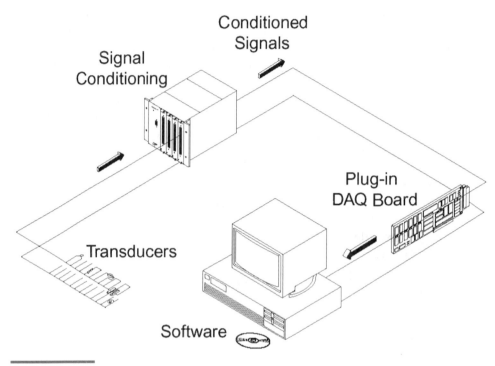

Figure 2.1
DAQ system.

LabVIEW can command DAQ boards to read analog input signals (A/D conversion), generate analog output signals (D/A conversion), read and write digital signals, and manipulate the on-board counters for frequency measurement, pulse generation, etc., to interface with the transducers. In the case of analog input, the voltage data from the sensor go into the plug-in DAQ board in the computer, which sends the data into computer memory for storage, processing, or other manipulation.

Signal conditioning modules "condition" the electrical signals generated by transducers so that they are in a form that the DAQ board can accept. For example, you would want to isolate a high-voltage input such as lightning, lest you fry both your board and your computer—a costly mistake! Signal conditioning modules can apply to many different types of conditioning: amplification, linearization, filtering, isolation, and so on. Not all applications will require signal conditioning, but many do, and you should pay attention to your specifications to avoid a potential disaster. In addition, information loss can be even worse than equipment loss! Noise, nonlinearity, overload, aliasing, etc. can hopelessly corrupt your data and LabVIEW

Figure 2.2
Many types of DAQ boards are available from National Instruments.

will not save you. Signal conditioning is often not optional—it's best to check before you start!

To acquire data in your lab using the virtual instrumentation approach, you will need a DAQ board, a computer configured with LabVIEW and DAQ driver software, and some method of connecting your transducer signal to the board, such as a connector block, breadboard, cable, or wire. You may also need signal conditioning equipment, depending on the specifications of your application.

For example, if you wanted to measure a temperature, then you would need to wire the temperature sensor to an analog input channel on the DAQ board in your computer (often via signal conditioning equipment, depending on the sensor). Then you would use LabVIEW's DAQ VIs to read the channel on the board, display the temperature on the screen, record it in a data file, and analyze it any way you need to.

The built-in LabVIEW data acquisition VIs only work with National Instruments' DAQ boards. If you are using a board from another vendor, you will have to get a driver from them (if they have one), or you will have to write your own driver code and call it from LabVIEW using code interface nodes or dynamic link libraries.

2.3   What Is a GPIB?

Hewlett Packard developed the *General Purpose Interface Bus*, or *GPIB*, in the late 1960s to facilitate communication between computers and instruments. A bus is simply the means by which computers and instruments transfer data, and GPIB provided a much-needed specification and protocol to govern this communication. The Institute of Electrical and Electronic Engineers (IEEE) standardized GPIB in 1975, and it became known as the IEEE 488 standard. GPIB's original purpose was to provide computer control of test and measurement instruments. However, its use has expanded beyond these applications into other areas, such as computer-to-computer communication and control of multimeters, scanners, and oscilloscopes.

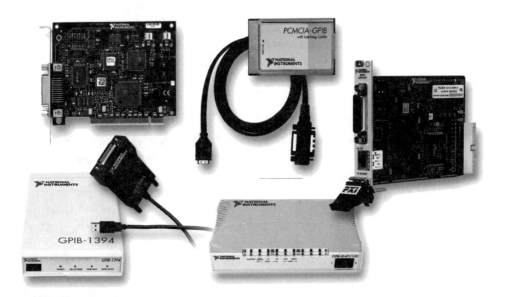

Figure 2.3
GPIB boards from National Instruments.

GPIB is a digital, 24-conductor parallel bus. It consists of eight data lines, five bus management lines (ATN, EOI, IFC, REN, and SRQ), three handshake lines, and eight ground lines. GPIB uses an eight-bit parallel, byte-serial, asynchronous data transfer scheme. In other words, whole bytes are sequentially moved across the bus at a speed determined by the slowest participant in the transfer. Because GPIB sends data in bytes (one byte=eight bits), the messages transferred are frequently encoded as ASCII character strings. Your computer can only perform GPIB communication if it has a GPIB board (or external GPIB box) and the proper drivers installed.

You can have many instruments and computers connected to the same GPIB bus. Every device, including the computer interface board, must have a unique GPIB address between 0 and 30, so that the data source and destinations can be specified by this number. Address 0 is normally assigned to the GPIB interface board. Instruments connected to the bus can use addresses 1 through 30. The GPIB has one Controller, usually your computer, that controls the bus management functions. To transfer instrument commands and data on the bus, the Controller addresses one Talker and one or more Listeners. The data strings are then sent across the bus from the Talker to the Listener(s). The LabVIEW GPIB VIs automatically handle the addressing and most other bus management functions, saving you the hassle of low-level programming. Figure 2.4 shows a typical GPIB system.

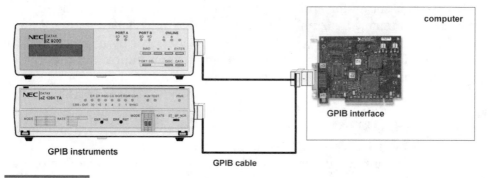

Figure 2.4
A typical GPIB system.

Although using GPIB is one way to bring data into a computer, it is fundamentally different from performing data acquisition, even though both use boards that plug into the computer. Using a special protocol, GPIB talks to another computer or instrument to bring in data acquired by that device, while data acquisition involves connecting a signal directly up to a DAQ board in the computer.

To use GPIB as part of your virtual instrumentation system, you need a GPIB board or external box, a GPIB cable, LabVIEW and a computer, and an IEEE 488-compatible instrument with which to communicate (or another computer containing a GPIB board). You also need to install the GPIB driver software on your computer as well, according to the directions that accompany LabVIEW or the board.

 LabVIEW's GPIB VIs communicate with National Instruments' GPIB boards, but not those from other manufacturers. If you have another vendor's board, you can either get driver software from them (if it's available) or write your own driver code and integrate it into LabVIEW. As with DAQ drivers, this is not an easy thing to do!

We'll talk more about DAQ and GPIB in Chapters 10 and 11.

2.4 Communication Using the Serial Port

Serial communication is another popular means of transmitting data between a computer and another computer or a peripheral device such as a programmable instrument. It uses the built-in *serial port* (either RS-232 or RS-422 standard) in your computer. Serial communication uses a transmitter to send data one bit at a time over a single communication line to a receiver. You can use this method when data transfer rates are low or when you must transfer data over long distances. It is slower and less reliable than communication by the GPIB, but you do not need a board in your computer and your instrument does not need to conform to the IEEE 488 standard. Figure 2.5 shows a typical serial communication system.

Serial communication is handy because most PCs have one or two serial ports built-in allowing you to send and receive data without buying any special hardware. Although most computers also now have USB (universal serial bus) ports built-in, USB is a more complex protocol that is oriented at peripherals, rather than communication. Serial communication (RS-232 or RS-485) is old compared to USB but is still widely used for many industrial devices.

Many GPIB instruments also have built-in serial ports. However, unlike GPIB, a serial port can communicate with only one device, which can be lim-

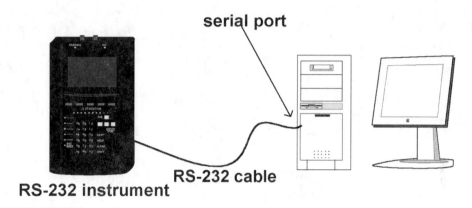

Figure 2.5
A typical serial communication system.

iting for some applications. Serial port communication is also painstakingly slow and has no built-in error checking capabilities. However, serial communication has its uses (it's certainly economical!), and the LabVIEW VISA Serial VIs contain ready-to-use functions for serial port operations. If you have a cable and a device to "talk" to, you are all set to try out serial communication!

2.5 Real-World Applications: Why We Analyze

Once you get data into your computer, you may want to process your data somehow. Modem, high-speed floating-point digital signal processors have become increasingly important to real-time and analysis systems. A few of the many possible analysis applications for LabVIEW include biomedical data processing, speech synthesis and recognition, and digital audio and image processing.

The importance of integrating analysis libraries into laboratory stations is obvious: The raw data collected from your DAQ board or GPIB instrument do not always immediately convey useful information. Often you must transform the signal, remove noise perturbations, correct for data corrupted by faulty equipment, or compensate for environmental effects such as temperature and humidity. Figure 2.6 shows data that epitomize the need for the analysis functions.

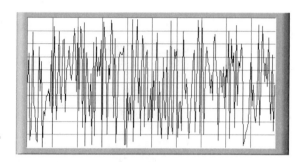

Figure 2.6

By analyzing and processing the digital data, you can extract the useful information from the noise and present it in a form more comprehensible than the raw data. The processed data looks more like that shown in Figure 2.7:

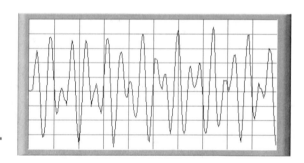

Figure 2.7

The LabVIEW block diagram programming method and the extensive set of LabVIEW analysis VIs simplify the development of analysis applications. The sample block diagram in Figure 2.8 illustrates the LabVIEW programming concept.

Figure 2.8

Because the LabVIEW analysis functions give you popular data analysis techniques in discrete VIs, you can wire them together, as shown in Figure 2.8, to analyze data. Instead of worrying about implementation details for analysis routines as you do in most programming languages, you can concentrate on solving your data analysis problems. LabVIEW's analysis VIs are powerful enough for experts to build sophisticated analysis applications using digital signal processing (DSP), digital filters, statistics, or numerical analysis. At the same time, they are simple enough for novices to perform sophisticated calculations.

The LabVIEW analysis VIs efficiently process blocks of information represented in digital form. They cover the following major processing areas:

- Pattern generation
- Digital signal processing
- Digital filtering
- Smoothing windows
- Statistical analysis
- Curve fitting
- Linear algebra
- Numerical analysis
- Measurement-based analysis

2.6 A Little Bit about PXI and VXI

Two other hardware platforms you should know about are PXI and VXI.

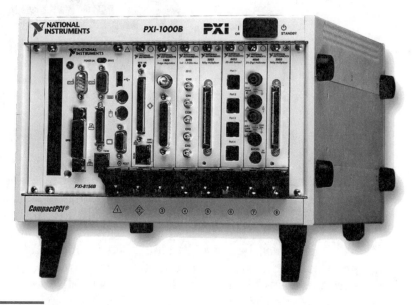

Figure 2.9
PXI chassis.

PXI, an acronym for *compactPCI extensions for Instrumentation*, defines a modular hardware platform based on Microsoft Windows and the Compact-PCI bus (a version of the PCI bus). A typical configuration involves a PXI chassis, which holds its own PC computer (called a *controller*) and slots for all types of measurement modules: analog input, imaging, sound, relays, GPIB and VXI interfaces, and more. The compact, ruggedized, and expandable system makes it an attractive platform for many applications. LabVIEW supports PXI systems from National Instruments. In addition, you can use the real-time version of LabVIEW (LabVIEW RT) on a PXI controller for a more robust system.

Figure 2.10
PXI systems.

The *VXIbus*, an acronym for *VMEbus eXtensions for Instrumentation*, is another instrumentation standard for instrument-on-a-card systems. First introduced in 1987 and based on the VMEbus (IEEE 1014) standard, the VXIbus is a higher-end and usually more costly system than PXI. VXI consists of a mainframe chassis with slots holding modular instruments on plug-in boards. A variety of instrument and mainframe sizes are available from numerous vendors, and you can also use VME modules in VXI systems. VXI has a wide array of uses in traditional test and measurement and ATE (automated test equipment) applications. VXI is also popular in data acquisition and analysis for research and industrial control applications that require high number of channels (hundreds or thousands).

VXIplug&play is a name used in conjunction with VXI products that have additional standardized features beyond the scope of the baseline specifications. VXIplug&play-compatible instruments include standardized software, which provides soft front panels, instrument drivers, and installation routines to take full advantage of instrument capabilities and to make your programming task as easy as possible. LabVIEW software for VXI is fully compatible with VXIplug&play specifications.

2.7 Connectivity

In some applications, you will want to share data with other programs, perhaps locally or across your local network. In many cases you may want to share the data over the Internet and allow other people to view or control your system over the Web.

LabVIEW has built-in features (such as a Web server and Web publishing tool) and functions that simplify this process. These VIs facilitate communication over a network or over the Internet. LabVIEW can use the DataSocket protocol to share data across networks, call and create dynamic link libraries (DLLs) or external code, and support ActiveX automation. Using the add-on Enterprise Connectivity Toolset, LabVIEW can also communicate with most SQL (structured query language) databases, such as Oracle, SQL Server, and Access.

2.7.1 Internet Connectivity

LabVIEW has several built-in features that make it very easy to share your VIs and data over the Internet. Using LabVIEW's Web server, you can allow people to remotely view images of the front panel of your VIs (without any additional programming).

With the *Enterprise Connectivity Toolset*, you can also use LabVIEW to e-mail, ftp, and telnet into remote systems, as well as provide enhanced Web serving capabilities.

The *LabVIEW Player* is a free utility that anyone can download at http://ni.com/labview/. With the LabVIEW Player, anyone can "play" (that is, view and run, but not edit or create) VIs created in LabVIEW,

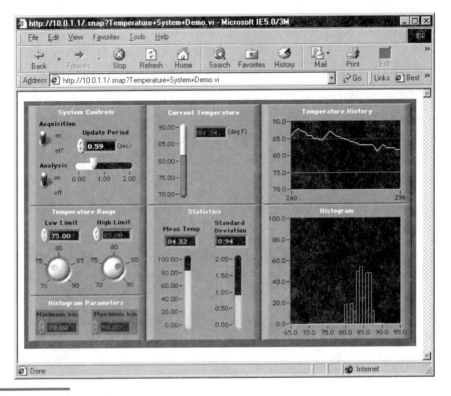

Figure 2.11
A VI's front panel image viewed in Internet Explorer, generated by LabVIEW's Web server.

without the need for installing or purchasing the full version. The LabVIEW Player is a convenient way to share VIs over the Internet, in a manner akin to which Adobe pdf files can be shared with anyone who has installed the Adobe Reader.

2.7.2 Networking

For our purposes, networking refers to communication among multiple processes that usually (but not necessarily) run on separate computers. This communication can occur on a closed, local network area (LAN) or over the Internet. One main use for networking in software applications is to allow one or more applications to use the services of another application. In addition to the Web publishing features of LabVIEW, you can use some of the networking functionality to communicate with other software or another LabVIEW program.

For communication among processes to work, the processes must use a common communications language, referred to as a protocol. LabVIEW supports the following protocols:

- *DataSocket*—this is a proprietary protocol from National Instruments for sharing instrumentation data. It has the advantage that it is very easy to use.
- *TCP/IP* (the basic protocol of most networks, including the Internet)
- *UDP*

In addition to these operating system–independent network protocols, LabVIEW provides support for some older and lesser-used protocols, such as *DDE* (Windows) and *AppleEvents* and *PPC* (MacOS).

This book will talk more about Internet capabilities and networking in Chapter 14.

2.7.3 ActiveX

ActiveX is a technology from Microsoft that defines a component-based architecture for building applications that can communicate with each other. ActiveX builds upon previous technologies such as OLE. With ActiveX, one application can share a piece of code (a *component*) with a completely different application. For example, since Microsoft Word is an ActiveX component, you can control and embed a Word document inside another ActiveX-capable application, such as LabVIEW VI. LabVIEW supports ActiveX automation and can contain ActiveX components. If you don't understand what we're talking about, don't worry. (ActiveX is a fairly complicated advanced topic. It is discussed in more detail in Chapter 14.)

2.7.4 DLLs and CINs

For increased flexibility, LabVIEW can both call and create external code routines or *dynamic link libraries (DLLs)* and integrate these routines into program execution. In case you were wondering, a dynamic link library is a library of shared functions that an application can link to at runtime, instead of at compile time. LabVIEW uses a special block diagram structure called a *code interface node (CIN)* to link conventional, text-based code to a VI. LabVIEW calls the executable code when the node executes, passing input data from the block diagram to the executable code, and returning data from the executable code to the block diagram. Similarly, you can use the Call Library function to call a DLL if you are running under Windows. You can also tell LabVIEW to compile its VIs as a DLL that other types of code (e.g., C++) can use.

Most applications never require the use of a CIN or DLL. Although the LabVIEW compiler can usually generate code that is fast enough for most tasks, CINs and DLLs are useful for some tasks that are time-critical, require a great deal of data manipulation, or that you've already written specific code for. They are also useful for tasks that you cannot perform directly from the diagram, such as calling system routines for which LabVIEW functions do not exist.

2.8   LabVIEW Add-on Toolkits

You can use the following special add-on toolkits with LabVIEW to increase your flexibility and capabilities. For more information about them, see Appendix A. Some of the more common toolkits are:

- Application Builder
- Enterprise Connectivity Toolset
- Internet Toolkit
- Database Connectivity Toolset
- SPC (Statistical Process Control) Toolkit
- Motion Control Toolkit
- Sound & Vibration Analysis Toolset
- OverVIEW Project Management Toolkit

Some toolkits are sold by National Instruments; others are available from third-party companies, often National Instruments Alliance Members, that have made add-ons to LabVIEW that do all sorts of things. If you have a specific task and you want to know if someone's already done it, we suggest posting to the info-LabVIEW user forum (see Appendix A for details).

2.9   Wrap It Up!

LabVIEW's built-in functions facilitate hardware communication with external devices so that you don't have to write involved programs. LabVIEW virtual instruments can work with several types of hardware to gather or exchange data: plug-in DAQ boards, GPIB boards, your computer's built-in serial port, or PXI and VXI hardware. You can use National Instruments DAQ boards, managed by LabVIEW, to read and generate analog input, analog output, and digital signals and also to perform counter/timer operations. LabVIEW can also control communication over the GPIB (assuming you have a GPIB board) or command a PXI or VXI instrumentation system. If you don't have any special hardware, LabVIEW can communicate with other devices through your computer's serial port.

LabVIEW analysis VIs make it easy for you to process and manipulate data once you have brought it into your computer. Rather than working through tricky algorithms by hand or trying to write your own low-level code, you can simply access the built-in LabVIEW functions that suit your needs.

You can use LabVIEW's built-in Internet capabilities to publish images of your VIs to the Web, share VIs with anyone using the LabVIEW Player, and communicate with other programs and computers linked by a network using protocols such as DataSocket or TCP/IP. LabVIEW supports ActiveX to communicate with other programs and can both call and create DLLs.

If you want to expand LabVIEW's considerable functionality, you can buy add-on toolkits to accomplish specific tasks. Toolkits are available to build standalone LabVIEW applications, enhance its Internet capabilities, design and analyze certain types of signals, perform statistical process and PID control, communicate with a SQL database, manage large LabVIEW projects, and much more.

The next chapters will teach you the fundamentals of LabVIEW programming. So get ready to write some code!

OVERVIEW

In this chapter, you will investigate the LabVIEW environment and learn how its three parts—the front panel, block diagram, and icon/connector—work together. When all three main components are properly developed, you have a VI that can stand alone or be used as a subVI in another program. You will also learn about the LabVIEW environment: pull-down and pop-up menus, floating palettes and subpalettes, the Toolbar, and how to get help. To finish up, we will discuss the power of subVIs and why you should use them.

GOALS

- Understand and practice using the front panel, block diagram, and icon/connector
- Learn the difference between controls and indicators
- Be able to recognize the difference between the block diagram terminals of controls and indicators
- Understand the principle of dataflow programming
- Become familiar with LabVIEW menus, both pop-up and pull-down
- Learn about the capabilities and uses of the Toolbar, **Tools** palette, **Controls** palette, **Functions** palette, and subpalettes
- Learn why the **Help** windows can be your most valuable ally
- Understand what a subVI is and why it's useful
- Work through the activities to get a feel for how LabVIEW works

KEY TERMS

- Control
- Indicator
- Wire
- SubVI
- Terminal
- Node
- Dataflow
- Pop-up menus
- Toolbar
- Palette
- Subpalette
- Help window

The LabVIEW Environment: Building Your Own Workbench

3

3.1 Front Panels

Simply put, the *front panel* is the window through which the user interacts with the program. When you run a VI, you must have the front panel open so that you can input data to the executing program. You will also find the front panel indispensable because that's where you see your program's output. Figure 3.1 shows an example of a LabVIEW front panel.

3.1.1 Controls and Indicators

The front panel is primarily a combination of *controls* and *indicators*. **Controls** simulate typical input objects you might find on a conventional instrument, such as knobs and switches. Controls allow the user to input values; they supply data to the block diagram of the VI. **Indicators** show output values produced by the program. Consider this simple way to think about controls and indicators:

Controls = Inputs from the User = Source Terminals

Indicators = Outputs to the User = Destinations or "Sinks"

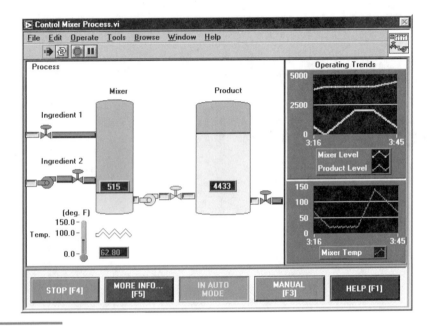

Figure 3.1
LabVIEW front panel.

They are generally not interchangeable, so make sure you understand the difference.

You "drop" controls and indicators onto the front panel by selecting them from a *subpalette* of the floating **Controls** palette window and placing them in a desired spot. Once an object is on the front panel, you can easily adjust its size, shape, position, color, and other attributes.

3.2 Block Diagrams

The *block diagram* window holds the graphical source code of a LabVIEW VI. LabVIEW's block diagram corresponds to the lines of text found in a more conventional language like C or BASIC—it is the actual executable code. You construct the block diagram by wiring together objects that perform specific functions. In this section, we will discuss the various components of a block diagram: *terminals, nodes,* and *wires.*

The simple VI shown in Figure 3.2 computes the sum of two numbers. Its diagram in Figure 3.3 shows examples of terminals, nodes, and wires.

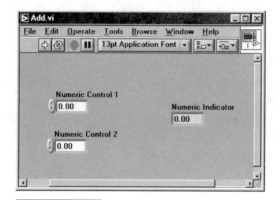

Figure 3.2

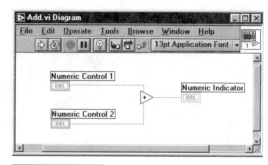

Figure 3.3

3.2.1 Terminals

When you place a control or indicator on the front panel, LabVIEW automatically creates a corresponding *terminal* on the block diagram. By default, you cannot delete a block diagram terminal that belongs to a control or indicator, although you may try to your heart's content. The terminal disappears only when you delete its corresponding control or indicator on the front panel.

Control terminals have thick borders, while indicator terminal borders are thin. It is very important to distinguish between the two since they are not functionally equivalent (Control = Input, Indicator = Output, and so they are not interchangeable).

Figure 3.4

You can think of terminals as entry and exit ports in the block diagram, or as sources and destinations. Data that you enter into Numeric Control 1 (shown in the Figure 3.3) exits the front panel and enters the block diagram through the Numeric Control 1 terminal on the diagram. The data from Numeric Control 1 follows the wire and enters the **Add** function input terminal. When the **Add** function completes its internal calculations, it produces new data values at its exit terminal. The data flows to the Numeric Indicator terminal and reenters the front panel, where it is displayed for the user.

3.2.2 Nodes

A *node* is just a fancy word for a program execution element. Nodes are analogous to statements, operators, functions, and subroutines in standard programming languages. The **Add** and **Subtract** functions represent one type of node. A structure is another type of node. Structures can execute code repeatedly or conditionally, similar to loops and Case statements in traditional programming languages. LabVIEW also has special nodes, called Formula Nodes, which are useful for evaluating mathematical formulas or expressions.

3.2.3 Wires

A LabVIEW VI is held together by *wires* connecting nodes and terminals. Wires are the data paths between source and destination terminals; they deliver data from one source terminal to one or more destination terminals. If you connect more than one source or no source at all to a wire, LabVIEW disagrees with what you're doing, and the wire will appear broken.

This principle of wires connecting source and destination terminals explains why controls and indicators are not interchangeable. Controls are source terminals, whereas indicators are destinations, or "sinks."

Each wire has a different style or color, depending on the data type that flows through the wire. The block diagram shown in Figure 3.3 depicts the wire style for a numeric scalar value—a thin, solid line. The chart in Figure 3.5 shows a few wires and corresponding types.

	Scalar	1D Array	2D Array	Color
Floating-point number	————	━━━━	═══	Orange
Integer number	————	━━━━	═══	Blue
Boolean	··········	wwwwwww	◊◊◊◊◊◊◊◊◊	Green
String	~~~~~~~	◻◻◻◻◻◻◻◻	░░░░░░░░░	Pink

Figure 3.5
Basic wire styles used in block diagrams.

To avoid confusing your data types, simply match up the colors and styles!

3.2.4 Dataflow Programming—Going with the Flow

Since LabVIEW is not a text-based language, its code cannot execute "line by line." The principle that governs LabVIEW program execution is called *dataflow*. Stated simply, a node executes only when data arrives at all its input terminals; the node supplies data to all of its output terminals when it finishes executing; and the data pass immediately from source to destination terminals. Dataflow contrasts strikingly with the control flow method of executing a text-based program, in which instructions are executed in the sequence in which they are written. This difference may take some getting used to. Whereas traditional execution flow is instruction driven, dataflow execution is data driven or *data dependent*.

3.3 The Icon and the Connector

When your VI operates as a *subVI,* its controls and indicators receive data from and return data to the VI that calls it. A VI's *icon* represents it as a subVI in the block diagram of another VI. An icon can include a pictorial representation or a small textual description of the VI, or a combination of both.

The VI's *connector* functions much like the parameter list of a C or Pascal function call; the connector terminals act like little graphical parameters to pass data to and from the subVI. Each terminal corresponds to its very own control or indicator on the front panel. During the subVI call, the input parameter terminals are copied to the connected controls, and the subVI executes. At completion, the indicator values are copied to the output parameter terminals.

Icon   Connector

Figure 3.6
An icon and its underlying connector.

Every VI has a default icon, which is displayed in the icon pane in the upper-right corner of the panel and diagram windows. The default icon is depicted in Figure 3.7.

Icon pane

Figure 3.7

A VI's connector is hidden under the icon; access it by choosing **Show Connector** from the front panel icon pane pop-up menu (we'll talk more about pop-up menus later). When you show the connector for the first time, LabVIEW helpfully suggests a connector pattern that has one terminal for each control and indicator currently on the front panel. You can select a different pattern if you desire, and you can assign up to 28 terminals before you run out of real estate on the connector.

3.3.1 Activity 3-1: Getting Started

Okay, you've read enough for now. It's time to get some hands-on experience. Go ahead and launch LabVIEW. You will step through the creation of a simple LabVIEW VI that generates a random number and plots its value on a waveform chart. You'll learn more in the next chapter about the steps you'll be taking; for now, just get a feel for the environment.

If you are using the full version of LabVIEW, just launch it and you'll be ready to start building your first VI.

If you are using the evaluation version of LabVIEW, you can still do these activities, since the evaluation version of LabVIEW has almost no restrictions on creating and editing VIs. Just be aware that your VI cannot run more than 5 minutes, and after 30 days your evaluation version will quit working.

If you are not comfortable working through the activities in this chapter without more background information or if you have trouble getting them to work, read Chapter 4 and then come back and try again.

1. At the LabVIEW dialog box during launch, click **New VI**. You should have an "Untitled 1" front panel on your screen.

Go to the floating **Controls** palette and click on the **Graph** button to access the Graph subpalette. If the **Controls** palette isn't visible, select **Show Controls Palette** from the **Windows** menu. Also make sure the front panel window is active, or you will see the **Functions** palette instead of the **Controls** palette. On the **Graph** subpalette, select **Waveform Chart** by releasing the mouse button. You will notice that, as you run the cursor over the icons in the Controls palette and subpalettes, the selected button or icon's name appears at the top of the palette, as shown in the Figures 3.8 and 3.9.

Figure 3.8

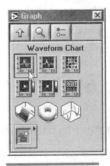

Figure 3.9

Positioning
Tool

You will see the outline of a chart with the cursor "holding" it. Position the cursor in a desirable spot on your front panel and click. The chart magically appears exactly where you placed it. If you want to move it, select the **Positioning** tool from the **Tools** palette, and then drag the chart to its new home. If the **Tools** palette isn't visible, select **Show Tools Palette** from the **Windows** menu.

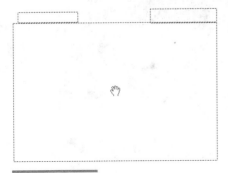

Figure 3.10

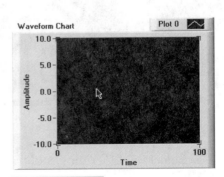

Figure 3.11

2. Go back to the floating **Controls** palette, by clicking on the "Up to Owning Palette" arrow on the **Graph** subpalette (this arrow is at the top-left corner of all control palettes). From the **Controls** palette, select the **Boolean** subpalette, and choose **Vertical Toggle Switch**.

Figure 3.12

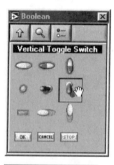

Figure 3.13

Place it next to the chart as shown in Figure 3.14.

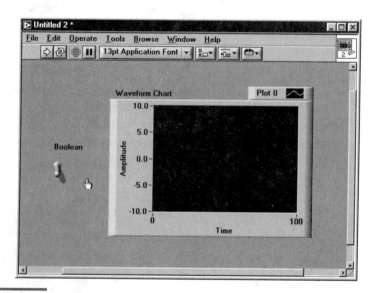

Figure 3.14

Operating Tool

3. Select the Operating tool from the floating **Tools** palette.

Figure 3.15

Enter Button

Now change the scale on the chart. Highlight the number "10" by click-dragging or by double-clicking on it with the Operating tool. Now type in `1.0` and click on the enter button that appears in the Toolbar at the top of the window.

4. Switch to the block diagram by selecting **Show Diagram** from the **Windows** menu. You should see two terminals already there (Figure 3.16).

Figure 3.16

5. Now you will put the terminals inside a While Loop to repeat execution of a segment of your program. Go to the **Structures** subpalette of the floating **Functions** palette and select the **While Loop**. Make sure the block diagram window is active, or you will see the **Controls** palette instead of the **Functions** palette.

Figure 3.17

Figure 3.18

Your cursor will change to a little loop icon. Now enclose the DBL and TF terminals: Click and hold down the mouse button while you drag the cursor from the upper-left to the lower-right corners of the objects you wish to enclose.

Figure 3.19

When you release the mouse button, the dashed line that is drawn as you drag will change into the While Loop border. Make sure to leave some extra room inside the loop.

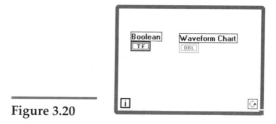

Figure 3.20

6. Go to the **Functions** palette and select **Random Number (0–1)** from the **Numeric** subpalette. Place it inside the While Loop.

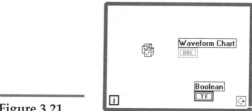

Figure 3.21

The While Loop is a special LabVIEW structure that repeats the code inside its borders until it reads a FALSE value. It is the equivalent of a Do-While Loop in a more conventional language. You'll learn more about loops in Chapter 6.

Positioning
Tool

7. Select the Positioning tool from the floating Tools palette and arrange your diagram objects so that they look like the previously shown block diagram.

Wiring Tool

8. Now select the Wiring tool from the **Tools** palette. Click once on the **Random Number (0–1)** icon, drag the mouse over to the DBL terminal, and click again.

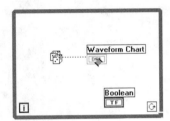

Figure 3.22

You should now have a solid orange wire connecting the two icons. If you mess up, you can select the wire or wire fragment with the Positioning tool and then hit the <delete> key to get rid of it. Now wire the Boolean TF terminal to the conditional terminal of the While Loop. The loop will execute while the switch on the front panel is TRUE (in the "up" position) and stop when the switch becomes FALSE.

Figure 3.23

Operating Tool

Run Button

9. You should be about ready to run your VI. First, switch back to the front panel by selecting **Show Panel** from the **Windows** menu. Using the Operating tool, flip the switch to the "up" position. Now click on the run button to run your VI. You will see a series of random

numbers plotted continuously across the chart. When you want to stop, click on the switch to flip it to the down position.

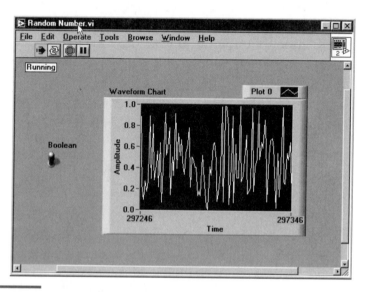

Figure 3.24

10. Create a directory or folder called MYWORK in your LabVIEW directory. Save your VI in your MYWORK directory or folder by selecting **Save** from the **File** menu and pointing out the proper location to save to. Name it **Random Number.vi**.

Save all of your subsequent activities in MYWORK so you can find them easily!

Remember, if you get stuck or just want to compare your work, the solutions to every activity in this book can be found in the EVERYONE directory or folder on the accompanying CD. You can view them in the sample version or the full version of LabVIEW.

Congratulate yourself—you've just written your first LabVIEW program! Don't worry that it doesn't actually do much—your programs will be more powerful and have more of a purpose soon enough!

3.4 Pull-Down Menus

Keep in mind that LabVIEW's capabilities are many and varied. This book by no means provides an exhaustive list of all of LabVIEW's ins and outs (it would be several thousand pages long if that were the case); instead, we try to get you up to speed comfortably and give you an overview of what you can do. If you want to know everything there is to know about a subject, we'd recommend looking it up in one of LabVIEW's many manuals, attending a seminar, or going to ni.com/labview. See Appendix A for an exhaustive list of other resources. Feel free to skim through this section and some of the subsequent ones, but remember that they're here if you need a reference.

LabVIEW has two main types of menus: pull-down and pop-up. You used some of them in the last activity, and you will use both extensively in all of your program development henceforth. Now you will learn more about what they can do. We'll cover pull-down menu items very briefly in this section. You might find it helpful to look through the menus on your computer as we explain them, and maybe experiment a little.

The menu bar at the top of a VI window contains several pull-down menus. When you click on a menu bar item, a menu appears below the bar. The pull-down menus contain items common to many applications, such as **Open**, **Save**, **Copy**, and **Paste**, and many other functions particular to LabVIEW. We'll discuss some basic pull-down menu functions here. You'll learn more about the advanced capabilities later.

Many menus also list shortcut keyboard combinations for you to use if you choose. To use keyboard shortcuts, press the appropriate key in conjunction with the <control> key on PCs, the <command> key on Macs, the <meta> key on Suns, and the <alt> key on HP machines.

Many of the menu items show keyboard shortcuts to the right of their corresponding commands. You may want to use the shortcuts instead of the menus.

Figure 3.25

Figure 3.26

Figure 3.27

File Menu

Pull down the **File** menu, which contains commands common to many applications such as **Save** and **Print**. You can also create new VIs or open existing ones from the **File** menu.

Edit Menu

Take a look at the **Edit** menu. It has some universal commands, like **Undo, Cut**, **Copy**, and **Paste**, that let you edit your window. You can also search for objects with the **Find...** command and remove bad wires from the block diagram.

Operate Menu

You can run or stop your program from the **Operate** menu (although you'll usually use Toolbar buttons). You can also change a VI's default values, control 'print and log at completion' features, and switch between run mode and edit mode.

Tools Menu

The **Tools** menu lets you access built-in and add-on tools and utilities that work with LabVIEW, such as the **Measurement & Automation Explorer**, where you configure your DAQ devices, or the **Web Publishing Tool** for

Figure 3.28

Figure 3.29

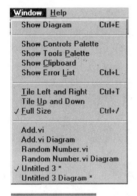

Figure 3.30

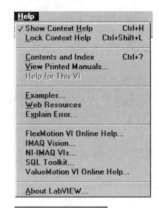

Figure 3.31

creating HTML pages from LabVIEW. You can view and change the myriad LabVIEW **Options....**

Browse Menu

The **Browse** menu contains features to simplify navigation among large sets of VIs. You can see VI hierarchy, determine all of a VI's subVIs, and view debugging breakpoints.

Windows Menu

Pull down the **Windows** menu. Here you can toggle between the panel and diagram windows, show the error list and the clipboard, "tile" both windows so you can see them at the same time, and switch between open VIs. You can also bring up floating palettes if you've closed them. In addition, you can show VI information and development history from this menu.

Help Menu

You can show, hide, or lock the contents of the Help window using the **Help** menu. You can also access LabVIEW's online reference information and view the About LabVIEW information window.

3.5 Floating Palettes

LabVIEW has three often-used floating palettes that you can place in a convenient spot on your screen: the **Tools** palette, the **Controls** palette, and the **Functions** palette. You can move them around by clicking on their title bar and dragging. Close them just like you would close any window in your operating system. If you decide you want them back, use the **Show... Palette** function in the **Windows** menu.

3.5.1 Controls and Functions Palettes

You will be using the **Controls** palette a lot, since that's where you select the controls and indicators that you want on your front panel. You will probably use the **Functions** palette even more often, since it contains the functions and structures used to build a VI.

The Controls and Functions palettes are unique in several ways. *Most importantly, the Controls palette is only visible when the front panel window is active, and the Functions palette is only visible when the block diagram window is active.* Both palettes have *subpalettes* containing the objects you need to access. As you pass the cursor over each subpalette button in the Controls and Functions palettes, you will notice that the subpalette's name appears at the top of the window.

Figure 3.32

If you click on a button the associated subpalette appears and replaces the previous active palette. To select an object in the subpalette, click the mouse button over the object, and then click on the front panel or block diagram to place it where you want it. Like palette button names, subpalette object names appear when you run the cursor over them. To return to the previous ("owning") palette, just click on the top-left arrow on each palette. You can search for a specific item in a palette by clicking on the spyglass icon, and you can edit your own palettes by clicking the options button.

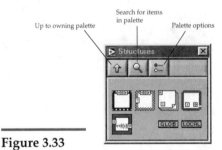

Figure 3.33

There is another way to navigate palettes that some people find a little easier. Instead of having each subpalette replace the current palette, you can pass through subpalettes in a hierarchical manner without them replacing their parent palettes. You can do this by right-clicking (Windows) or command-clicking (MacOS) the buttons on palettes.

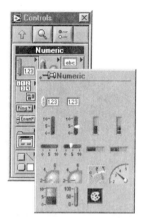

Figure 3.34

Note that some subpalettes have subpalettes containing more objects; these are denoted by a little triangle in the upper-right corner of the icon and a raised appearance. We'll discuss specific subpalettes and their objects in the next chapter.

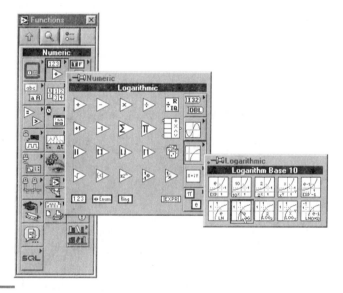

Figure 3.35

The Controls and Functions palettes can also be accessed by popping up in an empty area of the front panel or block diagram. "Popping up" is defined as right-mouse-clicking on the PC, Sun, and HP machines, and <command>-clicking on the Mac. You can also pop up with the soon-to-be-discussed Pop-up Tool.

3.5.2 The Thumbtack

If you use a subpalette frequently, you may want to "tear it off" by releasing the mouse button over the *thumbtack* located at the upper left of the palette. This thumbtack is available when you navigate hierarchically through palettes by right-clicking (<command>-clicking on the Mac), as we just described. You now have a stand-alone window that you can position anywhere and then close when you're done with it. You can leave open as many subpalettes as you like.

3.5.3 Customizable Palettes

If LabVIEW's default organization of the **Controls** and **Functions** palettes doesn't fit your needs, you can customize them according to your whim. Access the menu editor by clicking on the "options" icon for a palette. From here, you can create your own palettes and customize existing views by adding new subpalettes, hiding items, or moving them from one palette to another. For example, if you create a VI using trigonometric functions, you can place it in the existing **Trigonometric** subpalette for easy access.   Editing the palettes is handy for placing your most frequently used functions at the top level for easy access and burying those pesky functions you never want to see again at the bottom of a subpalette. You can also choose whether to show the icons, the text, or both on palettes.

You'll learn more about how to customize palettes in Chapter 4. You can also use the built-in "Data Acquisition" or "Test and Measurement" sets if those configurations are more convenient for you.

3.5.4 Tools Palette

A *tool* is a special operating mode of the mouse cursor. You use tools to perform specific editing and operation functions, similar to how you would use them in a standard paint program.

Figure 3.36

Like the **Controls** and **Functions** palettes, the **Tools** palette window can be relocated or closed. To select a tool, click the appropriate button on the **Tools** palette and your mouse cursor will change accordingly. If you're not sure which tool is which, hold your cursor over the button until a tip *strip* appears describing the tool.

Operating Tool

The Operating tool lets you change values of front panel controls and indicators. You can operate knobs, switches, and other objects with the Operating tool—hence the name. It is the only front panel tool available when your VI is running or in run mode (described shortly).

Positioning
Tool

The Positioning tool selects, moves, and resizes objects.

Labeling Tool

The Labeling tool creates and edits text labels.

Wiring Tool

The Wiring tool wires objects together on the block diagram. It is also used to assign controls and indicators on the front panel to terminals on the VI's connector.

Color Tool

The Color tool brightens objects and backgrounds by allowing you to choose from a multitude of hues. You can set both foreground and background colors by clicking on the appropriate color area in the Tools palette. If you pop up on an object with the Color tool, you can choose a hue from the color palette that appears.

Pop-up Tool

The Pop-up tool opens an object's pop-up menu when you click on the object with it. You can use it to access pop-up menus instead of the standard method for popping up (right-clicking under Windows and Unix and <command>-clicking on MacOS).

Scroll Tool

The Scroll tool lets you scroll in the active window.

Breakpoint
Tool

The Breakpoint tool sets breakpoints on VI diagrams to help you debug your code. It causes execution to suspend so that you can see what is going on and change input values if you need to.

Probe Tool

The Probe tool creates probes on wires so that you can view the data traveling through them while your VI is running.

Color Copy
Tool

Use the Color Copy tool to pick up a color from an existing object, and then use the Color tool to paste that color onto other objects. This technique is very useful if you need to duplicate an exact shade but can't remember which one it was. You can also access the Color Copy tool when the Color tool is active by holding down the <control> key on Windows, <option> on MacOS, <meta> on Sun, and <alt> on Linux and HP-UX.

You can use the <tab> key to tab through the Tools palette instead of clicking on the appropriate tool button to access a particular tool. Or press the space bar to toggle between the Operating tool and the Positioning tool when the panel window is active and between the Wiring tool and the Positioning tool when the diagram window is active. The <tab> and space bar shortcuts cycle through the most frequently used tool for your convenience—try using them, and see how they save you time!

You can also access a temporary copy of the Tools palette by pop-up clicking (<shift>-right click for Windows and Unix and <command-shift>-click on MacOS).

3.6 The Toolbar

The *Toolbar*, located at the top of LabVIEW windows, contains buttons you will use to control the execution of your VI, as well as text configuration options and commands to control the alignment and distribution of objects. You'll notice that the Toolbar has a few more options in the block diagram than in the front panel and that a few editing-related options disappear when you run your VI. If you're not sure what a button does, hold the cursor over it until a tip strip appears describing its function.

Figure 3.37

Run Button

The Run button, which looks like an arrow, starts VI execution when you click on it. It changes appearance when a VI is actually running. When a VI won't compile, the run button is broken.

Run Button (active)

Run Button (broken)

Continuous Run Button

The Continuous Run button causes the VI to execute over and over until you hit the stop button. It's kind of like a GOTO statement (sort of a programming "no-no"), so use it sparingly.

 Abort Button

The Abort button, easily recognizable because it looks like a tiny stop sign, becomes active when a VI begins to execute; otherwise the Abort button is grayed out. You can click on this button to halt the VI.

Using the Abort button is like pulling the power cord on your computer. Your program will stop immediately rather than coming to a graceful end, and data integrity can be lost this way. You should always code a more appropriate stopping mechanism into your program, as we will demonstrate later.

 Pause Button

The Pause button pauses the VI so that you can use single-step debugging options such as step into, step over, and step out. Hit the pause button again to continue execution.

 Step Into Button

The single-step buttons, Step Into, Step Over, and Step Out, force your VI to execute one step at a time so you can troubleshoot. We'll talk more about how to use them in Chapter 5.

 Step Over Button

 Step Out Button

 Execution Highlight Button

The Execution Highlight button causes the VI to highlight the flow of data as it passes through the diagram. When execution highlighting is on, you can see intermediate data values in your block diagram that would not otherwise appear.

Warning Button

The Warning button appears if you have configured your VI to show warnings and you have any warnings outstanding. You can list the warnings by clicking on the button. A warning is not an error; it just alerts you that you are doing something you may not have intended (for example, you have a front panel control with nothing wired to it).

You can change the font, size, style, justification, and color of LabVIEW text from the Font ring on the Toolbar.

Figure 3.38　　　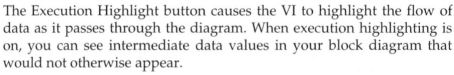

LabVIEW has an automatic alignment mechanism to help you line up and evenly space your icons. Select the objects you want to align by dragging around them with the Positioning tool, and then go to the Alignment ring on the Toolbar and choose how you want to align them (top edges flush, left edges flush, vertical centers, etc.). If you want to set uniform spacing between objects, use the Distribution ring in a similar fashion.

Alignment ring   Distribution ring

Figure 3.39

In a similar fashion, LabVIEW lets you group objects together to treat them as one control for graphical editing purposes, as well as set the depth order of objects, so that you can specify which objects should go in front of or behind others. You can do this with the Reorder ring.

Figure 3.40

Run Mode and Edit Mode

When you open a VI, it opens in *edit mode* so that you can make changes to it. When you run a VI, it automatically goes into *run mode* and you can no longer edit. Only the **Operating** tool is available on the front panel when the VI is in run mode. When your VI completes execution, your VI reverts to edit mode (unless you manually switched it to run mode before you ran it—then it stays in run mode). You can switch to run mode by selecting **Change to Run Mode** from the **Operate** menu; switch back to edit mode by choosing **Change to Edit Mode**. To draw a parallel with text-based languages, if a VI is in run mode, it has been successfully compiled and awaits your command to execute. Most of the time, you will not need to concern yourself with run and edit modes. But if you accidentally find that you suddenly have only the Operating tool, and you can't make any changes, at least now you'll know why.

If you prefer to open VIs in run mode (perhaps so uninvited users can't make changes), select Options... from the Tool menu. Go to the Miscellaneous options and choose "Open VIs in Run Mode."

3.7 Pop-Up Menus

As if pull-down menus didn't give you enough to learn about, we will now discuss the other type of LabVIEW menu, the pop-up menu. You will probably use pop-up menus more often than any other LabVIEW menu. To pop up, position the cursor over the object whose menu you desire; then click the right mouse button on Windows and UNIX machines, or hold down the <command> key and click on the Mac. You can also click on the object with the Pop-up tool. A pop-up menu will appear.

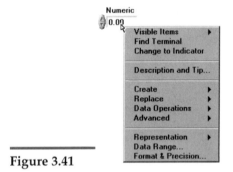

Figure 3.41

Virtually every LabVIEW object has a pop-up menu of options and commands. Options available in this pop-up menu depend on the kind of object, and they are different when the VI is in edit mode or run mode. For example, a numeric control will have a very different pop-up menu than a graph indicator. If you pop up on empty space in a front panel or block diagram, you will get the **Controls** or **Functions** palette, respectively.

You will find that instructions throughout this book guide you to select a command or option from an object pop-up menu—so try popping up now!

How to Pop Up
Windows and UNIX: right mouse click on the object
Mac: <command>-click on the object
All Platforms: Click on the object with the Pop-up tool

Pop-up menus are ever-present in LabVIEW. They contain most configuration options for an object. So remember, when in doubt about how to do something, try popping up!

3.7.1 Pop-Up Menu Features to Keep in Mind

Many pop-up menu items expand into submenus called hierarchical menus, denoted by a right arrowhead (see Figure 3.42). Hierarchical menus some-times have a selection of mutually exclusive options. The currently selected option is denoted by a check mark for text-displayed options or is sur-rounded by a box for graphical options.

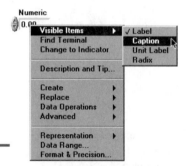

Figure 3.42

Some menu items pop up dialog boxes containing options for you to con-figure. Menu items leading to dialog boxes are denoted by ellipses (...).

Menu items without right arrowheads or ellipses are usually commands that execute immediately upon selection. A command usually appears in verb form, such as **Change to Indicator**. When selected, some commands are replaced in the menu by their inverse commands. For example, after you choose **Change to Indicator**, the menu selection becomes **Change to Control**.

Sometimes different parts of an object have different pop-up menus. For example, if you pop up on an object's label, the menu contains only a Size to Text option. Popping up elsewhere on the object gives you a full menu of options. So if you pop up and don't see the menu you want, try popping up elsewhere on the object.

3.7.2 Pop-Up Features Described

Pop-up menus allow you to specify many traits of an object. The following options appear in numerous pop-up menus, and we thought they were important enough to describe them individually. We'll let you figure out the other options, since we would put you to sleep detailing them all. Feel free to skim over this section and refer back to it when necessary.

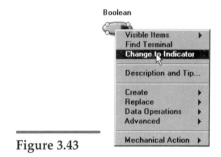

Figure 3.43

Visible Items

Many items have **Visible Items** menus with which you can show or hide certain cosmetic features like labels, captions, scrollbars, or wiring terminals. If you select **Visible Items**, you will get another menu off to the side, listing options of what can be shown (this list varies depending on the object). If an option has a check next to it, that option is currently visible; if it has no check, it is hidden. Release the mouse on an option to toggle its status.

Find Terminal and Find Control/Indicator

If you select **Find Terminal** from a front panel pop-up menu, LabVIEW will locate and highlight its corresponding block diagram terminal. If you select **Find Control/Indicator** from a block diagram pop-up menu, LabVIEW will show you its corresponding front panel object.

Change to Control and Change to Indicator

By selecting **Change to Indicator**, you can turn an existing control (an input object) into an indicator (an output object), or vice versa if you select **Change** to **Control**. When an object is a control, its pop-up menu contains the option to **Change to Indicator**. When it is an indicator, the pop-up menu reads **Change to Control**.

 *Since **Change to Control/Indicator** is an option in the pop-up menu, it is easy to accidentally select it without realizing what you've done. Because controls and indicators are not functionally interchangeable in a block diagram, the resulting errors may befuddle you.*

A control terminal in the block diagram has a thicker border than an indicator terminal. Always pay attention to whether your objects are controls or indicators to avoid confusion!

Description and Tip

Selecting this option will allow you to enter a description and a "tip." The description will appear in the Help window for that control, and the tip will show up when you place the mouse cursor over this control (this type of help is sometimes called tool-tip or hesitation help).

Create...

The **Create...** option is an easy way for you to create an property node, local variable, or reference for a given object (these advanced topics will be covered in detail in Chapter 12.)

Replace

The **Replace** option is extremely useful. It gives you access to the **Controls** or **Functions** palette (depending on whether you're in the front panel or block diagram) and allows you to replace the object you popped up on with one of your choice. Where possible, wires will remain intact.

Data Operations

The Data Operations pop-up menu has several handy options to let you manipulate the data in a control or indicator:

- **Reinitialize to Default** returns an object to its default value, while **Make Current Value Default** sets the default value to whatever data are currently there.
- Use **Cut Data**, **Copy Data**, and **Paste Data** to take data out of or put data into a control or indicator.
- **DataSocket Connection...** brings up a dialog box where you can configure this control to be connected to a DataSocket URL. We'll talk more about DataSocket in Chapter 14.

Advanced

The Advanced pop-up option gives you access to less-frequently used features that let you fine-tune the appearance and behavior of the control or indicator:

- Use **Key Navigation...** to associate a keyboard key combination with a front panel object. When a user enters that key combination while a VI is running, LabVIEW acts as if the user had clicked on that object, and the object becomes the key focus (key focus means the cursor is active in that field).
- **Synchronous Display** is a selectable option that forces LabVIEW to refresh the display of this control or indicator on every update. This adds significant overhead, so you shouldn't use this option unless you have a good reason.
- **Customize...** will bring up the *Control Editor* to allow you to customize the graphical appearance of the control. We'll talk about creating your own custom controls in Chapter 15.
- **Hide Control/Indicator**. You can choose to hide a front panel object using this option, which comes in handy when you don't want the user to see the front panel object but still need it in the diagram. If you need to show the front panel object again, you must select Show Control/Indicator on the pop-up menu of the block diagram terminal.
- **Enabled State** allows you to set a control's state as **enabled**, **disabled**, or **disabled & grayed**. This comes in handy if you still want to show a control or indicator on the front panel, but you don't want the user to use it.

There are some other pop-up options that are specific to different types of controls (numeric, Boolean, etc.), but we'll leave them for later.

Don't worry about memorizing all of these features right now—you'll come across them as you work with LabVIEW and they'll make a lot more sense!

 The same object will have a different pop-up menu in run mode than it will in edit mode. If you cannot find a certain pop-up option, it may not present for that object, you may need to switch modes, or you should pop up elsewhere on the object.

3.8 Help!

3.8.1 The Context Help Window

The LabVIEW *Context Help window* offers indispensable help information for functions, constants, subVIs, and controls and indicators. To display the window, choose **Show Context Help** from the **Help** menu or use the keyboard shortcut: <control-H> on Windows, <command-H> on the Mac, <meta-H> on the Sun, and <alt-H> on HP-UX and Linux. If your keyboard has a <help> key, you can press that instead. You can resize the Help window and move it anywhere on your screen to keep it out of the way.

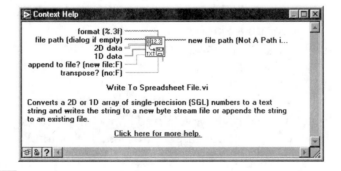

Figure 3.44
Help window.

When you hold the cursor over a function, a subVI node, or a VI icon (including the icon of the VI you have open, at the top-right corner of the VI window), the Help window shows the icon for the function or subVI with wires of the appropriate data type attached to each terminal. *Input wires point to the left, and output wires point to the right.* Terminal names appear beside each wire. If the VI has a description associated with it, this description is also displayed.

Lock Button

For some subVIs or functions, the Help window will show the names of required inputs in bold, with default values shown in parentheses. In some cases, the default value can be used and you do not need to wire an input at all. You can lock the Help window so that its contents do not change when you move the mouse by selecting **Lock Context Help** from the **Help** menu or by pressing the Lock button in the Help window.

If you position the Wiring tool over a specific node on a function or subVI, the Help window will flash the labeled corresponding node so that you can make sure you are wiring to the right place. Sometimes you may need to use the scrollbar to see all of the text in the Help window.

For VIs and functions with large numbers of inputs and outputs, the Help window can be overwhelming, so LabVIEW gives you the choice between simple or detailed views. You can use the simple view to emphasize the important connections and deemphasize less commonly used connections.

Simple/ Detailed Help Button

Switch between views by pressing the Simple/Detailed Diagram Help button on the lower-left corner of the Help window. In simple help view, required connections appear in bold text; recommended connections appear in plain text; and optional connections are not shown. Wire stubs appear in the place of inputs and outputs that are not displayed, to inform you that additional connections exist (and you can see them in the detailed help view).

In detailed help view, required connections appear in bold text; recommended connections appear in plain text; and optional connections appear as disabled text.

If a function input does not need to be wired, the default value often appears in parentheses next to the input name. If the function can accept multiple data types, the Help window shows the most common type.

Online Help

Online Help
Button

LabVIEW's Help window provides a quick reference to functions, VIs, controls, and indicators. However, there are times when you'd prefer to look at a more detailed, indexed description for information on using a VI or function. LabVIEW has extensive online help that you can access by selecting **Contents and Index...** from the **Help** menu or by pressing the Online Help button in the Help window.

You can type in a key word to search for, view an extensive keyword index, or choose from a variety of topics to browse through. You can also set your own links to online help documents, which we'll talk about in Chapter 15.

> *Currently, not all LabVIEW VIs link to online help; if this is the case the online help menu item and online help button will be grayed out*

3.9 A Word about SubVIs

If you want to take full advantage of LabVIEW's abilities, you must understand and use the hierarchical nature of the VI. A *subVI* is simply a stand-alone program that is used by another program. After you create a VI, you can use it as a subVI in the block diagram of a higher-level VI as long as you give it an icon and define its connector. A LabVIEW subVI is analogous to a subroutine in C or another text-based language. Just as there is no limit to the number of subroutines you can use in a C program, there is no limit to the number of subVIs you can use in a LabVIEW program (memory permitting, of course).

If a block diagram has a large number of icons, you can group them into a subVI to maintain the simplicity of the block diagram. You can also use one subVI to accomplish a function common to several different top-level VIs. This modular approach makes applications easy to debug, understand, and modify. We'll talk more about how to build subVIs later, but it's such an important part of the LabVIEW programming environment that we want you to keep it in mind as you're learning the basics.

3.10 Activity 3-2: Front Panel and Block Diagram Basics

In this activity, you will practice some simple exercises to get a feel for the LabVIEW environment. Try to do the following basic things on your own. If you have any trouble, glance back through the chapter for clues.

1. Open a new VI and toggle between the front panel and block diagram.

Use the keyboard shortcuts listed in the pull-down menus!

2. Resize the windows so that both front panel and block diagram are visible simultaneously. You may need to move them around.

Do this using the standard resizing technique for your platform. Or try the Tile function!

3. Drop a digital control, a string control, and a Boolean indicator on the front panel by selecting them from the Controls palette.

   To get the digital control, click on the **Numeric** palette button in the **Controls** palette and select **Digital Control** from the subpalette that appears.

Figure 3.45

Now click your mouse on the front panel in the location where you want your digital control to appear. Voilà—there it is! Now create the string control and Boolean indicator in the same fashion.

Figure 3.46

Notice how LabVIEW creates corresponding terminals on the block diagram when you create a front panel object. Also notice that floating-point numeric terminals are orange (integer numerics will be blue), strings are pink, and Booleans are green. This color-coding makes it easier for you to distinguish among data types.

4. Now pop up on the digital numeric control (by right-mouse-clicking on Windows and UNIX platforms or <command>-clicking on Mac) and select **Change to Indicator** from the pop-up menu. Notice how the appearance of the numeric's front panel changes (the little arrows go away). Also notice how the terminal on the block diagram changes (the border is much thinner for indicators). Switch the object back and forth between control and indicator until you can easily recognize the differences on both front panel and block diagram. Note that for some objects (like a few Booleans), front panel indicators and controls can look the same, but their block diagram terminals will always be different.

Positioning
Tool

5. Choose the Positioning tool from the floating **Tools** palette, and then select an object on the front panel. Hit the <delete> key to remove it. Delete all front panel objects so that you have an empty front panel and block diagram.

Enter Button

6. Drop another digital control from the **Numeric** subpalette of the **Controls** palette onto the front panel. If you don't click on anything first, you should see a little box above the control. Type Number 1, and you will see this text appear in the box. Click the Enter button on the Toolbar to enter the text. You have just created a label. Now create another digital control labeled Number 2, a digital indicator labeled N1+N2, and a digital indicator labeled N1-N2.

Operating Tool

Use the Operating tool to click on the increment arrow of Number 1 until it contains the value "4.00." Give Number 2 a value of "3.00."

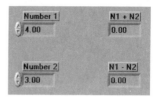

Figure 3.47

7. Switch back to the diagram. Drop an **Add** function from the **Numeric** subpalette of the **Functions** palette in the block diagram (this works just like creating front panel objects). Now repeat the process and drop a **Subtract** function.

8. Pop up on the **Add** function and select the **Visible Items>>Terminals** option (you'll notice that before you select it, the option is not checked, indicating that terminals are not currently shown). Once you show them, observe how the input and output terminals are arranged; then redisplay the standard icon by again selecting **Visible Items>>Terminals** (this time the option appears with a check mark next to it, indicating that terminals are currently shown).

9. Bring up the Help window by using either the keyboard shortcut or the **Show Context Help** command from the **Help** menu. Position the cursor over the **Add** function. The Help window provides valuable information about the function's use and wiring pattern. Now move the cursor over the **Subtract** function and watch the Help window change.

Wiring Tool

10. You may have to use the Positioning tool to reposition some of the terminals as shown in Figure 3.47. Then use the Wiring tool to wire the terminals together. First select it from the **Tools** palette; then click once on the DBL terminal and once on the appropriate terminal on the **Add** function to draw a wire. A solid orange line should appear. If you mess up and get a dashed black line instead of a solid orange one, select the wire fragment with the Positioning tool and hit the <delete> key; then try again. Click once and release to start the wire, click any time you want to add a new segment (which turns a corner), and click on a destination to finish the wire.

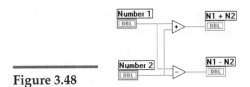
Figure 3.48

Notice that when you pass the Wiring tool over the **Add** and **Subtract** functions, little wire stubs appear showing where the terminals are located. In addition, as you pass the cursor over a terminal, its name appears in a tip strip. Like learning to type, wiring can be kind of tricky until you get the hang of it, so don't worry if it feels a little awkward right now.

11. Switch back to the front panel and pop up on the icon pane (the little window in the upper-right corner). Select **Show Connector** from the menu. Observe the connector that appears. If you can't get the pop-up menu to appear, you are probably trying to pop up in the icon pane of the block diagram.

Figure 3.49

Now pop up again on the connector and look at its menu to see the configuration options you have. The connector defines the input and output parameters of a VI so that you can use it as a subVI and pass data to it. You can choose different patterns for your connectors depending on how many parameters you need to pass. Show the icon again by selecting **Show Icon**. Remember, the icon is just the pictorial representation of a VI; when you use a VI as a subVI, you will wire to this icon in the block diagram of the top-level VI just like you wired to the **Add** function.

Run Button

12. Run the VI by clicking on the Run button. The N1+N2 indicator should display a value of "7.00" and N1-N2 should be "1.00." Feel free to change the input values and run it over and over.

13. Save the VI by selecting **Save** from the **File** menu. Call it **Add.vi** and place it in your MYWORK directory or VI library.

Congratulations! You have now mastered several important basic Lab-VIEW skills!

3.11 Wrap It Up!

The LabVIEW environment has three main parts: the *front panel, the block diagram,* and the *icon/connector.* The front panel is the user interface of the program—you can input data through *controls* and observe output data through *indicators.* When you place an object on the front panel using the **Controls** palette, a corresponding terminal appears in the block diagram, making the front panel data available for use by the program. Wires carry data between *nodes,* which are LabVIEW program execution elements. A node will execute only when all input data are available to it, a principle called *dataflow.*

A VI should also have an *icon* and a *connector.* When you use a VI as a subVI, its icon represents it in the block diagram of the VI you use it in. Its connector, usually hidden under the icon, defines the input and output parameters of the subVI.

LabVIEW has two types of menus: pull-down and pop-up. *Pull-down* menus are located in the usual menu spot at the top of your window or screen, while *pop-up* menus can be accessed by "popping up" on an object. To pop up, right-mouse-click on Windows and UNIX machines and <command>-click on the Mac, or click with the Pop-up tool. Pull-down menus tend to have more universal commands, whereas pop-up menu commands affect only the object you pop up on. Remember, when in doubt about how to do something, pop up to see its menu options!

The **Tools** palette gives you access to the special operating modes of the mouse cursor. You use these tools to perform specific editing and operation functions, similar to how you would use them in a standard paint program. You will find front panel control and indicator graphics located in the **Controls** palette and block diagram constants, functions, and structures in the **Functions** palette. These palettes often have objects nestled several layers down in *subpalettes,* so make sure you don't give up your search for an object too soon.

The Help window provides priceless information about functions and how to wire them up; you can access it from the **Help** menu. LabVIEW also contains extensive online help that you can call up from the **Help** menu or by pressing the online help button in the Help window. Between these two features, your questions should never go unanswered!

You can easily turn any VI into a subVI by creating its icon and connector and placing it in the block diagram of another VI. Completely stand-alone and modular, subVIs offer many advantages: They facilitate debugging, allow many VIs to call the same function without duplicating code, and offer an alternative to huge messy diagrams.

Don't worry if this seems like a lot to remember. It will all become natural to you as you work your way through the book.

OVERVIEW

Get ready to learn about LabVIEW's basic principles in this chapter. You will learn how to use different data types and how to build, change, wire, and run your own VIs. You will also learn some helpful shortcuts to speed your development. Make sure you understand these fundamentals before you proceed, because they are integral to all developments you will achieve in LabVIEW.

GOALS

- Become comfortable with LabVIEW's editing techniques
- Learn the different types of controls and indicators, and the special options available for each
- Master the basics of creating a VI, such as wiring and editing
- Create and run a simple VI

KEY TERMS

- Preferences
- Numeric
- String
- Boolean
- Path

- Ring control
- Format and precision
- Numeric representation
- Label
- Caption

LabVIEW
Foundations

4

4.1 Creating VIs — It's Your Turn Now!

We've gone over a few basics of the LabVIEW environment, and now we're going to show you exactly how to build your own VIs. Since people remember things better if they actually *do* them, you might step through these instructions on your computer as you read them so that you can learn the techniques more quickly.

4.1.1 Placing Items on the Front Panel

You will usually want to start your programs by "dropping" controls and indicators on a front panel to define your user inputs and program outputs. You've done this once or twice before in activities, but we'll mention it here for reference (and as a good way to start this interactive section). As you run the cursor over the **Controls** palette, you will see the names of subpalettes appear at the top of the palette. Right-click (<command>-click on the Mac) and hold down the mouse button on one of these buttons to access the corresponding subpalette. Choose the desired object from the subpalette by releasing the

75

mouse; again, you will notice that objects' names appear at the top of the sub-palette when the mouse is over them.

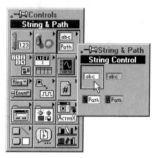

Figure 4.1

Now click on the front panel, in the place you want your object to appear . . . and there it is!

Figure 4.2

Figure 4.3

 *You can also access the **Controls** palette by popping up in an empty area of the front panel.*

Now create a new VI and drop a digital control on your front panel.

Remember, when you drop an item on the front panel, its corresponding terminal appears on the block diagram. You might find it helpful to select **Tile Left and Right** from the **Windows** menu so that you can see the front panel and block diagram windows at the same time.

4.1.2 Labeling Items

Labels are blocks of text that are the names of the specific components on front panels and block diagrams. An object first appears in the front panel window with a default name for a label (e.g, "Numeric," "String," etc.). If you want to rename the label at this time, start entering text from the keyboard. After you enter text into a label, any one of the following actions completes the entry:

- Press <enter> on the numeric keypad.
- Click on the enter button in the **Tools** palette.
- Click somewhere outside the label on the front panel or block diagram.
- Press <shift-enter> on Windows and HP or <shift-return> on Mac and Sun, from the alphanumeric keyboard.

The label appears on the corresponding block diagram terminal as well as the front panel object.

LabVIEW has two kinds of labels: owned labels and free labels. Owned labels belong to and move with a particular object; they annotate that object only. When you create a control or indicator on the front panel, a blank owned label accompanies it, awaiting input. A front panel object and corresponding block diagram terminal will have the same owned label. A free label is not associated with any particular object and can be created and deleted at will.

You can select **Visible Items>> Label** from the owning object's pop-up menu to create or change a label that isn't currently visible. You can hide owned labels, but you cannot copy or delete them independently of their owners. Structures and functions come with a default label that is hidden until you show it. You may want to edit this label to reflect the object's function in your program. You can also show labels of subVIs (which are really just their names), but you cannot edit them.

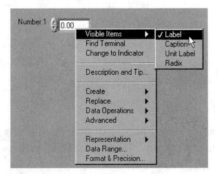

Figure 4.4

Now label the digital control you just created My Label.

Captions

In addition to labels, front panel objects can have a *caption*. A caption is very much like a label— some text that describes the control or indicator. To enter a caption for a control or indicator, pop up on it and select **Visible Items>>Caption**.

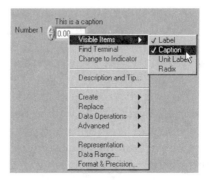

Figure 4.5

So why would you need or want a caption *and* a label? Most of the time you probably don't. In certain advanced programming situations (which we'll touch on later), you need to use a control's label to reference it, and you may want to have a separate caption. With captions, you can keep labels short ("Temperature") but enter a longer description for the caption ("This displays the current temperature in degrees Celsius"). Think of captions as a convenient way to link a comment to a front panel object.

Creating Free Labels

Free labels are not attached to any object, and you can create, move, or dispose of them independently. Use them to annotate your panels and diagrams. Use the Labeling tool to create free labels and to edit virtually any visible text.

Labeling Tool

To create a free label, select the Labeling tool from the Tools palette and click anywhere in empty space. A small, bordered box appears with a text cursor at the left margin ready to accept typed input. Type the text you want to appear in the label and enter it in one of the four ways previously described. If you do not type any text in the label, the label disappears as soon as you click somewhere else. Create a free label on the front panel that says `hippopotamus`.

4.1.3 Changing Font, Style, Size, and Color of Text

You can change text attributes in LabVIEW using the options in the Font ring on the Toolbar. Select objects with the Positioning tool or highlight text with the Labeling or Operating tools; then make a selection from the **Font** ring. The changes apply to everything selected or highlighted. If nothing is selected, the changes apply to the default font and will affect future instances of text.

Figure 4.6   `13pt Application Font   ▾`

Change your `hippopotamus` label so that it uses 18-point font.

If you select **Font Dialog...** from the menu, a dialog box appears; you can change multiple font attributes at the same time using this dialog box.

LabVIEW uses System, Application, and Dialog fonts for specific portions of its interface. These fonts are predefined by LabVIEW, and changes to them affect all controls that use them.

- The Application font is the default font, used for the **Controls** palette, the **Functions** palette, and text in new controls.
- The System font is used for menus.
- LabVIEW uses the Dialog font for text in dialog boxes.

4.1.4 Placing Items on the Block Diagram

A user interface isn't much good if there's no program to support it. You create the actual program by placing functions, subVIs, and structures on the block diagram. To do this, access the **Functions** palette just like you did the **Controls** palette. Then select the item you want from a subpalette, and click on the diagram to place it.

Drop an **Add** function from the **Numeric** subpalette of the **Functions** palette onto the block diagram.

4.1.5 Editing Techniques

Once you have objects in your windows, you will want to be able to move them around, copy them, delete them, etc. Read on to learn how.

Selecting Objects

Positioning
Tool

You must select an item before you can move it. To select something, click the mouse button while the Positioning tool is on the object. When you select an object, LabVIEW surrounds it with a moving dotted outline called a marquee, shown in Figure 4.7.

Figure 4.7
Marquee.

To select more than one object, <shift>-click on each additional object. You can also deselect a selected object by <shift>-clicking on it.

Another way to select single or multiple objects is to drag a selection rectangle around them. To do this, click in an open area with the Positioning tool and drag diagonally until all the objects you want to select lie within or are touched by the selection rectangle that appears. When you release the mouse button, the selection rectangle disappears and a marquee surrounds each selected object. The marquee is sometimes referred to as "marching ants," for obvious reasons. Once you have selected the desired objects, you can move, copy, or delete them at will.

You cannot select a front panel object and a block diagram object at the same time. However, you can select multiple objects on the same front panel or block diagram.

Clicking on an unselected object or clicking in an open area deselects everything currently selected. <shift>-clicking on an object selects or deselects it without affecting other selected objects.

Now select the digital control you created earlier.

Moving Objects

You can move an object by selecting and dragging it to the desired location.
If you hold down the <shift> key and then drag an object, LabVIEW restricts
the direction of movement horizontally or vertically (depending on which
direction you first move the object). You can also move selected objects in
small, precise increments by pressing the appropriate arrow key; hold down
the <shift> key at the same time to make the arrow keys move objects by a
larger amount.

If you change your mind about moving an object while you are dragging
it, drag the cursor outside all open windows and the dotted outline will disappear. Then release the mouse button, and the object will remain in its original location. If the object lingers at the window edge, the window will
autoscroll.

Move your digital control to the other side of the screen.

Duplicating Objects

Positioning
Tool
You can duplicate LabVIEW objects after you have selected them. From the
Edit menu, select the **Copy** option, click the cursor where you want the new
object, and then select the **Paste** option. You can also clone an object by
using the Positioning tool to <control>-click on the object if you use Windows, <option>-click if you use a Mac, <meta>-click on a Sun, and <alt>-
click on Linux machines.

Then drag the cursor away while still holding down the mouse. You will
drag away the new copy, displayed as a dotted line, while the original stays
in place. You can also duplicate front panel and block diagram objects from
one VI to another. For example, if you select a block of code on one diagram
and drag it to another, the appropriate wires will stay in place and any necessary front panel objects will be created.

*You cannot duplicate control and indicator terminals on the block diagram—you must
copy the items on the front panel.*

Copy your digital control using both methods. You should now have three digital controls labeled `Number 1`, `Number 2`, and `Number 3` on your front panel and three corresponding terminals on the block diagram. Notice how LabVIEW automatically changes the label numbers for you. To find out which one belongs to which, pop up on a control or on a terminal and select **Find Terminal** or **Find Control**. LabVIEW will find and highlight the object's counterpart.

Deleting Objects

To delete an object, select it and then choose **Clear** from the **Edit** menu or press <delete>.

By default, you can only delete controls and indicators from the front panel. If you try to delete their terminals on the block diagram, the deletes are ignored.

Although you can delete most objects, you cannot delete control or indicator components such as labels and digital displays. Instead, you can hide these components by selecting **Visible Items** from the pop-up menu and then deselecting the appropriate option.

Delete one of your digital controls.

Resizing Objects

Positioning
Tool

You can change the size of most objects. When you move the Positioning tool over a resizable object, resizing handles appear at the corners of the object, as in Figure 4.8.

Figure 4.8

When you pass the Positioning tool over a resizing handle, the cursor changes to the Resizing tool. Click and drag this cursor until the dotted border outlines the size you want (Figure 4.9).

Figure 4.9

To cancel a resizing operation, continue dragging the frame corner outside the window until the dotted frame disappears. Then release the mouse
button. The object maintains its original size.

If you hold down the <shift> key while you resize, the object will change
size only horizontally, vertically, or in the same proportions in both directions, depending on the object and in which direction you drag first.

Resize one of your digital controls.

 *Some controls and indicators don't allow certain sizing operations. For example, digital
controls can only grow horizontally (however, you can use a larger font to make the object bigger horizontally and vertically).*

Moving, Grouping, and Locking Objects

Objects can sit on top of and often hide other objects, either because you
placed them there or through some wicked twist of fate. LabVIEW has several commands in the Edit menu that move them relative to each other. You
may find these commands very useful for finding "lost" objects in your programs. If you see an object surrounded by a shadow, chances are it's sitting
on top of something. In Figure 4.10, the string control is not actually inside
the loop, it is sitting on it.

Figure 4.10

Move To Front moves the selected object to the top of a stack of objects.
Move Forward moves the selected object one position higher in the stack.
Move To Back and **Move Backward** work similarly to **Move To Front** and
Move Forward except that they move items down the stack rather than up.

Figure 4.11

On the front panel, you can also group two or more objects together. You do this by selecting the objects you want to group and choosing **Group** from the **Reorder** ring. Grouping objects will make them behave as one object when you move them, resize them, or delete them.

Figure 4.12
Two controls that are grouped.

Ungroup will break up the group back into the individual objects.

Lock objects will fix an object's size and position so it cannot be resized, moved, or deleted. This is handy if you are editing a front panel with a lot of objects and don't want to accidentally edit certain controls.

Coloring Objects

You can change the color of most LabVIEW objects, including controls, indicators, the front panel background, labels, and some block diagram elements. Not all elements can have their color changed. For example, block diagram terminals of front panel objects and wires use color codes for the type of data they carry, so you cannot change them.

Color Tool

To change the color of an object or the background window, pop up on it with the Color tool. The palette in Figure 4.13 appears in color.

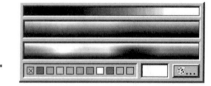

Figure 4.13

As you move through the palette while depressing the mouse button, the object or background you are coloring redraws with the color currently touched by the cursor. This gives you a preview of the object in the new color. If you release the mouse button on a color, the object retains the selected color. To cancel the coloring operation, move the cursor out of the color palette before releasing the mouse button. Selecting the option from the color palette calls up a dialog box for picking custom colors.

Color one of your digital controls by popping up and selecting the color. Then color another control using the **Tools** palette method.

Matching Colors

Color Copy
Tool

Sometimes it's hard to match a shade you've used. So you can also duplicate the color of one object and transfer it to a second object without going through the color palette. You can use the Color Copy tool on the **Tools** palette, which looks like an eye dropper (some call it the "sucker" tool), to set the active colors. Simply click with it on an object displaying the colors you want to pick up; then switch to the Color tool to color other things.

You can also access the Color Copy tool by <control>-clicking under Windows, <option>-clicking on the Mac, <meta>-clicking on Sun, and <alt>-clicking on Linux machines with the Color tool on the object whose color you want to duplicate. Then you can release the keystroke and click on another object with the Color tool; that object assumes the color you chose.

Transparency

If you select the box with a "T" in it from the color palette and color an item, LabVIEW makes the object transparent. You can use this feature to layer objects. For instance, you can place invisible controls on top of indicators, or you can create numeric controls without the standard three-dimensional container. Transparency affects only the appearance of an object. The object responds to mouse and key operations as usual.

Object Alignment and Distribution

Sometimes you want to make your VIs look just perfect, and you need a way to evenly line up and space your objects. LabVIEW's alignment and distribution functions make this easy. To align objects, select them with the Positioning tool (it's usually easiest just to drag a rectangle around them all, rather than <shift>-clicking on each one individually), then go to the Align ring, located in the Toolbar right next to the Font ring, and choose how you want them lined up. The Distribute ring works similarly to space objects evenly.

Figure 4.15
Align ring.

Figure 4.16
Distribute ring.

Be careful when you use these functions, because sometimes you'll end up with all of your objects on top of each other and you'll wonder what happened. For example, if you have three buttons in a row, and you align by Left Edges, all left edges will be flush, and all the objects will be stacked on top of each other. If this happens, select **Undo** from the **Edit** menu or use the Positioning tool to pick them off one by one.

Activity 4-1: Editing Practice

In this activity, you will practice some of the editing techniques you've just learned. Remember, just as the **Controls** palette is visible only when the front panel window is active, the **Functions** palette can only be seen when the block diagram window is up.

Positioning
Tool

1. Open **Editing Exercise.vi**, found in CH4.LLB in the EVERYONE directory or folder. The front panel of the Editing Exercise VI contains a number of LabVIEW objects. Your objective is to change the front panel of the VI shown.

2. First, you will reposition the digital control. Choose the Positioning tool from the **Tools** palette. Click on the digital control and drag it to another location. Notice that the label follows the control—the control *owns* the label. Now, click on a blank space on the panel to deselect the control; then click on the label and drag it to another location. Notice that the control does not follow. An owned label can be positioned anywhere relative to the control, but when the control moves, the label will follow.

3. Reposition the three slide switches as a group. Using the Positioning tool, click in an open area near the three switches, hold down the mouse button, and drag until all the switches lie within the selection rectangle. Click on the selected switches and drag them to a different location.

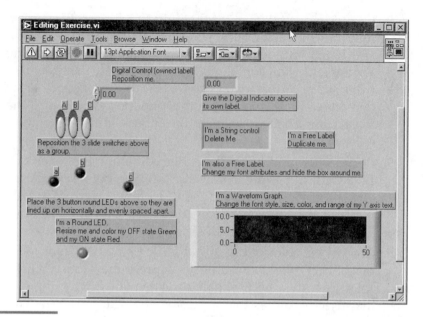

Figure 4.17

4. Delete the string control by selecting it with the Positioning tool, and then pressing <delete> or selecting **Clear** from the **Edit** menu.

5. Duplicate the free label. Hold down the <control> key on a computer running Windows, the <option> key on the Mac, the <meta> key on the Sun, or the <alt> key under Linux, then click on the free label and drag the duplicate to a new location.

Labeling Tool

Color Tool

Pop-up Tool

6. Change the font style of the free label. Select the text by using the Labeling tool. You can double-click on the text or click and drag the cursor across the text to select it. Modify the selected text using the options from the **Font** ring. Then hide the box around the label by popping up on the box with the Color tool and selecting the T (for transparent) from the color palette. Remember, use the right mouse button on Windows, Sun, and Linux, and <command>-click on Macintosh, to pop up on an item. Or you can use the Pop-up tool from the **Tools** palette and simply click on the object to access its pop-up menu.

7. Now use the **Font** ring again to change the font style, size, and color of the Y-axis text on the Waveform Graph.

Enter Button

8. Create an owned label for the digital indicator. Pop up on the digital indicator by clicking the right mouse button under Windows, Sun, and Linux or by clicking the mouse button while holding down the <command> key on the Mac; then choose **Visible Items>>Label** from the pop-up menu. Type `Digital Indicator` inside the bordered box. Press <enter> on the numeric keypad, click the enter button on the Toolbar, or click the mouse button outside the label to enter the text.

Positioning
Tool

9. Resize the round LED. Place the Positioning tool over a corner of the LED until the tool becomes the resizing cursor. Click and drag the cursor outward to enlarge the LED. If you want to maintain the current ratio of horizontal to vertical size of the LED, hold down the <shift> key while you resize

Color Tool

10. Change the color of the round LED. Using the Color tool, pop up on the LED. While continuing to depress the mouse button, choose a color from the selection palette. When you release the mouse button, the object assumes the last color you selected. Now click with the Operating tool on the LED to change its state to ON and then color the new state.

Vertical Centers
Axis

Horizontal
Centers Axis

11. Place the three LED indicators so that they are aligned horizontally and evenly spaced. Using the Positioning tool, click in an open area near the LEDs and drag a rectangle around them. Align them horizontally by choosing the Vertical Centers axis from the **Align** ring in the Toolbar. Then space the LEDs evenly by choosing Horizontal Centers axis from the **Distribute** ring.

12. Your panel should now look something like the one shown in Figure 4.18.

13. Close the VI by selecting **Close** from the **File** menu. Do not save any changes. Pat yourself on the back—you've mastered LabVIEW's editing techniques!

4.2 Basic Controls and Indicators and the Fun Stuff They Do

We're now going to talk a bit about the goodies contained in palettes of the **Controls** palette. LabVIEW has four types of simple controls and indicators: *numeric, Boolean, string,* and the less frequently used *path.* You will also en-

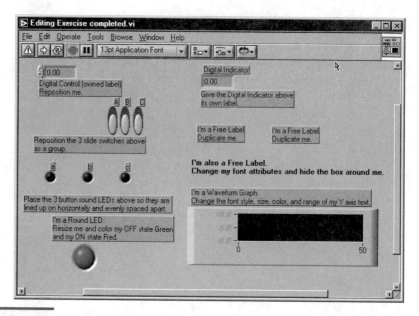

Figure 4.18

counter a few more complex data types such as arrays, clusters, tables, charts, and graphs that we will expand on later.

*The **Controls** palette is visible only when the front panel window is active, not when the block diagram is up. You can drive yourself crazy looking for it if you don't keep this in mind.*

When you need to enter numeric or text values into any controls or indicators, you can use the Operating or Labeling tool. New or changed text is not registered until you press the <enter> key on the numeric keypad, click the enter button on the Toolbar, or click outside the object to terminate the editing session.

Hitting the <enter> key (Windows and Linux) or <return> key (Macintosh and Sun) on the alphanumeric keyboard (not the one on the numeric keypad) enters a carriage return and does not register your change (unless you've configured your system otherwise). You must use the <enter> key on the numeric keypad to enter text to LabVIEW. If you must use the alphanumeric keyboard, hit <shift-enter> or <shift-return> to enter text.

4.2.1 Numeric Controls and Indicators

Numeric controls allow you to enter numeric values into your VIs; numeric indicators display numeric values you wish to see. LabVIEW has many types of numeric objects: knobs, slides, tanks, thermometers, and, of course, the simple digital display. To use numerics, select them from the **Numeric** palette of the **Controls** subpalette. All numerics can be either controls or indicators, although each type defaults to one or the other. For example, a thermometer defaults to an indicator because you will most likely use it as one. By contrast, a knob appears on the front panel as a control because knobs are usually input devices.

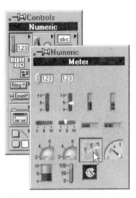

Figure 4.19

Representation

The appearance of numeric terminals on the block diagram depends on the *representation* of the data. The different representations provide alternative methods of storing data, to help use memory more effectively. Different numeric representations may use a different number of bytes of memory to store data, or they may view data as *signed* (having the capacity for negative values) or *unsigned* (having only zero or positive values). Block diagram terminals are blue for integer data and orange for floating-point data (integer data have no digits to the right of the decimal point). The terminals contain a few letters describing the data type, such as "DBL" for double-precision floating-point data.

The numeric data representations available in LabVIEW are shown in Table 4.1, along with their size in bytes and a picture of a digital control terminal with that representation.

Table 4.1

Representation	Abbreviation	Terminal	Size (bytes)
byte	I8		1
unsigned byte	U8		1
word	I16		2
unsigned word	U16		2
long	I32		4
unsigned long	U32		4
single precision	SGL		4
double precision	DBL		8
extended precision	EXT		$10^a/12^b/16^c$
complex single	CSG		8
complex double	CDB		16
complex extended	CXT		$20^a/24^b/32^c$

[a]Windows.
[b]MacOS.
[c]Unix.

You can change the representation of numeric constants, controls, and indicators by popping up on the object and selecting **Representation>>**. Remember, you pop up on a numeric control or indicator by right-mouse-clicking on machines running Windows or UNIX or <command>-clicking on MacOS. You can then choose from the palette shown in Figure 4.20.

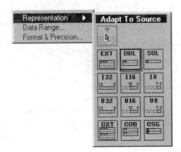

Figure 4.20

If you are concerned about memory requirements, you will want to use the smallest representation that will hold your data without losing information, especially if you are using larger structures such as arrays. **Adapt To Source** automatically assigns the representation of the source data to your indicator—a good habit to get into. LabVIEW also contains functions that convert one data type to another, which will be covered in detail in Chapters 9 and 12.

Format and Precision

LabVIEW lets you select whether your digital displays are formatted for numeric values or for time and date. If numeric, you can choose whether the notation is floating point, scientific, engineering, or relative time in seconds; you can also choose the *precision* of the display, which refers to the number of digits to the right of the decimal point, from 0 through 20. The precision affects only the display of the value; the internal accuracy still depends on the representation.

You can specify the format and precision by selecting **Format & Precision...** from an object's pop-up menu. The dialog box in Figure 4.21 appears. If you'd rather show time and date, choose **Time & Date** from the **Format** ring and your dialog box will change accordingly (Figure 4.22).

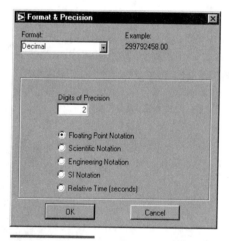

Figure 4.21

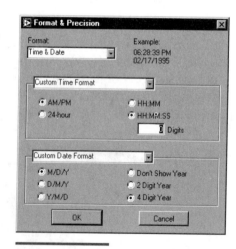

Figure 4.22

Sometimes it can be hard to discern exact values from graphical controls and indicators like graphs and thermometers. Use the pop-up option to **Visible Items>>Digital Display** to bring up a digital window next to the object and display the precise numeric value. This digital display is part of the object itself and will not have a block diagram terminal.

Numeric Range Checking

LabVIEW gives you the option to enforce a certain valid range of numeric values and increments of data. For example, you might only want an input between 0 and 100, in increments of 2. You can set range checking by popping up on the appropriate numeric value and selecting **Data Range....**

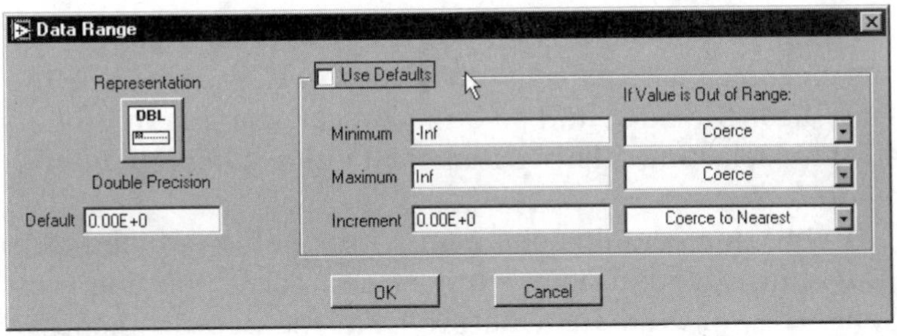

Figure 4.23

From the dialog box that appears (Figure 4.23), you can leave the default representation (by leaving the box checked), or you can change numeric representation, input maximum and minimum acceptable values, set the increments you want, and change the default value for that object, as well as select a course of action to follow if values are out of range.

- If you choose to **Ignore** out-of-range values, LabVIEW does not change or flag them. Clicking on the increment or decrement arrows of a control will change the value by the increment you set, up to the maximum values (or down to the minimum). However, you can still type in or pass a parameter out of the limits.

- If you choose to **Coerce** your data, LabVIEW will set all values below the minimum to equal the minimum and all values above the maximum to the maximum. Values in improper increments will be rounded.

Rings

Rings are special numeric objects that associate unsigned 16-bit integers with strings, pictures, or both. You can find them in the **Ring & Enum** sub-palette of the **Controls** palette. They are particularly useful for selecting mutually exclusive options such as modes of operation, calculator function, etc.

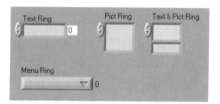

Figure 4.24

When you create a ring, you enter text or paste a picture into the ring that becomes associated with a certain number (0 for the first text message, 1 for the next, and so on). You can see this number (shown in Figure 4.24) by selecting **Visible Items>>Digital Display** from the ring's pop-up menu.

A new ring contains one item with a value of 0 and an empty display. If you want to add another number and corresponding message, select **Add Item After** or **Add Item Before** from the pop-up menu and a blank entry window will appear. You can then type in text with the Labeling tool or import a picture.

If you click on a ring with the Operating tool, you will see a list of all possible messages or pictures, with the current one checked. Rings are useful if you want a user to select an option that will then correspond to a numeric value in the block diagram. Try dropping a ring on the front panel; then show the digital display and add a few items.

4.2.2 Booleans

Booleans are named for George Boole, an English logician and mathematician whose work forms the basis for Boolean algebra. For our purposes, you can think of Boolean as just a fancy word for "on or off." Boolean data can have one of two states: true or false. LabVIEW provides a myriad switches, LEDs, and buttons for your Boolean controls and indicators, all accessible from the

Boolean subpalette of the **Controls** palette. You can change the state of a Boolean by clicking on it with the Operating tool. Like numeric controls and indicators, each type of Boolean has a default type based on its probable use (i.e., switches appear as controls, LEDs as indicators).

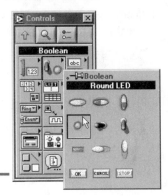

Figure 4.25

Boolean
Terminal

Boolean terminals appear green on the block diagram and contain the letters "TF."

 Control terminals have thick borders, whereas indicator borders are thin. It is very important to distinguish between the two since they are not functionally equivalent (Control = Input = data source and Indicator = Output = data sink, so they are not interchangeable).

Labeled Buttons

LabVIEW has three buttons with text messages built into them: the OK, Cancel, and Stop buttons.

Not just these three, but all Booleans also have a **Visible Items>>Boolean Text** option that will display the word "ON" or "OFF," depending on their states. This text is merely informative for the user. Each labeled button can contain two text messages: one for the TRUE state and one for the FALSE state. When you first drop buttons, the TRUE state says "OK" and the FALSE state says "OFF." You can then use the Labeling tool to change each message.

Mechanical Action

A Boolean control has a handy pop-up option called **Mechanical Action**, which lets you determine how the Boolean behaves when you click on it (e.g., whether the value switches when you press the mouse button, switches when you release it, or changes just long enough for one value to be read and then returns to its original state). Mechanical Action is covered in more detail in Chapter 8.

Customizing Your Boolean with Imported Pictures

You can design your own Boolean style by importing pictures for the TRUE and FALSE state of any of the Boolean controls or indicators. You can learn more about how to do this in Chapter 15.

4.2.3 Strings

Simply put, string controls and indicators display text data. Strings most often contain data in ASCII format, the standard way to store alphanumeric characters. String terminals and wires carrying string data appear pink on the diagram. The terminals contain the letters "abc." You can find strings in the **String & Path** subpalette of the **Controls** palette.

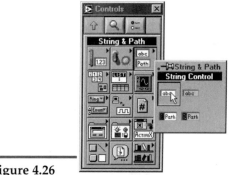

Figure 4.26

Although string controls and indicators can contain numeric characters, they do not contain numeric data. You cannot do any numerical processing on string data; that is, you can no more add an ASCII "9" character than you can an "A." If you need to use numeric information that is stored in string format (to perform arithmetic, for example), you must first convert it to numeric format using the appropriate functions (see Chapter 9).

String controls and indicators are fairly simple. Their pop-up menus contain few special options. We'll talk more about strings and their more complex cousin, the table, in Chapter 9.

4.2.4 Paths

You use *path* controls and indicators to display paths to files, folders, or directories. If a function that is supposed to return a path fails, it will return <Not A Path> in the path indicator. Paths are a separate, platform-independent data type especially for file paths, and their terminals and wires appear bluish-green on the block diagram. A path is specified by drivename followed by directory or folder names and then finally the filename itself. On a computer running Windows, directory and filenames are separated by a backslash (\); on a Mac, folder and filenames are separated by a colon (:); on UNIX machines, a forward slash (/) separates files and directories.

Figure 4.27
Windows path.

Figure 4.28
Mac path.

4.2.5 Decorations

Just for fun, you can use LabVIEW's special **Decorations** subpalette of the **Controls** palette to enhance your front panel's appearance. These decorations have a solely aesthetic function—they are the only objects from the **Controls** palette that do not have corresponding block diagram terminals.

4.2.6 Custom Controls and Indicators

To make programming even more fun, LabVIEW lets you create your own custom controls and indicators. So if LabVIEW doesn't provide exactly the one you want, make your own! You'll learn how to do that in Chapter 16.

4.2.7 Summary of Basic Controls and Indicators

Just to make sure you get your data types straight, we'll recap the four types of simple controls and indicators:

Numerics contain standard numeric values.

Booleans can have one of two states: on or off (true or false, one or zero).

Strings contain text data. Although they can contain numeric characters (0–9), you must convert string data to numeric data before you can perform any arithmetic on it.

Paths give you a platform-independent data type especially for file paths.

4.3 Wiring Up

Your neatly arranged front panel full of sharp-looking controls and indicators won't do you much good if you don't connect the wires in your diagram to create some action in your program. The following sections detail everything you need to know about wiring techniques.

Wiring Tool   You use the Wiring tool to connect terminals. The cursor point or "hot spot" of the tool is the tip of the unwound wire segment, as shown.

Wiring Cursor Hot Spot ⟶

To wire from one terminal to another, click the Wiring tool on the first terminal, move the tool to the second terminal, and then click on the second terminal. It does not matter which terminal you click on first. The terminal area blinks when the hot spot of the Wiring tool is correctly positioned on the terminal. Clicking connects a wire to that terminal.

Figure 4.28

Once you have made the first connection, LabVIEW draws a wire as you move the cursor across the diagram, as if the wire were reeling off the spool. You do not need to hold down the mouse button.

To wire from an existing wire, perform the operation described above, starting or ending the operation on the existing wire. The wire blinks when the Wiring tool is correctly positioned to fasten a new wire to the existing wire.

4.3.1 Automatic Wiring

Another way you can wire functions is to use LabVIEW's automatic wiring feature. When you select a function from the **Controls** palette, you will notice that as you drag it over the block diagram, LabVIEW draws temporary wires (they look like little "whiskers" protruding from the function) to show you valid connections. If you drag the control near a terminal or other object that has a valid input or output, you'll notice LabVIEW connects the two. Releasing the mouse button at this point "snaps" the wiring into place.

Figure 4.29
Automatic wiring.

For auto-wiring to work, you will need to drag the function very, very close to the other object you are trying to wire it to. If the automatic wiring seems awkward or doesn't work for you, don't worry about it—just stick to "manual" wiring.

You can wire directly from a terminal outside a structure to a terminal within the structure using the basic wiring operation (you'll learn more about structures in Chapter 6). LabVIEW creates a tunnel where the wire crosses the structure boundary, as shown in Figures 4.30 and 4.31. Figure 4.30 shows what the tunnel looks like as you are drawing the wire; Figure 4.31 depicts a finished tunnel.

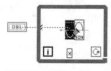

Figure 4.30

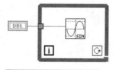

Figure 4.31

4.3.2 Wiring Complicated Objects

When you are wiring a complicated built-in node or subVI, it helps to pay attention to the wire "whiskers" and tip strips that appear as the Wiring tool approaches the icon. Wire whiskers, the truncated wires shown around the VI icon in Figure 4.32, indicate the data type needed at that terminal by their style, thickness, and color. Dots at the end of stubs indicate inputs, while outputs have no dots. Whiskers are drawn in the suggested direction to use if you want to wire clean diagrams, and also can be used for the automatic wiring feature we just described.

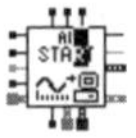

Figure 4.32

You may also want to take advantage of the Help window feature that highlights each connector pane terminal. When you pass the Wiring tool over a terminal, the corresponding Help window terminal will blink so that you can be sure you are wiring to the right spot. You can also use the Help window to determine which connections are recommended, required, or optional.

4.3.3 Bad Wires

When you make a wiring mistake, a broken wire—a black dotted line—appears instead of the usual colored wire pattern. Until all such bad wires have been vanquished, your run button will appear broken and the VI won't compile. You can remove a bad wire by selecting and deleting it. A better method is to obliterate all bad wires at once by selecting **Remove Broken Wires** from the **Edit** menu or by using the keyboard shortcut, <control-B> under Windows, or <command-B> on the Mac.

 Sometimes bad wires are mere fragments, hidden under something or so small you can't even see them. In some cases, all you need to do to fix a broken run arrow is Remove Broken Wires.

If you don't know why a wire is broken, click on the broken run button or pop up on the broken wire and choose **List Errors**. A dialog box will appear describing your problem(s).

This procedure is for removing moderately bad wires. If you have very bad and totally evil wires, then you must reboot your computer (just kidding!).

4.3.4 Wiring Tips

The following tips may make wiring a little easier for you:

- You can make a 90-degree turn in your wire, or "elbow" it, only once without clicking.
- Click the mouse to tack the wire and change direction.
- Change the direction from which the wire leaves a tack point by pressing the space bar.
- Double-click with the wiring tool to begin or terminate a wire in an open area.
- When wires cross, a small gap appears in the first wire drawn, as if it were underneath the second wire, (see Figure 4.33). You can also select **Tools>>Options...** and go to the **Block Diagram** menu, and then check the box to **Show dots at wire junctions**.
- Right-mouse-click to delete a wire while you're wiring, or <command>-click on the Mac.
- Use the Help window for more information about an object and to help you wire to the right terminal.

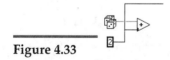

Figure 4.33

Create two numeric controls on a front panel and wire them to the inputs of the Add function. Don't wire the output just yet.

4.3.5 Wire Stretching

You can move wired objects individually or in groups by dragging the selected objects to the new location using the Positioning tool. Wires connected to the selected objects stretch automatically. If you duplicate the selected objects or move them from one diagram or subdiagram into another (for example, from the block diagram into a structure subdiagram such as a While Loop), LabVIEW leaves behind the connecting wires, unless you select them as well.

Wire stretching occasionally creates wire stubs or loose ends. You must remove these by using the **Remove Broken Wires** command from the Edit menu (or the keyboard shortcut) before the VI will execute.

Now move the **Add** function with the Positioning tool and watch how the attached wires adjust.

4.3.6 Selecting and Deleting Wires

Positioning
Tool

A wire segment is a single horizontal or vertical piece of wire. The point at which three or four wire segments join is a *junction*. A *bend* in a wire is where two segments join. A wire branch contains all the wire segments from junction to junction, terminal to junction, or terminal to terminal if there are no junctions in between. One mouse click with the Positioning tool on a wire selects a segment. A double-click selects a branch. A triple-click selects an entire wire. Press the <delete> or <backspace> key to remove the selected portion of wire.

Select and delete one of your wires; then rewire it.

4.3.7 Moving Wires

Positioning
Tool

You can reposition one or more segments by selecting and dragging them with the Positioning tool. For fine tuning, you can also move selected segments one pixel at a time by pressing the arrow keys on the keyboard. LabVIEW stretches adjacent, unselected segments to accommodate the change. You can select and drag multiple wire segments, even discontinuous segments, simultaneously. When you move a tunnel, LabVIEW normally maintains a wire connection between the tunnel and the wired node.

Move a wire segment first using the Positioning tool and then using the arrow keys.

4.3.8 Wiring to Off-Screen Areas

If a block diagram is too large to fit on the screen, you can use the scroll bars to move to an off-screen area and drag whatever objects you need to that area. Dragging the Wiring tool slightly past the edge of the diagram window while you are wiring automatically scrolls the diagram. You can also click in empty space with the Positioning tool and drag outside the block diagram, and more space will be created.

4.3.9 Adding Constants, Controls, and Indicators Automatically

Instead of creating a constant, control, or indicator by selecting it from a palette and then wiring it manually to a terminal, you can pop up on the terminal and choose **Create>> Constant**, **Create>>Control**, or **Create>>Indicator** to automatically create an object with an appropriate data type for that terminal. The new object will be automatically wired for you, assuming that makes sense. Remember this feature as you develop your programs, because it's amazingly convenient!

Create an indicator to display the results of your **Add** by popping up on the function and selecting **Create>>Indicator**. LabVIEW will create an indicator terminal wired to the **Add** output on the block diagram as well as a corresponding front panel indicator, saving you the effort.

4.4 Running Your VI

Run Button
You can run a VI using the **Run** command from the **Operate** menu, the associated keyboard shortcut, or by clicking on the Run button. While the VI is executing, the Run button changes appearance.

Run Button
(active)
The VI is currently running at its top level if the Run button is black and looks like it's "moving."

Run Button
(subVI)
The VI is executing as a subVI, called by another VI, if the Run button has a tiny arrow inside the larger arrow.

Continuous
Run Button

If you want to run a VI continuously, press the Continuous Run button, but be careful—this is not a good programming habit to get into. You can accidentally catch your program in an endless loop and have to reboot to get out. If you do get stuck, try hitting the keyboard shortcut for the Abort command: <control-.> under Windows, <command-.> on Macs, <meta-.> on Suns, and <alt-.> under Linux.

Abort Button

Press the Abort button to abort execution of the top-level VI. If a VI is used by more than one running top-level VI, it is grayed out. Using the Abort button causes an immediate halt of execution and is not good programming practice, as your data may be invalid. You should code a "soft halt" into your programs that gracefully wraps up execution. You will learn how very soon.

Pause Button

The Pause button pauses execution when you press it, and then resumes execution when you press it again.

You can run multiple VIs at the same time. After you start the first one, switch to the panel or diagram window of the next one and start it as previously described. Notice that if you run a subVI as a top-level VI, all VIs that call it as a subVI are broken until the subVI completes. You cannot run a subVI as a top-level VI and as a subVI at the same time.

4.4.1　Activity 4-2: Building a Thermometer

Now you're going to put together a VI that actually does something! This program will take a voltage reading from a channel on your data acquisition (DAQ) board if you happen to have one, or from a simulation if you don't, and display it in a thermometer on the front panel. You should have channel 0 of your board connected to a temperature sensor or similar voltage (preferably between 0 and 1 volt). If you do have a DAQ board, read the instructions that accompany it for information on how to set it up. You will also learn more about DAQ setup in Chapters 10 and 11.

Make sure you save this activity, because you will be adding to it later. If you didn't save it, you can find our version of **Thermometer.vi** *in* EVERYONE\CH4.LLB *when you need it.*

Operating Tool

Labeling Tool

1. Open a new front panel.

2. Drop a thermometer on the panel by selecting it from the **Numeric** palette of the **Controls** menu. Label it `Temperature` by typing inside the owned label box as soon as the thermometer appears on the panel. Rescale the thermometer by dragging over the "10.0" with the Operating tool or the Labeling tool and entering `100`.

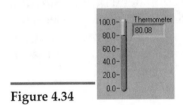

Figure 4.34

3. Build one of the block diagrams shown in Figures 4.35 and 4.36. You might find it helpful to select **Tile Left and Right** from the **Windows** menu so that you can see both the front panel and the block diagram at the same time. Build diagram A if you are using a DAQ board or diagram B if you need to use simulated data. *You do not need to build both unless you really want to practice!*

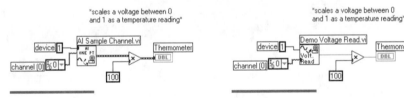

Figure 4.35
Diagram A. Using a DAQ board.

Figure 4.36
Diagram B. Using simulated data.

You will find **AI Sample Channel.vi**, used in diagram A, under **Functions>>Data Acquisition>>Analog Input**. If you do not have a board, use the **Demo Voltage Read**, located under **Functions>>Tutorial**, shown in diagram B. Remember to use the Help window to assist you in wiring to the correct terminals! Notice how, when you place the wiring tool over a node of a function, the corresponding terminal in the Help window will blink to let you see that you are wiring to the right place.

You can create a constant by popping up on the appropriate terminal of **Demo Voltage Read** or **AI Sample Channel** and selecting **Create Constant**. Just make sure you pop up on the terminal you want to wire to; otherwise you will create a constant for the wrong terminal.

Other block diagram components are described in the following list:

DAQ Channel
I/O constant

You will need a DAQ Channel I/O constant, a purple-colored icon located in the **Data Acquisition** subpalette of the **Functions** palette, to specify the analog input channel you want to use, which may or may not be channel 0 as shown. In the **Demo Voltage Read**, this value is ignored, but we left it there to simulate the real thing. *Even though the DAQ Channel I/O constant contains a numeric character, you must use a constant of DAQ Channel I/O data type rather than numeric data type, or you will not be able to wire it up.*

Operating Tool

Labeling Tool

When you drop a DAQ Channel I/O constant on the block diagram, it contains no data. You can choose a value by selecting it with the operating tool (if you've predefined a channel name in the Measurement & Automation Explorer), or you can type a value in it by choosing the Labeling tool.

If (A) you haven't configured your DAQ board in the Measurement & Automation Explorer, or (B) you are using simulated data, you will need to first pop up on the DAQ Channel I/O constant and check "Allow Undefined Names."

□

This numeric constant, located in the **Numeric** subpalette of the **Functions** palette, specifies the device number of your board, which may or may not be 1. In the **Demo Voltage Read**, this value is ignored.

When you drop a numeric constant on the block diagram, it contains the highlighted value "0." You can type and enter a new number immediately (if you don't click anywhere else first).

*Under Windows, the device number is assigned in the **Measurement & Automation Explorer** (it is probably 1 unless you have multiple boards). On UNIX machines, the device number is assigned through a configuration file. On a Mac, the device number is the slot number of your DAQ board; you can see this by opening up the **NI-DAQ** control panel. For more information on device number see the manual accompanying your DAQ board, or look in Chapter 10.*

Diagram A assumes you have a temperature sensor wired to channel 0 of your DAQ board (another voltage source is fine but will probably not provide a very accurate "temperature" unless it reads around 0.8 volts).

This numeric constant simply scales your voltage (somewhat artificially) into a "valid" temperature. If your voltage input is not between 0 and 1.0, you might want to change the value of this constant to make your output "temperature" more appropriate.

4. Use the Help window, found under the Help menu, to display the wiring pattern of the function terminals. Pay special attention to color coding to avoid broken wires. Remember, numeric data types are blue or orange, strings are pink, and Booleans are green.

Run Button

5. Run the VI by clicking on the Run button. You will see the thermometer display the voltage brought in from your board or from the simulation function. If you can't get your VI to compile, read Chapter 5, which explains debugging techniques. Then try again. If you're having trouble getting the data acquisition to work, don't worry for now and just try the VI with the simulation function (hint: use the **Replace** feature to drop in **Demo Voltage Read.vi**). We just want to give you a good sample of LabVIEW programming, not drag you through a data acquisition troubleshooting session.

6. Save the VI in your MYWORK directory by selecting **Save** from the **File** menu. Name it **Thermometer.vi**. You will be using this VI as a subVI later on in the book.

As you do more and more programming in LabVIEW, you'll find some of these shortcuts very useful for putting together your VI more quickly. Look over all of these, and refer back to them later as reminders; you're sure to find some that will make you say, "I wish I'd known that!"

Keyboard Shortcuts

Many LabVIEW menu options have keyboard shortcuts. For example, to create a new front panel window, you can select the **New** option from the **File** menu or press the keyboard equivalent <control-N> (for Windows) or <command-N> (for MacOS).

In general the keyboard auxiliary keys, <control> on Windows or <command> on MacOS, have the equivalent of <meta> key for the Sun and the <alt> key for Linux and HP-UX.

Examples

Glance through the examples that ship in LabVIEW's EXAMPLES directory. You can use these programs as is or modify them to suit your application. You can browse the examples by selecting **Examples...** from the **Help** menu.

4.5 Useful Tips

Changing Tools

When LabVIEW is in edit mode, pressing <tab> toggles through the tools. If the front panel is active, LabVIEW rotates from the Operating tool to the Positioning tool to the Labeling tool to the Color tool. If the block diagram is active, LabVIEW toggles through the tools in the same order, except that it selects the Wiring tool instead of the Color tool.

You can also press the space bar to alternate between the Operating and Positioning tools in the front panel, and between the Wiring and Positioning tools in the block diagram.

Changing the Direction of a Wire

Pressing the space bar while you wire changes the direction that the current wire branch leaves the last tack point. Thus, if you accidentally move horizontally from a tack point but want the wire to move down initially, pressing the space bar changes the initial orientation from horizontal to vertical.

Canceling a Wiring Operation

To delete a wire as you are wiring under Windows and UNIX, click the right mouse button. On a Mac, wire off the screen and click.

Removing the Last Tack Point

Clicking while wiring tacks a wire down. <control>-clicking (Windows) or <command>-clicking (MacOS) while wiring removes the last tack point, and <control>-clicking or <command>-clicking again removes the next-to-the-last tack point. If the last tack point is the terminal, <control>-clicking or <command>-clicking removes the wire.

Inserting an Object into Existing Wires

You can insert an object, such as an arithmetic or logic function, into an existing wiring scheme without breaking the wire and rewiring the objects. Pop up on the wire where you wish to insert the object, and choose **Insert>>**; then go ahead and choose the object you want to insert from the **Functions** palette that appears.

Moving an Object Precisely

You can move selected objects very small distances by pressing the arrow keys on the keyboard once for each pixel you want the objects to move. Hold down the arrow keys to repeat the action. To move the object in larger increments, hold down the <shift> key while pressing the arrow key.

Incrementing Digital Controls More Quickly

If you press the <shift> key while clicking on the increment or decrement buttons of a digital control, the display increments or decrements very quickly. The size of the increment increases by successively higher orders of magnitude; for example, by 1, then by 10, then by 100, and so on. As the range limit approaches, the increment decreases by orders of magnitude, slowing down to normal as the value reaches the limit.

Entering Items in a Ring Control

To add items quickly to ring controls, press <shift-enter> or <shift-return> after typing the item name to accept the item and position the cursor to add the next item.

Cloning an Object

To clone objects, select the objects to be copied, hold down the <control> key (Windows) or the <option> key (MacOS), and drag the duplicates to the new position. The original objects remain where they are. You can clone objects into another VI window as well.

Moving an Object in Only One Direction

If you hold down the <shift> key while moving or cloning objects, LabVIEW restricts the direction of movement horizontally or vertically, depending on which direction you move the mouse first.

Matching the Color

To pick a color from an object, click on the object with the Color Copy tool. Then color other objects by clicking on them using the Color tool.

Replacing Objects

You can easily replace a front panel or block diagram object by popping up on it and selecting **Replace>>**. A **Controls** or **Functions** palette will appear (depending on which window you're in), and you can choose a new object or function. The new one will replace the old, and any wires that are still legal will remain intact.

Making Space

To add more working space to your panel or diagram window, control-drag out a region with the Positioning tool that goes beyond the window border. The front panel or block diagram will scroll and you will see a rectangle marked by a dotted line, which defines your new space.

Designing Custom Palettes

If you're using a certain front panel object or function very often in your application, take advantage of the "sticky" palettes and keep the palette that contains your object open. Just keep the palette open by releasing the mouse button on its thumbtack or by left-clicking on the palette button instead of popping it up at the parent palette. LabVIEW also gives you the option of adding VIs or custom controls you've created to the standard palettes you're using for quick access. To create a custom palette, follow these steps:

1. Open the palette you wish to customize
2. Click on the "Options" button for that palette (top row, third button on every palette)
3. Select "Edit Palettes..." from the dialog box.
4. When you're finished, save the new palettes by giving them a name. Later you can switch between your custom palettes, the default palettes, and others by choosing the Options from the palette.

Configuring Your Preferences

LabVIEW has many preferences and options you can configure to suit your taste and convenience by selecting **Options...** from the **Tools** menu. You can select which preference menu to view from the menu ring at the top of the **Options** dialog window.

Select **Options...** from the **Tools** menu and browse through the different options available to you. If you want to know more about options, look in the LabVIEW manuals or online help.

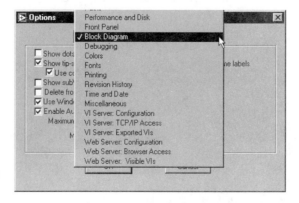

Figure 4.37

4.6   Wrap It Up!

LabVIEW has special editing tools and techniques fitting to its graphical environment. The Operating tool changes an object's value. The Positioning tool selects, deletes, and moves objects. The Wiring tool creates the wires that connect diagram objects. The Labeling tool creates and changes owned and free labels. Owned labels belong to a particular object and cannot be deleted or moved independently, while free labels have no such restrictions.

LabVIEW has four types of simple controls and indicators: *numeric, Boolean, string,* and *path.* Each holds a separate data type and has special pop-up options. Control and indicator terminals and wires on the block diagram are color coded according to data type: floating-point numbers are orange, integer numbers are blue, Booleans are green, strings are pink, and paths are bluish-green.

You place objects on the front panel or block diagram using the **Controls** or **Functions** palette, respectively. You can also access these palettes by popping up in an empty section of the panel or diagram.

To run your VI, click on the run button or select **Run** from the **Operate** menu. If your run button is broken, it means something is wrong in your VI. Read the next chapter to learn good debugging techniques.

4.7   Additional Activities

Activity 4-3: Comparison Practice

Build a VI that compares two input numbers. If they are equal, an LED on the front panel turns on. Name it **Comparison Practice.vi**.

Figure 4.38

Activity 4-4: Very Simple Calculator

Build a VI that adds, subtracts, multiplies, and divides two input numbers and displays the result on the front panel. Use the front panel in Figure 4.39 to get started. Call it **Very Simple Calculator.vi**.

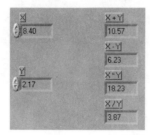

Figure 4.39

If you get stuck, you can find the answers in EVERYONE \CH4.LLB.

OVERVIEW

In this chapter, you will learn more fundamentals of the LabVIEW environment. We'll talk about load and save basics, LabVIEW's special library files, debugging features, subVI implementation, and documentation procedures.

GOALS

- Be able to load and save your VIs (and then save regularly!)
- Learn how to make LabVIEW's powerful debugging features work for you
- Create your very first subVI and understand how it's used
- Document your achievements for all the world to see

KEY TERMS

- VI library
- Broken VI
- Single-step mode
- Node
- Execution highlighting

- Probe
- Breakpoint
- SubVI
- Icon editor
- Required, recommended, and optional inputs

Yet More
Foundations

5

5.1 Loading and Saving VIs

Obviously, you will be loading and saving VIs quite a bit during your development. LabVIEW has many features that can accommodate your file storage needs, and this section discusses how you can make them work for you.

You can load a VI by selecting **Open** from the **File** menu, and then choosing the VI from the dialog box that appears. As the VI loads, you will see a status window that describes the VIs that are currently being loaded and allows you to cancel the loading process. You can load a specific VI and launch LabVIEW at the same time by double-clicking on the VI's icon, or, on Windows and Macintosh operating systems, by dragging the icon on top of the LabVIEW icon.

Save VIs by selecting **Save** (or a similar option) from the **File** menu. LabVIEW then pops up a file dialog box so you can choose where you want to save. If you save VIs as individual files, they must conform to the file naming restrictions of your operating system (e.g., MacOS 9.x and earlier limits filenames to 31 characters). Some LabVIEW projects require many VIs (hundreds) that may take up a good deal of space. To avoid these restrictions, you can save VIs in a compressed form in a special LabVIEW file called a *VI library*, described later in this chapter.

Keep in mind that LabVIEW references VIs by name. You cannot have two VIs with the same name in memory at one time. When searching for a VI of a given name, LabVIEW will load the first VI it finds, which may not be the one you intended.

Note that an asterisk (*) marks the titles of VIs that you have modified but have not yet saved. We're sure you already know about saving your work constantly, but we'd like to stress the importance of saving frequently and backing up everything you do on a computer—you never know when lightning will strike (literally)!

Figure 5.1

Never save your VIs in the vi.lib directory. This directory is updated by National Instruments during new version releases of LabVIEW, and if you put anything in there you may lose your work.

5.1.1 Save Options

You can save VIs with one of four save options in the **File** menu.

Select the **Save** option to save a new VI and then specify a name for the VI and its destination in the disk hierarchy; or use this option to save changes to an existing VI in a previously specified location.

Select the **Save As...** option to rename the VI in memory and to save a copy of the VI to disk under the new name. If you enter a new name for the VI, LabVIEW does not overwrite the disk version of the original VI. In addition, all VIs currently in memory that call the old VI now point to the new VI. If you do not change the name of the VI in the dialog box, LabVIEW prompts you to verify that you want to overwrite the original file.

Save with Options... brings up a dialog box in which you can choose to save the VI for application or development distribution, save the entire VI hierarchy, save to a previous version of LabVIEW, or do a custom save. You also have the option to save VIs without block diagrams or to password-protect your diagrams, but make sure you keep an extra copy somewhere that retains the block diagram and/or remember your password, in case you ever need to modify it! To save a specified VI or VIs to a single new location without being interrupted by multiple prompts, save **To new location—single prompt**.

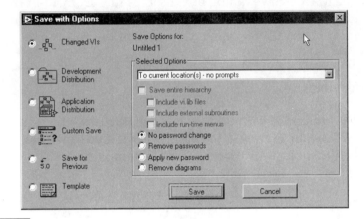

Figure 5.2

You cannot edit a VI after you save it without a block diagram. Always make a copy of the original VI before saving it without a block diagram.

5.1.2 Revert

You can use the **Revert...** option in the **File** menu to return to the last saved version of the VI you are working on. A dialog box will appear to confirm whether you want to discard any changes in the VI.

5.1.3 Save and Load Dialogs

LabVIEW supports the File dialog box format that your system uses. When you open or save a VI, your system File dialog box appears. If you click on a VI library (a special LabVIEW file storage structure), LabVIEW replaces the system dialog box with its own File dialog box so that you can select files within the library.

Because the interface for selecting and saving into a VI library is somewhat awkward with the system dialog box, you may prefer to set the LabVIEW options to use the standard LabVIEW dialog box if you commonly use VI libraries. Select **Options...** from the **Tools** menu, and then go to the **Miscellaneous** menu and uncheck the **Use native file dialogs** box.

The MacOS Save dialog box disables files from VI libraries. You will need to click the **Use LLBs** button in order to save into a VI library, or use the standard LabVIEW dialog box instead of the native one.

5.1.4 Filter Rings

At the bottom of your Save or Load dialog box, you will see a filter ring that allows you to **View All**, view only **VIs & Controls**, **VIs**, **Templates,** or **Controls** files, or just see those with a **Custom Pattern** that you specify.

 The Custom Pattern option is not available on some native-style dialog boxes.

If you choose **Custom Pattern**, another box appears in which you can specify a pattern. Only files matching that pattern will show up in the dialog box. Notice that an asterisk automatically appears inside the box; this is the "wild card character" and is considered a match with any character or characters.

5.2 VI Libraries

VI libraries are special LabVIEW files that have the same load, save, and open capabilities as directories and folders within the LabVIEW environment. You can group several VIs together and save them as a VI library. VI libraries offer several advantages and disadvantages; for example, they can contain only compressed versions of VIs, not data or other files. In addition, your operating system sees VI libraries as single files, and you can access their contents only from LabVIEW.

Read on to determine if VI libraries or individual files best meet your storage needs.

5.2.1 Reasons for Using VI Libraries

- You can use up to 255 characters to name your files.
- You can transfer a VI library to other platforms more easily than you can transfer multiple individual VIs.
- You can slightly reduce the file size of your project because VI libraries are compressed to reduce disk space requirements.

5.2.2 Reasons to Save VIs as Individual Files

- You can use the file system to manage the individual files (e.g., copy, move, rename, backup) without having to go through LabVIEW.
- You can use subdirectories.
- You can store VIs and controls in individual files more robustly than you can store your entire project in the same file.
- You can use the Professional Development System built-in source code control tools or third-party source code control tools.

Note that many of the VIs shipped with LabVIEW are kept in VI libraries so that they are stored in consistent locations on all platforms. For the activities in this book, we've asked you to save your work in a MYWORK directory so that you can access individual files more easily.

5.2.3 How to Use VI Libraries

Create a VI library from the **Save** or **Save As...** dialog box by clicking on the New VI Library button under Windows or the New... button on MacOS. If you're on a Mac configured to use native dialog boxes, you will have to click on the Use LLBs button from the save dialog box, and then select New... from the dialog box that appears.

Enter the name of the new library in the dialog box that appears, shown in Figure 5.3, figure, and append a .llb extension. Then click on the VI Library button and the library is created. If you do not include the .llb extension, LabVIEW adds it.

Figure 5.3

Usually you will create a VI library when you are saving a VI, so after the library is created, a dialog box appears to let you name your VI and save it in your new library.

Once you've created a VI library, you can save VIs in it and access them through LabVIEW much like a directory or folder, but you cannot see the individual VIs from your operating system. Remember that on Macs configured to use native dialogs, you'll have to select Use LLBs from the Save dialog box in order to access them.

5.2.4 The Edit VI Library Dialog

Since you can't edit the contents of a VI library through your operating system, you must use the **Edit VI Library** dialog box. The **Edit VI Library** dialog box, available from the **Tools** menu and shown in Figure 5.4, initially displays a list of the files in the VI library. As you move through the list, the creation and last modification dates for the selected file are shown at the bottom of the dialog box.

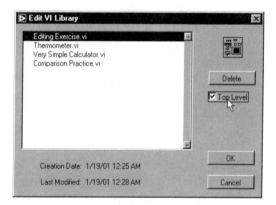

Figure 5.4

If you mark a VI as **Top Level**, it will load automatically when you open the VI library. You can have more than one top-level VI in a library. Top-level VI names will also appear in a separate section at the top of the Load dialog, making it easier for you to determine which VIs are main VIs and which are subVIs.

5.2.5 The VI Library Manager

You can use the **VI Library Manager** (from the **Tools** menu) to simplify copying, renaming, and deleting files within VI libraries as well as your file system. You also can use this tool to create new VI libraries and directories and convert VI libraries to and from directories. Creating new VI libraries and directories and converting VI libraries to and from directories is important if you need to manage your VIs with source code control tools.

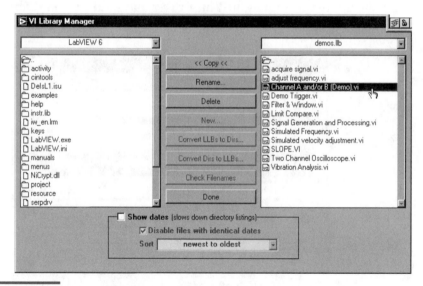

Figure 5.5

5.3 Debugging Techniques

Have you *ever* written a program with no errors in it? LabVIEW has many built-in debugging features to help you develop your VIs. This section explains how to use these conveniences to your best advantage.

5.3.1 Fixing a Broken VI

Run button
(broken)
A *broken VI* is a VI that cannot compile or run. The Run button appears as a broken arrow to indicate that the VI has a problem. It's perfectly normal for a VI to be broken while you are creating or editing it, until you finish wiring all the icons in the diagram. Sometimes you may need to **Remove Broken Wires** (found in the **Edit** menu) to clean up loose wires, but be careful not to delete wires you want!

To find out why a VI is broken, click on the broken **Run** button or select **Show Error List** from the **Windows** menu. An information box titled "Error List" appears listing all errors for the VI. You can choose to see the error list for other open VIs using a menu ring at the top of the window. To find out more about a particular error, click on it. The Error List window will display more information. To locate a particular error in your VI, double-click on the error in the list or highlight it and press the Find button. LabVIEW brings the relevant window to the front and highlights the object causing the error.

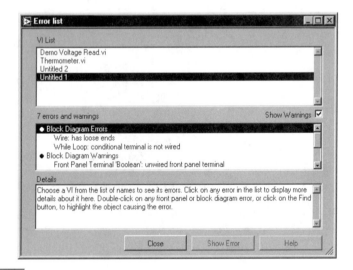

Figure 5.6

5.3.2 Warnings

Warning
Button
If you want extra debugging help, you can choose to **Show Warnings** in the Error List window by clicking in the appropriate box. A warning is something that's not illegal and won't cause a broken run arrow but does not make sense

to LabVIEW, such as a control terminal that is not wired to anything. If you have **Show Warnings** checked and have any outstanding warnings, you will see the Warning button on the Toolbar. You can click on the Warning button to see the Error List window, which will describe the warning.

You can also configure LabVIEW's options to show warnings by default. Go to the Debugging menu in the **Options** dialog box (accessed by selecting **Tools>>Options...**) and check the **Show warnings in error box by default** box.

5.3.3 Most Common Mistakes

Certain mistakes are made more frequently than others, so we thought we'd list them to make your life easier. If your run button is broken, one of these might describe your problem.

- A function terminal requiring an input is unwired. You cannot leave unwired functions on the diagram while you run a VI to try out different algorithms.
- The block diagram contains a bad wire due to a data-type mismatch or a loose, unconnected end, which may be hidden under something or so tiny that you can't see it. The **Remove Broken Wires** command from the **Edit** menu eliminates the bad wires, but you might have to look a little harder to find a data-type conflict.
- A subVI is broken, or you edited its connector after placing its icon on the diagram. Use the **Replace** or **Relink to subVI** pop-up option to re-link to the subVI.
- You have a problem with an object that is disabled, invisible, or altered using a property node (which we'll talk more about in Chapter 12).
- You have unwittingly wired two controls together or wired two controls to the same indicator. The Error List window will bear the message, "Signal: has multiple sources," for this problem. You can often solve it by changing one of those controls to an indicator.

5.3.4 Single-Stepping through a VI

Ⅱ
Pause Button

For debugging purposes, you may want to execute a block diagram node by node. *Nodes* include subVIs, functions, structures, code interface nodes

(CINs), formula nodes, and property nodes. To begin single-stepping, you can start a VI by clicking on one of the single-step buttons (instead of the Run button), pause a VI by setting a breakpoint, or click on the Pause button. To resume normal execution, hit the Pause button again.

You may want to use execution highlighting (described next) as you single-step through your VI, so you can visually follow data as it flows through the nodes.

While in *single-step mode,* press any of the three step buttons that are active to proceed to the next step. The step button you press determines how the next step will be executed.

Step Into
Button

Press the *Step Into* button to execute the first step of a subVI or structure and then pause at the next step of the subVI or structure. Or use the keyboard shortcut: down arrow key in conjunction with <control> under Windows and <command> on Macs (<meta> on Suns and <alt> on HP workstations).

Step Over
Button

Press the *Step Over* button to execute a structure (sequence, loop, etc.) or a subVI and then pause at the next node. Or use the keyboard shortcut: right arrow key in conjunction with <control> under Windows and <command> on Macs.

Step Out Button

Press the *Step Out* button to finish executing the current block diagram, structure, or VI and then pause. Or use the keyboard shortcut: up arrow key in conjunction with <control> under Windows and <command> on Macs.

5.3.5 Execution Highlighting

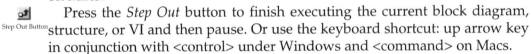
Execution
Highlighting
Button

Sometimes it's nice to see exactly where your data is and what's happening to it. In LabVIEW, you can view an animation of VI block diagram execution. To enable this mode, click on the *Execution Highlighting* button in the Toolbar.

As data pass from one node to another, the movement of data is marked by bubbles moving along the wires. You will notice that highlighting greatly reduces the performance of a VI. Click again on the Execution Highlighting button to resume normal execution. Figure 5.7 shows a VI running with execution highlighting enabled.

Node values are automatically shown during execution highlighting, if you select **Auto probe** during execution highlighting from the **Debugging** menu of the **Options** dialog.

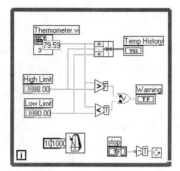

Figure 5.7

You commonly use execution highlighting in conjunction with single-step mode to gain an understanding of how data flow through nodes. When these two modes are used together, execution glyphs on the subVI's icon indicate which VIs are running and which are waiting to run.

5.3.6 Using the Probe

Use the *probe* to check intermediate values in a VI that executes but produces questionable or unexpected results. For instance, assume you have a diagram with a series of operations, any one of which may be the cause of incorrect output data. To fix it, you could create an indicator to display the intermediate results on a wire, or you can leave the VI running and simply use a probe. To access the probe, select the Probe tool from the **Tools** palette and click its cursor on a wire, or pop up on the wire and select Probe. The probe display, which is a floating window, first appears empty if your VI is not running. When you run the VI, the probe display shows the value carried by its associated wire.

You can use the probe with execution highlighting and single-step mode to view values more easily. Each probe and the wire it references are automatically numbered uniquely by LabVIEW to help you keep track of them. A probe's number will not be visible if the name of the object probed is longer than the Probe window itself; if you lose track of which probe goes

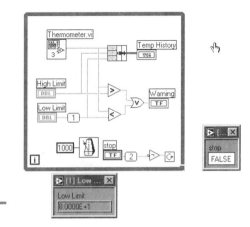

Figure 5.8

with which wire, you can pop up on a probe or a wire and select **Find Wire** or **Find Probe**, respectively, to highlight the corresponding object.

You cannot change data with the probe.

You can also probe using any conventional indicator by selecting **Custom Probe>** from the wire's pop-up menu, and then choosing the desired indicator with which to probe. For example, you could use a chart to show the progress of a variable in a loop, since it displays past values as well as current ones. LabVIEW won't let you select an indicator of a different data type than the wire.

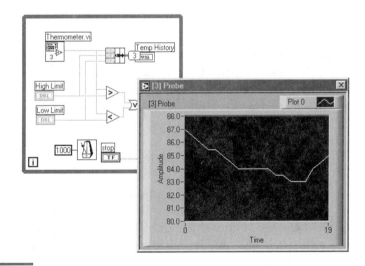

Figure 5.9

5.3.7 Setting Breakpoints

Don't panic—*breakpoints* do not "break" a VI; they only suspend its execution so that you can debug it. Breakpoints are handy if you want to inspect the inputs to a VI, node, or wire during execution. When the diagram reaches a breakpoint, it activates the pause button; you can single-step through execution, probe wires to see their data, change values of front panel objects, or simply continue running by pressing the Pause button or the Run button.

Breakpoint
Tool

To set a breakpoint, click on a block diagram object with the Breakpoint tool from the **Tools** palette. Click again on the object to clear the breakpoint. The appearance of the breakpoint cursor indicates whether a breakpoint will be set or cleared.

Set Breakpoint Cursor   Clear Breakpoint Cursor

Depending on where they are placed, breakpoints behave differently.

- If the breakpoint is set on a *block diagram,* a red border appears around the diagram and the pause will occur when the block diagram completes.
- If the breakpoint is set on a *node,* a red border frames the node and execution pauses just before the node executes.
- If the breakpoint is set on a *wire,* a red bullet appears on the wire and any attached probe will be surrounded by a red border. The pause will occur after data have passed through the wire.

When a VI pauses because of a breakpoint, the block diagram comes to the front, and the object causing the break is highlighted with a marquee.

Breakpoints are saved with a VI but only become active during execution.

5.3.8 Suspending Execution

You can also enable and disable breakpoints with the **Suspend when Called** option, found in the **Execution** menu of **VI Properties . . .** (which you access from the icon pane pop-up menu in a VI's front panel). **Suspend when Called** causes the breakpoint to occur at all calls to the VI on which it's set. If a subVI is called from two locations in a block diagram, the subVI break-point suspends execution at both calls.

If you want a breakpoint to suspend execution only at a particular call to the subVI, set the breakpoint using the **SubVI Node Setup...** option. Pop up on the subVI icon (in the block diagram of the calling VI) to access this option. You will learn more about VI setup options in Chapter 13.

5.3.9 Activity 5-1: Debugging Challenge

In this activity, you will troubleshoot and fix a broken VI. Then you will practice using other debugging features, including execution highlighting, single-step mode, and the probe.

1. Open up the VI called **Debug Exercise.vi**, located in EVERYONE\
   CH5.LLB.

2. Switch to the block diagram. Notice that the run arrow is broken. You must find out why and rectify the situation so the VI will run.

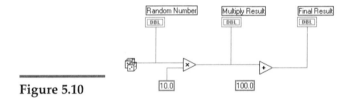

Figure 5.10

3. Click on the broken run arrow. An Error List dialog box (Figure 5.11) appears describing the errors in the VI.

4. Click on the "Add: contains unwired or bad terminal" error. The Error List window will give you a more detailed description of the error. Now double-click on the error, or click the Find button. Lab-VIEW will highlight the offending function in the block diagram to help you locate the mistake.

5. Draw in the missing wire. The Run button should appear solid. If it doesn't, try to **Remove Broken Wires**.

If you can't find the missing wire, think about how many inputs the Add function should have.

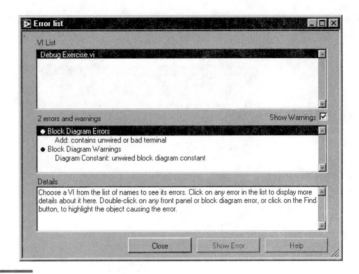

Figure 5.11

6. Switch back to the front panel and run the VI a few times.

Execution Highlighting Button

7. Tile the front panel and block diagram (using the **Tile** command under the **Windows** menu) so you can see both at the same time. Enable execution highlighting and run the VI in single-step mode by pressing the appropriate buttons on the Toolbar in the block diagram.

Step Into Button

Step Over Button

Step Out Button

8. Click the Step Over button each time you wish to execute a node (or click the Step Out button to finish the block diagram). Notice that the data appear on the front panel as you step through the program. First, the VI generates a random number and then multiplies it by 10.0. Finally, the VI adds 100.0 to the multiplication result. Notice how each of these front panel indicators is updated as new data reach their block diagram terminals, a perfect example of dataflow programming. Remember, you can exit single-step mode and complete the VI by pressing the pause button. Also notice that the tip strips describing the single-step buttons change text to give you an exact description of what they will do when you click them, given the context of where you are.

9. Now enable the probe by popping up on any wire segment and selecting **Probe**.

10. Step through the VI again and note how the probe displays the data carried by their corresponding wire.

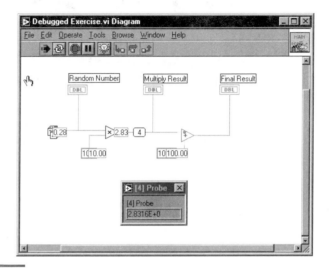

Figure 5.12

11. Turn off execution highlighting by clicking its button. You're almost done.

12. Save your finished VI in your MYWORK directory by selecting the **Save As . . .** option from the **File** menu so you won't overwrite the original. Name it **Debugged Exercise.vi**. If you're feeling adventuresome, or you think VI libraries will suit your needs, try creating a VI library and saving your work again in it for practice.

13. Close the VI by selecting **Close** from the **File** menu. Good job!

5.4 Creating SubVIs

Much of LabVIEW's power and convenience stems from its modularity. You can build the parts of your program one complete module at a time by creating subVIs. A subVI is simply a VI used in (or called by) another VI. A subVI node (comprised of icon/connector in a calling VI block diagram) is analogous to a subroutine call in a main program. A block diagram can contain several identical subVI nodes that call the same subVI several times.

You can use any VI as a subVI in the block diagram of another VI, provided its icon has been created and its connector assigned. Drop existing VIs on a block diagram to use as subVIs with the **Select a VI...** button in the **Functions** palette. Choosing this option produces a file dialog box from which you can select any VI in the system; its icon will appear on your diagram.

Figure 5.13

 A VI can't call itself directly by calling itself as a subVI. If you really need to do this, and implement recursion, you can have a VI call itself indirectly using a VI reference, which we'll talk about in Chapter 12.

5.4.1 Creating a SubVI from a VI

Before you use a VI as a subVI, you must supply a way for the VI to receive data from and pass data to the calling VI. To do this, you need to assign the VI controls and indicators to terminals on its connector pane, and you must create an icon to represent the VI.

Designing the Icon

Every subVI must have an icon to represent it in the block diagram of a calling VI; the icon is its graphical symbol. You can create the icon by selecting

Edit Icon from the pop-up menu of the icon pane in the upper right-hand corner of the front panel. You must be in edit mode to get this menu.

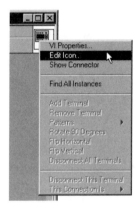

Figure 5.14

You can also access the *Icon Editor* by double clicking on the icon in the icon pane. The Icon Editor window, shown in Figure 5.15, will appear. Use its tools to design the icon of your choice.

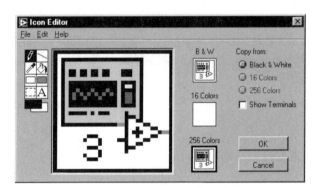

Figure 5.15

	Pencil	Draws and erases pixel by pixel.
	Line	Draws straight lines. Press <shift> to restrict drawing to horizontal, vertical, and diagonal lines.
	Dropper	Copies the foreground color from an element in the icon. Use the <shift> key to select the background color with the dropper.

⬧	Fill bucket	Fills an outlined area with the foreground color.
▫	Rectangle	Draws a rectangle in the foreground color. Double-click on this tool to frame the icon in the foreground color. Use the <shift> key to constrain the rectangle to a square shape.
▪	Filled Rectangle	Draws a rectangle bordered with the foreground color and filled with the background color. Double-click to frame the icon in the foreground color and fill it with the background color.
⬚	Selects	Selects an area of the icon for moving, cloning, or other changes.
A	Text	Enters text into the icon design. Double-click on this tool to change the font attributes.
▬	Foreground/ Background	Displays the current foreground and background colors. Click on each to get a palette from which you can choose new colors.

The buttons at the right of the editing screen perform the following functions:

- *Undo:* Cancels the last icon edit operation you performed.
- *OK:* Saves your drawing as the VI icon and returns to the front panel window.
- *Cancel:* Returns to the front panel window without saving any changes.

Although rarely done any more, you can design a separate icon for display in monochrome, 16-color, or 256-color mode (this feature is an inheritance from the old days of LabVIEW where some people would need to run it on monochrome or 16-color monitors). You can design and save each icon version separately; you can also copy an icon from color to black and white (or vice versa) using the **Copy from...** buttons. Your VIs should always have at least a black and white icon, because color icons do not show up in a palette menu, and they don't show up on black and white screens. If no black and white icon exists, LabVIEW will show a blank icon.

Assigning the Connector

Before you can use a VI as a subVI, you will need to assign connector terminals, just like you must define parameters for a subroutine in a conventional language. The connector is LabVIEW's way of passing data into and out of a subVI. The connector of a VI assigns the VI's control and indicators to input and output terminals. To define your connector, pop up in the icon pane and select **Show Connector** (when you want to see the icon again, pop up on the connector and select **Show Icon**). LabVIEW chooses a default connector based on the number of controls and indicators on the front panel. If you want a different one, choose it from the Patterns menu, obtained by popping up on the connector. You can also rotate and flip your connector if it doesn't have a convenient orientation, using commands in the connector pop-up menu.

Follow these steps to assign a terminal to a control or indicator:

1. Click on a terminal in the connector. The cursor automatically changes to the Wiring tool, and the terminal turns black as shown in Figure 5.16.

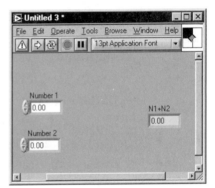

Figure 5.16

2. Click on the control or indicator you want that terminal to represent. A moving dotted line frames the control or indicator as shown in Figure 5.17.

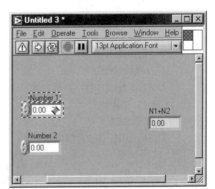

Figure 5.17

3. Click in an open area on the front panel. The dotted line disappears and the selected terminal dims as pictured in Figure 5.18, indicating that you have assigned the control or indicator to that terminal.

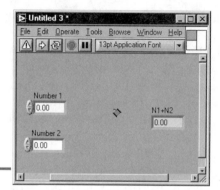

Figure 5.18

If the terminal is white or black, you have not made the connection correctly. Repeat the previous steps if necessary. You can have up to 28 connector terminals for each VI.

You can reverse the order of the first two steps.

If you make a mistake, you can **Disconnect** a particular terminal or **Disconnect All** by selecting the appropriate action from the connector's pop-up menu.

5.4.2 Creating SubVIs from a Block Diagram Selection

Sometimes you won't realize you should have used a subVI for a certain section of code until you've already built it into the main program. Fortunately, you can also create subVIs by converting a part of the code in an existing VI. Use the Positioning tool to select the section of the VI that you want to replace with a subVI, choose **Create SubVI** from the **Edit** menu, and watch LabVIEW replace that section with a subVI, complete with correct wiring and an icon. You can double-click on your new subVI to view its front panel, edit its icon, look at its connector, and save it under its new name. Use **Create SubVI** with caution, as you may cause some unexpected results. You will learn more about this handy feature in Chapter 13, where we'll describe the restrictions that apply as well as common pitfalls.

5.4.3 SubVI Help: Recommended, Required, and Optional Inputs

If you bring up the Help window on a subVI node in a block diagram, its description and wiring pattern will appear. Input labels appear on the left, while outputs appear on the right. You can also bring up the Help window on the current VI's icon pane to see its parameters and description. You will learn how to edit the description in the next section.

Built-in LabVIEW functions automatically detect if you have not wired a required input and break the VI until you do. You can also configure your subVIs to have the same types of required, recommended, and optional inputs as functions. When an input is *required*, you cannot run the VI as a subVI without wiring that input correctly. When an input or output is *recommended*, you can run the VI, but the Error List window will list a warning (if you have warnings enabled) that tells you a recommended input or output is unwired. When an input is *optional*, no restrictions are enforced, and the connection is often considered advanced.

To mark a connection as required, recommended, or optional (or see what state it's in currently), pop up on the assigned terminal in the connector pane and take a look at the **This Connection Is>>** pullout menu. A checkmark next to **Required, Recommended**, or **Optional** indicates its current state.

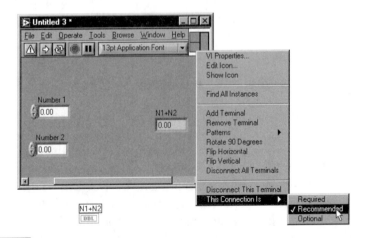

Figure 5.19

In the Help window, required connections appear bold, recommended connections are in plain text, and optional connections are grayed if you are using detailed help view. If the Help window is set to simple help view, optional connections are hidden.

5.5 Documenting Your Work

It's important to document your VIs so that others can understand them and so you don't forget why you did something or how it works. This section discusses a few ways you can document your work in LabVIEW.

5.5.1 Creating Descriptions and Tips for Individual Objects

If you want to enter a description of a LabVIEW object, such as a control, indicator, or function, choose **Description and Tip...** from the object's pop-up menu. Enter the description in the resulting Description dialog box, shown in Figure 5.20, and click OK to save it. You can also enter a tip in the Tip box.

LabVIEW displays the description text in the Help window whenever you pass the cursor over a front panel control or indicator. The tip is displayed as a "tool tip" whenever a cursor pauses over the front panel object if the VI is in run mode, regardless of whether the Help window is open.

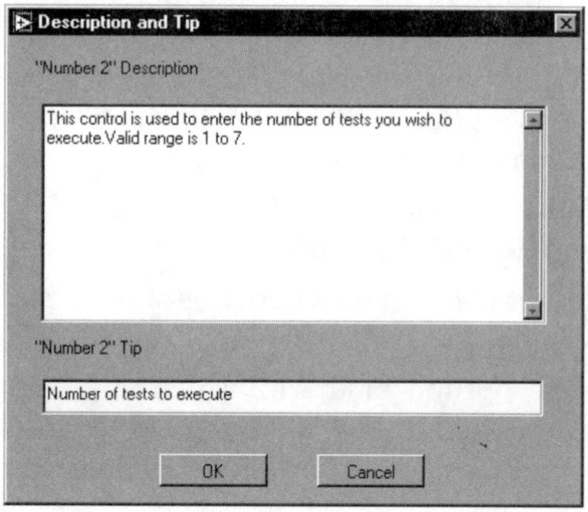

Figure 5.20

The best way to set up online help for your VIs is to enter tips and descriptions for all of their controls, indicators, and functions.

5.5.2 Documenting VIs in the VI Properties

LabVIEW also gives you an easy way to document an entire VI. Selecting **VI Properties...** from the **File** menu and choosing the **Documentation** dialog displays the VI Documentation dialog box for the current VI.

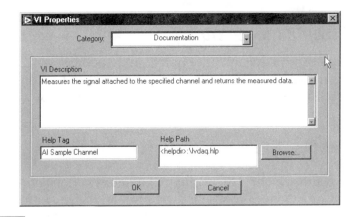

Figure 5.21

You can use the **VI Properties** dialog box to perform the following functions:

- Enter a description of the VI in the **Documentation** category. The description area has a scrollbar so that you can edit or view lengthy descriptions. When you use the VI as a subVI, the Help window will display this description when the cursor is over its block diagram icon. You can optionally enter a Help tag and a path to an external Help file.
- See a list of changes made to the VI since you last saved it by pressing the **Revision History...** button in the **General** category.
- View the path of the VI (i.e., where it is stored) in the General category
- See how much memory the VI uses in the **Memory Category**. The memory usage portion of the information box displays the disk and system memory used by the VI. The figure applies only to the amount of memory the VI is using and does not reflect the memory used by any of its subVIs.
- And perform other interesting things, which we'll leave for the Advanced section.

5.6 A Little about Printing

LabVIEW has three kinds of printing you can use when you want to make a hard copy of your work.

- You can use the **Print Window** option from the **File** menu to make a quick printout of the contents of the current window.

- You can make a comprehensive, custom printout of a VI, including information about the front panel, block diagram, subVIs, controls, VI history, and so on, by selecting the **Print...** option from the **File** menu. Choosing this option takes you through a print wizard where you can specify the format you want. You have the option of not only printing a hard copy but printing to HTML and RTF files as well. We'll talk more about this in Chapter 15.

- You can use the LabVIEW programmatic printing features to make VI front panels print under the control of your application. Select **Print at Completion** from the **Operate** menu to enable programmatic printing. LabVIEW will then print the contents of the front panel any time the VI finishes executing. If the VI is a subVI, LabVIEW prints when that subVI finishes, before returning to the caller. We'll also talk more about this option in Chapter 15.

5.7 Activity 5-2: Creating SubVIs — Practice Makes Perfect

Okay, it's time to go back to the computer again. You will turn the **Thermometer** VI you created in the last chapter into a subVI so that you can use it in the diagram of another VI.

1. Open the **Thermometer.vi** that you created in Activity 4-2. If you saved it in your MYWORK directory (or VI library) like we told you to, it should be easy to find. If you are using the sample software or can't find it, use the **Thermometer.vi** found in EVERYONE\CH4.LLB.

2. Create an icon for the VI. Pop up on the icon pane in the front panel (it doesn't work in the diagram) and select **Edit Icon...** from the menu to open the Icon Editor. Use the tools described earlier in this chapter (in Section 5.4.1) to create the icon; then click the OK button

to return to the main VI. Your icon should appear in the icon pane as shown in Figure 5.22.

Figure 5.22

3. Create the connector by popping up in the icon pane and selecting **Show Connector**. Since you have only one indicator on your front panel, your connector should have only one terminal and should appear as a white box, as shown in Figure 5.23.

Figure 5.23

4. Assign the terminal to the thermometer indicator. Using the Wiring tool (which is probably what the cursor will be automatically), click on the terminal in the connector. The terminal will turn black. Then click on the thermometer indicator. A moving dotted line will frame the indicator. Finally, click in an open area on the panel. The dotted line will disappear and the selected terminal will turn from black to gray, indicating that you have assigned the indicator to that terminal. Pop up and choose **Show Icon** to return to the icon.

5. Document the Temperature indicator by selecting **Description and Tip...** from the pop-up menu. Type in the description and tip as shown in Figure 5.24 and click OK when you're finished.

6. Document **Thermometer.vi** by selecting, from the **File** menu, **VI Properties... >>Documentation** and typing in a description of what it does, as shown in Figure 5.25. Click OK to return to the main VI.

7. Now bring up the Help window by choosing **Show Help** from the **Help** menu. When you place the cursor over the icon pane, you will see the VI's description and wiring pattern in the **Help** window. If your Temperature indicator is not labeled in the VI, it won't have a label in the Help window either.

8. If you have a printer connected to your computer, choose **Print Window** from the **File** menu to print the active window. You can decide whether you want to print the front panel or block diagram.

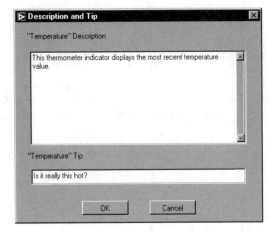

Figure 5.24

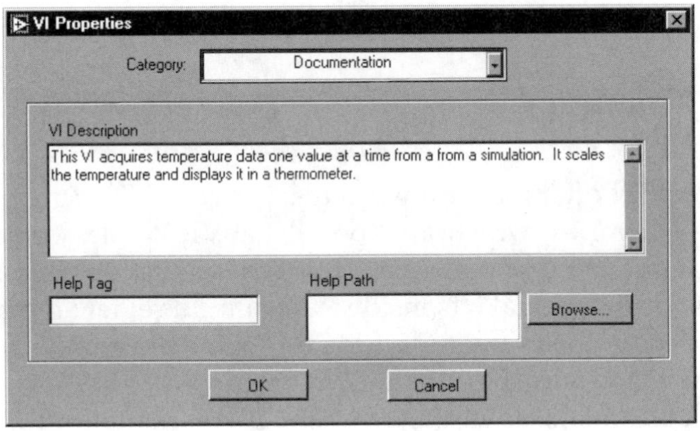

Figure 5.25

9. Save the changes by selecting **Save** from the **File** menu. Excellent work! Since you will use this VI as a subVI in the next chapter, make sure to put it in MYWORK so you can find it!

10. Just for fun, use the Positioning tool to select a portion of the block diagram as shown in Figure 5.26. Then choose **Create SubVI** from the **Edit** menu to automatically turn it into a subVI. Notice that the Temperature indicator remains part of the caller VI. Double-click on the new subVI to see its front panel.

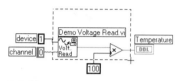
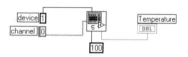

Figure 5.26
Selected portion of block diagram.

Figure 5.27
Created subVI.

11. Close both the new subVI and **Thermometer.vi**. This time, do not save any changes.

Frequently, people create an icon for a VI and forget about the connector. If you haven't assigned the connector, you will be unable to wire inputs and outputs to your VI when you try to use it as a subVI, and you may find the source of your wiring inability very difficult to locate.

5.8 Wrap It Up!

LabVIEW offers several ways to save your VIs. You may want to save them in VI libraries, which are special LabVIEW files containing groups of VIs. Your operating system sees VI libraries as single files; only LabVIEW can access the individual VIs inside. You should review the pros and cons of using VI libraries before you decide how you want to save your work. Regardless of how you save, be sure to do it frequently!

You can take advantage of LabVIEW's many useful debugging features if your VIs do not work right away. You can *single-step* through your diagram node by node, animate the diagram using *execution highlighting*, and suspend subVIs when they are called so that you can look at input and output values by setting a *breakpoint*. You can also use the *probe* to display the value a wire is carrying at any time. Each of these features allow you to take a closer look at your problem.

SubVIs are the LabVIEW equivalent of subroutines. All subVIs must have an icon and a connector. You can create subVIs from existing VIs or from a selected part of a block diagram. To prevent wiring mistakes with subVIs, use their online help and specify their inputs as *required, recommended,* or *optional.* Import an existing subVI into a calling VI's block diagram by choosing **Select a VI...** from the **Functions** palette, and then selecting the subVI you want from the dialog box. SubVIs represent one of LabVIEW's most powerful features. You will find it very easy to develop and debug reusable, low-level subVIs and then call them from higher-level VIs.

As with all programming, it is a good idea to document your work in Lab-VIEW. You can document an entire VI by entering a description under the **Documentation** dialog of the **VI Properties...** from the **File** menu. This description is also visible in the Help window if you pass the cursor over the VI's icon. You can document individual front panel objects and block diagram functions by selecting **Description and Tip...** from the pop-up menu and then entering your text in the resulting dialog box. Descriptions for front panel objects will also appear in the Help window when you pass the cursor over the object.

LabVIEW offers several options for printing VIs; you can print the active window, specify what parts of a VI you want to print (such as front panel, block diagram, or subVI information), or set the VI to print programmatically.

Congratulations! You've just covered LabVIEW's fundamental operations. Now you're ready to learn about some of LabVIEW's powerful structures and functions and how to write cool programs with them.

5.9 Additional Activities

Here's some more practice for you. If you get really stuck, look in EVERY-ONE\CH5.LLB for the solutions.

Activity 5-3: Find the Average

Create a subVI that averages three input numbers and outputs the result. Remember to create both the icon and connector for the subVI. Save the VI as **Find the Average.vi**.

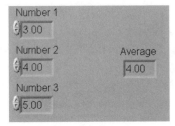

Figure 5.28

Activity 5-4: Divide by Zero (Who Says You Can't?)

Build a VI that generates a random number between 0 and 10, divides it by an input number, and displays the result on the front panel. If the input number is zero, the VI lights an LED to flag a "divide by zero" error. Save the VI as **Divide by Zero.vi**.

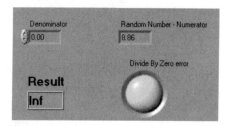

Figure 5.29

*Use the **Equal?** function found in the **Comparison** subpalette of the **Functions** palette.*

OVERVIEW

Structures, an important type of node, govern execution flow in a VI, just as control structures do in a standard programming language. This chapter introduces you to the four main structures in LabVIEW: the While Loop, the For Loop, the Case Structure, and the Sequence Structure. You will also learn how to implement lengthy formulas using the Formula Node, how to pop up a dialog box containing your very own message, and a few basics on how to control the timing of your programs. You might want to take a look at some examples of structures in EXAMPLES\GENERAL\ STRUCTS.LLB *in LabVIEW.*

GOALS

- Know the uses of the While Loop and the For Loop and understand the differences between them
- Recognize the necessity of shift registers in graphical programming
- Understand the different types of Case Structures — numeric, string, and Boolean
- Learn how to regulate execution order using Sequence Structures
- Use the Formula Node to implement long mathematical formulas
- Make LabVIEW pop up a dialog box that says anything you tell it to
- Understand how to use some of LabVIEW's simple timing functions

KEY TERMS

- For Loop
- While Loop
- Iteration terminal
- Conditional terminal
- Count terminal

- Tunnel
- Coercion dot
- Shift register
- Case Structure
- Selector terminal

- Dialog box
- Sequence Structure
- Formula Node

Controlling Program Execution with Structures

6

6.1 Two Loops

If you've ever programmed in any language, you've probably wanted to repeat a section of code. LabVIEW offers two loop structures to make this easy. You can use the *For Loop* and *While Loop* to control repetitive operations in a VI. A For Loop executes a specified number of times; a While Loop executes until a specified condition is no longer true. You can find both loops under the **Structures** subpalette of the **Functions** palette.

6.1.1 The For Loop

A *For Loop* executes the code inside its borders, called its *subdiagram*, for a total of *count* times, where the count equals the value contained in the *count terminal*. You can set the *count* by wiring a value from outside the loop to the count terminal. If you wire 0 to the count terminal, the loop does not execute.

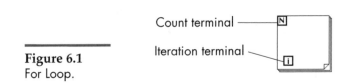

Figure 6.1
For Loop.

The *iteration terminal* contains the current number of completed loop iterations; 0 during the first iteration, 1 during the second, and so on, up to N–1 (where N is the number of times you want the loop to execute).

The For Loop is equivalent to the following pseudocode:

```
for i = 0 to N-1
        Execute subdiagram
```

6.1.2 The While Loop

The *While Loop* executes the subdiagram inside its borders until the Boolean value wired to its *conditional terminal* is FALSE. LabVIEW checks the conditional terminal value at the *end* of each iteration. If the value is TRUE, another iteration occurs. The default value of the conditional terminal is FALSE, so if you leave it unwired, the loop iterates only once.

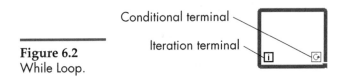

Figure 6.2
While Loop.

The While Loop's *iteration terminal* behaves exactly like the one in the For Loop.

The While Loop is equivalent to the following pseudocode:

```
Do
        Execute subdiagram
While condition is TRUE
```

You can also change the state that the conditional terminal of the While Loop checks, so that instead looping *while true*, you can have it loop *unless it's true*. To do this, you pop up on the conditional terminal, and select "**Stop if True.**" The While Loop will look like Figure 6.3.

Figure 6.3
While Loop with "Stop if True" conditional terminal.

The While Loop in Figure 6.3 is equivalent to the following pseudocode:

```
Do
        Execute subdiagram
While condition is NOT TRUE
```

6.1.3 Placing Objects inside Structures

When you first select a structure from the **Structures** subpalette of the **Functions** palette, the cursor appears as a miniature of the structure you've selected, for example, the For Loop or the While Loop. You can then click where you want one corner of your structure to be, and drag to define the borders of your structure. When you release the mouse button, the structure will appear containing all objects you captured in the borders.

Once you have the structure on the diagram, you can place other objects inside, either by dragging them in or by placing them inside when you select them from the **Functions** palette. To make it clear that you are dragging something *into* a structure, the structure's border will highlight as the object moves inside. When you drag an object *out* of a structure, the *block diagram's* border (or that of an outer structure) will highlight as the object moves outside.

You can resize an existing structure by grabbing and dragging a corner with the Positioning tool.

If you move an existing structure so that it overlaps another object, the overlapped object will be visible above the edge of the structure. If you drag an existing structure completely over another object, that object will display a thick shadow to warn you that the object is *over* or *under* rather than *inside* the structure. Both of these situations are shown in Figure 6.4.

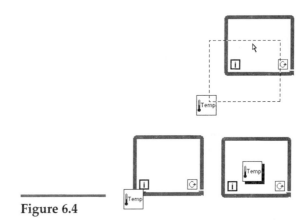

Figure 6.4

Terminals Inside Loops and Other Behavioral Issues

Data pass into and out of a loop through a little box on the loop border called a *tunnel*. Since LabVIEW operates according to dataflow principles, inputs to a loop must pass their data in before the loop executes. *Loop outputs pass data out only after tile loop completes all iterations.*

Also, according to dataflow, *you must place a terminal inside a loop if you want that terminal checked or updated on each loop iteration.* For example, the left While Loop in Figure 6.5 checks its Boolean control each time it loops. When the loop reads a FALSE value, it terminates.

Figure 6.5

If you place the terminal of the Boolean control outside the While Loop, as pictured in the right loop of Figure 6.5, you create an infinite loop or a loop that executes only once, depending on the Boolean's initial value. True to dataflow, LabVIEW reads the value of the Boolean *before* it enters the loop, not within the loop or after completion.

Similarly, the Digital Indicator in the loop in Figure 6.6 will update during each loop iteration. The Digital Indicator in the loop in Figure 6.7 will update only once, after the loop completes. It will contain the random number value from the last loop iteration.

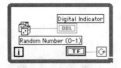

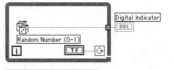

Figure 6.6   **Figure 6.7**

If you want to remove a loop without deleting its contents, pop up on its border and select **Remove While Loop** or **Remove For Loop**, respectively. If you simply highlight the loop with the Positioning tool and delete it, all of the objects inside will be deleted too.

You can generate arrays of data in a loop and store them on loop boundaries using LabVIEW's *auto-indexing* capability. We'll talk more about arrays and auto-indexing in the next chapter.

Remember, the first time through a For Loop or a While Loop, the iteration count is zero! If you want to show how many times the loop has actually executed, you must add one to the count!

6.1.4 Activity 6-1: Counting with Loops

In this activity, you get to build a For Loop that displays its count in a chart on the front panel. You will choose the Number of Iterations, and the loop will count from zero up to that number minus one (since everything is zero-based). You will then build a While Loop that counts until you stop it with a Boolean switch. Just for fun (and also to illustrate an important point), you will observe the effect of putting controls and indicators outside the While Loop.

1. Create a new panel by selecting **New VI** from the **File** menu or by clicking the **New VI** button in the LabVIEW initial dialog box.

2. Build the front panel and block diagram shown in Figures 6.8 and 6.9. The For Loop is located in the **Structures** subpalette of the **Functions** palette. You might use the **Tile Left and Right** command from the **Windows** menu so that you can see both the front panel and the block diagram at the same time.

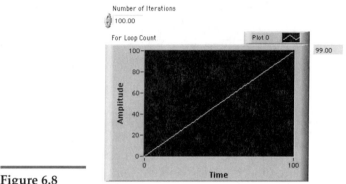

Figure 6.8

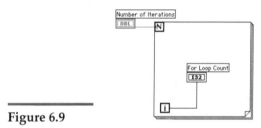

Figure 6.9

Drop a **Waveform Chart** from the **Graph** subpalette of the **Controls** palette onto your front panel. Label it <u>For Loop Count</u>. We'll talk more about charts and graphs in Chapter 8. Use a digital control from the **Numeric** subpalette for your <u>Number of Iterations</u> control.

3. Pop up on the Waveform Chart and select **AutoScale Y** from the **Y Scale** pull-out menu so that your chart will scale to fit the For Loop count. Then pop up on the chart and **Visible Items>>Digital Display**. Input a number to your <u>Number of Iterations</u> control and run the VI. Notice that the digital indicator counts from 0 to N–1, NOT 1 to N (where N is the number you specified)! Each time the loop executes, the chart plots the For Loop count on the Y axis against time on the X axis. In this case, each unit of time represents one loop iteration.

4. Notice the little gray dot present at the junction of the count terminal and the <u>Number of Iterations</u> wire. It's called a coercion dot, and we'll talk about it after this exercise. Pop up on the <u>Number of Iterations</u> control and choose **I32 Long** from the subpalette to make it go away.

5. You can save the VI if you want, but we won't be using it again. Open up another new window so you can try out the While Loop.

6. Build the VI shown in Figure 6.11. Remember, Booleans appear on the front panel in their default FALSE position.

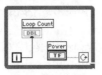

Figure 6.10   **Figure 6.11**

Operating Tool

7. Flip the switch up to its TRUE position by clicking on it with the Operating tool and run the VI. When you want to stop, click on the switch to flip it down to FALSE. Loop Count will update during each loop iteration.

8. With the switch still in the FALSE position, run the VI again. Notice that the While Loop executes once, but only once. Remember, the loop checks the conditional terminal at the *end* of an iteration, so it always executes at least once, even if nothing is wired to it.

9. Now go to the block diagram and move the Loop Count indicator outside the loop as shown in Figure 6.12. You will have to rewire the indicator; the tunnel is created automatically as the wire leaves the loop.

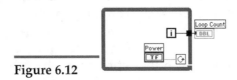

Figure 6.12

10. Make sure the switch is TRUE and run the VI. Notice that the indicator updates only after the loop has finished executing; it contains the final value of the iteration count, which is passed out after the loop completes. You will learn more about passing data out of loops in Chapter 7. *Until then, do not try to pass scalar data out of a For Loop like you just did in a While Loop, or you will get bad wires and you won't understand why.* It can easily be done, but you will have to learn a little about auto-indexing first.

11. Save the VI. Place it in your MYWORK directory and call it **Loop Count.vi**.

Abort Button

12. Now, just to demonstrate what *not* to do, drag the switch out of the loop (but leave it wired). Make sure the switch is TRUE, run the VI, and then hit the switch to stop it. It won't stop, will it? Once LabVIEW enters the loop, it will not check the value of controls outside of the loop (just like it didn't update the Loop Count indicator until the loop completed). Go ahead and hit the Abort button on the Toolbar to halt execution. If your switch had been FALSE when you started the loop, the loop would have only executed once instead of forever. Close the VI and do not save changes.

The Coercion Dot

Remember the little gray dot present at the junction of the For Loop's count terminal and the Number of Iterations wire in the last activity? It's the *coercion dot*, so named because LabVIEW is coercing one numeric representation to fit another. If you wire two terminals of different numeric representations together, LabVIEW converts one to the same representation as the other. In the previous exercise, the count terminal has a 32-bit integer representation, while the Number of Iterations control is by default a double-precision floating-point number until you change it. In this case, LabVIEW converts the double-precision floating-point number to a long integer. In doing so, LabVIEW makes a new copy of the number in memory, in the proper representation. This copy takes up space. Although the extra space is negligible for scalar numbers (single-valued data types), it can add up quickly if you are using arrays (which store multiple values). Try to minimize the appearance of the coercion dot on large arrays by changing the representation of your controls and indicators to exactly match the representation of the data they carry.

When a VI converts floating-point numbers to integers, it rounds to the nearest integer. A number with a decimal value of ".5" is rounded to the nearest even integer.

*An easy way to create a count terminal input with the correct data type and representation is to pop up on the count terminal and select **Create Constant** (for a block diagram constant) or **Create Control** (for a front panel control).*

6.2   Shift Registers

Shift registers, available for While Loops and For Loops, are a special type of variable used to transfer values from one iteration of a loop to the next. They are unique to and necessary for LabVIEW's graphical structure; we'll talk more about their uses in a little while. You create a shift register by popping up on the left or right loop border and selecting **Add Shift Register** from the pop-up menu.

Right
Terminal

Left
Terminal

A shift register comprises a pair of terminals directly opposite each other on the vertical sides of the loop border. The right terminal stores the data upon the completion of an iteration. These data are "shifted" at the end of the iteration and appear in the left terminal at the beginning of the next iteration, as shown in Figure 6.13. A shift register can hold any data type—numeric, Boolean, string, array, and so on. The shift register automatically adapts to the data type of the first object that you wire to it. It appears black when you first create it, but the shift register assumes the color of the data type wired to it.

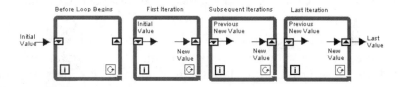

Figure 6.13
Shift registers.

You can configure the shift register to remember values from several previous iterations, as shown in Figure 6.14. This is a useful feature when you are averaging data values acquired in different iterations. To access values from previous iterations, create additional terminals by popping up on the *left* terminal and choosing **Add Element** from the pop-up menu.

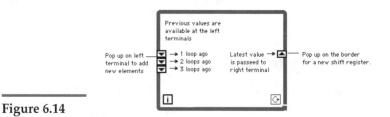

Figure 6.14

You can have many different shift registers storing many different variables on the same loop. Just pop up on the loop border and add them until you have as many pairs as you need. The left terminal will always stay parallel to its right terminal; if you move one, they both move. So if you have a lot of shift registers on your loop and can't tell exactly which ones are parallel, just select one and its partner will be automatically selected, or move one terminal a little and watch its mate follow.

Don't make the common mistake of confusing multiple variables stored in multiple shift registers with a single variable stored from multiple previous iterations in one shift register. Figure 6.15 shows the difference.

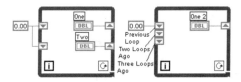

Figure 6.15
(*Left*) Two separate variables. (*Right*) Several loop values of one variable.

If you're still a little confused, don't worry. Shift registers are a completely new and different concept, unlike anything you may have encountered in a traditional programming language. Stepping through the next exercise should demonstrate them more clearly for you.

Make sure to wire directly to the shift register terminal so that you don't accidentally create an unrelated tunnel into or out of the loop.

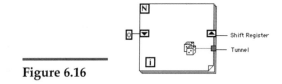

Figure 6.16

6.2.1 Activity 6-2: Shift Register Example

To give you an idea of how shift registers work, you will observe their use in accessing values from previous iterations of a loop. In this VI, you will be retrieving count values from previous loops.

1. Open **Shift Register Example.vi**, located in EVERYONE\CH6.LLB.

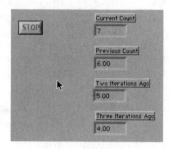

Figure 6.17

The front panel has four digital indicators. The <u>Current Count</u> indicator will display the current value of the loop count (it is wired to the iteration terminal). The <u>Previous Count</u> indicator will display the value of the loop count one iteration ago. The <u>Two Iterations Ago Count</u> indicator will display the value from two iterations ago, and so on.

2. Open the block diagram window by choosing **Show Diagram** from the **Windows** menu.

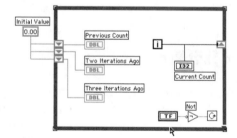

Figure 6.18

The zero wired to the left shift register terminals initializes the elements of the shift register to zero. At the beginning of the next iteration, the old <u>Current Count</u> value will shift to the top-left terminal to become <u>Previous Count</u>. <u>Previous Count</u> shifts down into <u>Two Iterations Ago Count</u>, and so on.

3. After examining the block diagram show both the panel and the diagram at the same time by choosing **Tile Left and Right** from the **Windows** menu.

Execution
Highlighting
Button

4. Enable the execution highlighting by clicking on the Execution Highlighting button.

Step Into
Button

5. Run the VI and carefully watch the bubbles. If the bubbles are moving too fast, stop the VI and click on the Step Into button to put the VI in single-step mode. Click on the button again to execute each step of the VI. Watch how the front panel indicator values change.

Notice that in each iteration of the While Loop, the VI "funnels" the previous values through the left terminals of the shift register using a first in, first out (FIFO) algorithm. Each iteration of the loop increments the count terminal wired to the right shift register terminal, <u>Current Count</u>, of the shift register. This value shifts to the left terminal, <u>Previous Count</u>, at the beginning of the next iteration. The rest of the shift register values at the left terminal funnel downward through the terminals. In this example, the VI retains only the last three values. To retain more values, add more elements to the left terminal of the shift register by popping up on it and selecting **Add Element**.

Stop the VI by pressing the <u>STOP</u> button on the front panel. If you are in single-step mode, keep pressing the step button until it completes.

6. Close the VI. Do not save any changes. Congratulations on another job well done!

6.2.2 Why You Need Shift Registers

Observe the example illustrated in Figure 6.19. In loop (A), you are creating a running sum of the iteration count. Each time through the loop, the new sum is saved in the shift register. At the end of the loop, the total sum of 45 is passed out to the numeric indicator. In loop (B), you have no shift registers, so you cannot save values between iterations. Instead, you add zero to the current "*i*" each time, and only the last value of 9 will be passed out of the loop.

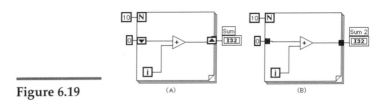

Figure 6.19

Or what about a case where you need to average values from successive loop iterations? Maybe you want to take a temperature reading once per second, and then average those values over an hour. Given LabVIEW's graphical nature, how could you wire a value produced in one loop iteration into the next iteration without using a shift register?

6.2.3 Initializing Shift Registers

To avoid unforeseen and possibly nasty behavior, you should *always initialize your shift registers* unless you have a specific reason not to and make a conscious decision to that effect. To initialize the shift register with a specific value, wire that value to the left terminal of the shift register from outside the loop, as shown in the left two loops in Figure 6.20. If you do not initialize it, the initial value will be the default value for the shift register data type the first time you run your program. In subsequent runs, the shift register will contain whatever values are left over from previous runs.

For example, if the shift register data type is Boolean, the initial value will be FALSE for the first run. Similarly, if the shift register data type is numeric, the initial value will be zero. The second time you run your VI, an uninitialized shift register will contain values left over from the first run! Study Figure 6.20 to make sure you understand what initialization does. The two loops in the left column show what happens when you run a program that contains an initialized shift register twice. The right column shows what happens if you run a program containing an uninitialized shift register two times. Note the initial values of the shift registers in the two bottom loops.

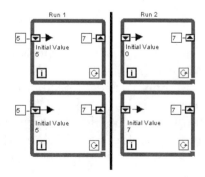

Figure 6.20

LabVIEW does not discard values stored in the shift register until you close the VI and remove it from memory. In other words, if you run a VI containing uninitialized shift registers, the initial values for the subsequent run will be the ones left over from the previous run. You seldom want this behavior, and the resulting problems can be very difficult to spot!

6.3 Case Structures

Selector
Terminal

That's enough about loops for now—let's move on to another powerful structure. A *Case Structure* is LabVIEW's method of executing conditional text, sort of like an "if-then-else" statement. You can find it in the **Structures** subpalette of the **Functions** palette. The Case Structure, shown in Figure 6.21, has two or more subdiagrams, or cases; only one of them executes, depending on the value of the Boolean, numeric, or string value you wire to the *selector terminal*.

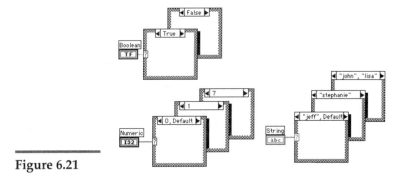

Figure 6.21

If a Boolean value is wired to the selector terminal, the structure has two cases, FALSE and TRUE.

If a numeric or string data type is wired to the selector, the structure can have from zero to almost unlimited cases. Initially only two cases are available, but you can easily add more. You can specify more than one value for a case, separated by commas, as shown in Figure 6.21. In addition, you always select a "Default" case that will execute if the value wired to the selector terminal doesn't match any of the other cases; this is very handy when you can't think of every possible case but want to specify a "catch-all" case.

When you first place it on the panel, the Case Structure appears in its Boolean form; it assumes numeric values as soon as you wire a numeric data type to its selector terminal.

Case Structures can have multiple subdiagrams, but *you can only see one case at a time*, sort of like a stacked deck of cards (unlike what appears in Figure 6.21, where we cheated and took several pictures). Clicking on the decrement (left) or increment (right) arrow at the top of the structure displays the previous or next subdiagram, respectively. You can also click on the display at the top of the structure for a pull-down menu listing all cases, and then highlight the one you want to go to. Yet another way to switch cases is to pop up on the structure border and select **Show Case>>**.

Decrement
Arrow

Increment
Arrow

If you wire a floating-point number to the selector, LabVIEW rounds that number to the nearest integer value. LabVIEW coerces negative numbers to 0 and reduces any value higher than the highest-numbered case to equal the number of that case.

You can position the selector terminal anywhere along the left border. You must always wire something to the selector terminal, and when you do, the selector automatically assumes that data type. If you change the data type wired to the selector from a numeric to a Boolean, cases 0 and 1 change to FALSE and TRUE. If other cases exist (2 through *n*), LabVIEW does not discard them, in case the change in data type is accidental. However, you must delete these extra cases before the structure can execute.

For string data types wired to case selectors, you should always specify the case values as strings between quotes. The only exception is the keyword `Default`, which should never be in quotes.

6.3.1 Wiring Inputs and Outputs

The data at all Case Structure input terminals (tunnels and selector terminal) are available to all cases. Cases are not required to use input data or to supply output data, *but if any one case outputs a value, all must output a value.* When you wire an output from one case, a little white tunnel appears in the same location on all cases. The run arrow will be broken until you wire data to this output tunnel from every case, at which time the tunnel will turn black and the run arrow will be whole again (provided you have no other errors). Make sure you wire *directly* to the existing output tunnel, or you might accidentally create more tunnels.

Why must you always assign outputs for each case, you ask? Because the Case Structure must supply a value to the next node regardless of which case executes. LabVIEW forces you to select the value you want rather than selecting one for you, as a matter of good programming practice.

6.3.2 Adding Cases

If you pop up on the Case Structure border, the resulting menu gives you options to **Add Case After** and **Add Case Before** the current case. You can also choose to copy the currently shown case by selecting Duplicate Case. You can delete the current case (and everything in it) by selecting **Remove Case**.

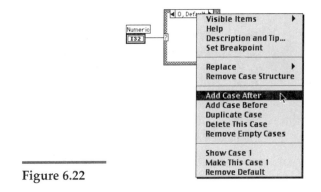

Figure 6.22

6.3.3 Dialog Boxes

Let us switch gears for a moment; we will digress from structures to tell you about dialog boxes so that you can use them in the next activity. The **One Button Dialog** and **Two Button Dialog** functions, shown in Figures 6.23 and 6.24, bring up a *dialog box* containing a message of your choice. You can find these functions in the **Time & Dialog** subpalette of the **Functions** palette. The **One Button Dialog** stays open until you click the OK button, while the **Two Button Dialog** box remains until you click either the OK or the Cancel button. You can also rename these buttons by inputting "button name" strings to the functions. These dialog boxes are *modal*; in other words, you can't activate any other LabVIEW window while they are open. They are very useful for delivering messages to or soliciting input from your program's operator.

Figure 6.23
One button dialog.

Figure 6.24
Two button dialog.

6.3.4 Activity 6-3: Square Roots

This activity will give you some practice with Case Structures and dialog boxes. You will build a VI that returns the square root of a positive input number. If the input number is negative, the VI pops up a dialog box and returns an error.

1. Open a new panel.
2. Build the front panel shown in Figure 6.25.

Figure 6.25

The Number digital control supplies the input number. The Square Root Value indicator will display the square root of the number.

3. Open the block diagram window. You will construct the code shown in Figures 6.26 and 6.27.

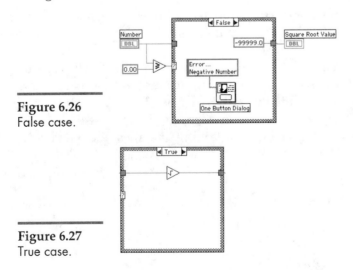

Figure 6.26
False case.

Figure 6.27
True case.

4. Place the Case Structure (**Structures** subpalette) in the block diagram window. Like you did with the For Loop and While Loop, click with the structure cursor and drag to define the boundaries you want.

The **Greater or Equal?** function returns a Boolean value, so the Case Structure remains in its default Boolean form.

Remember, you can display only one case at a time. To change cases, click on the arrows in the top border of the Case Structure. Note that Figures 6.26 and 6.27 show two cases from the same structure so you will know what to build. *Do not create two different Case Structures for this activity!*

5. Select the other diagram objects and wire them as shown in Figures 6.26 and 6.27. *Make sure to use the Help window to practice displaying terminal inputs and outputs!*

Greater or
Equal? Function

Greater or Equal? function (**Comparison** subpalette). In this activity, checks whether the number input is negative. The function returns a TRUE if the number input is greater than or equal to zero.

Square Root
Function

Square Root function (**Numeric** subpalette). Returns the square root of the input number.

Numeric
Constants

Numeric Constants (**Numeric** subpalette). "–99999.0" supplies the error case output, and "0" provides the basis for determining if the input number is negative.

One Button
Dialog Function

One Button Dialog function (Time & Dialog menu). In this exercise, it displays a dialog box that contains the message "Error . . . Negative Number."

Error...
Negative Number
String Constant

String Constant (**String** subpalette). Enter text inside the box with the Operating or Labeling tool. (You will study strings in detail in Chapter 9.)

In this exercise, the VI will execute either the TRUE case or the FALSE case of the Case Structure. If the input <u>Number</u> is greater than or equal to zero, the VI will execute the TRUE case, which returns the square root of the number. If <u>Number</u> is less than zero, the FALSE case outputs a –99999.00 and displays a dialog box containing the message "Error . . . Negative Number."

Remember that you must define the output tunnel for each case, which is why we bothered with the –99999.00 error case output. When you create an output tunnel in one case, tunnels appear at the same location in the other cases. Unwired tunnels look like hollow squares. Be sure to wire to the output tunnel for each unwired case, clicking on the tunnel itself each time, or you might accidentally create another tunnel.

6. Return to the front panel and run the VI. Try a number greater than zero and another less than zero.

7. Save and close the VI. Name it **Square Root.vi** and place it in your MYWORK directory or VI library.

Square Root VI Logic

```
If (Number >= 0) then
      Square Root Value = SQRT (Number)
Else
      Square Root Value = -99999.0
      Display Message "Error...Negative Number"
End If
```

6.3.5 The Select Function

In simple "if-then-else" cases, you might find it more convenient to use Lab-VIEW's **Select** function, which works much like a Case Structure.

Figure 6.28
Select function.

The **Select** function, found in the **Comparison** subpalette of the **Functions** palette, returns a value of **t** if the **s** input value is TRUE and returns a value of **f** if the **s** input is FALSE. This function could accomplish almost the same thing as the Case Structure in the last activity, with the exception of popping up the dialog box.

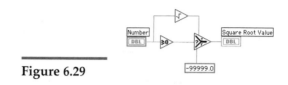

Figure 6.29

6.4 Sequence Structures

Determining the execution order of a program by arranging its elements in a certain sequence is called *control flow*. Visual Basic, C, and most other procedural programming languages have inherent control flow because statements execute in the order in which they appear in the program. LabVIEW uses the *Sequence Structure* to obtain control flow within a dataflow framework. A Sequence Structure executes frame 0, followed by frame 1, then frame 2, until the last frame executes. Only when the last frame completes do data leave the structure.

The Sequence Structure, shown in Figure 6.30, looks like a frame of film. It can be found in the **Structures** subpalette of the **Functions** palette. Like the Case Structure, only one frame is visible at a time—you must click the arrows at the top of the structure to see other frames; or you can click on the top display for a listing of existing frames, or pop up on the structure border and choose **Show Frame....**When you first drop a Sequence Structure on the block diagram, it has only one frame; thus, it has no arrows or numbers at the top of the structure to designate which frame is showing. Create new frames by popping up on the structure border and selecting **Add Frame After** or **Add Frame Before**.

Figure 6.30
Sequence structure.

You use the Sequence Structure to control the order of execution of nodes that are not data dependent on each other. Within each frame, as in the rest of the block diagram, data dependency determines the execution order of nodes. You will learn about another way to control execution order called artificial data dependency in Chapter 16.

Output tunnels of Sequence Structures can have only one data source, unlike Case Structures, whose outputs must have one data source per case. The output can originate from any frame, but keep in mind that data are passed out of the structure only when the structure completes execution entirely, not when the individual frames finish. Data at input tunnels are available to all frames.

6.4.1 Sequence Locals

To pass data from one frame to any subsequent frame, you must use a terminal called a *sequence local*. To obtain a sequence local, choose **Add Sequence Local** from the *structure border* pop-up menu. This option is not available if you pop up too close to another sequence local or over the subdiagram display window. You can drag the sequence local terminal to any unoccupied location on the border. Use the **Remove** command from the sequence local pop-up menu to remove a terminal, or just select and delete it.

When it first arrives on the diagram, a sequence local terminal is just a small yellow box. Figures 6.31 to 6.34 show the sequence local terminal in its various forms. When you wire source data to the sequence local, an outward-pointing arrow appears in the terminal of the frame containing the data source. The terminals in subsequent frames contain an inward-pointing arrow, indicating that the terminal is a data source for that frame. In frames before the source frame, you cannot use the sequence local (after all, it hasn't been assigned a value), and it appears as a dimmed rectangle.

Figure 6.31
Sequence local cannot be used.

Figure 6.32
Sequence local is sink of data.

Figure 6.33
Sequence local is source of data.

Figure 6.34
Sequence local that is a source; you cannot wire an input value to it.

6.4.2 Timing

Sometimes you will find it useful to control or monitor the timing of your VI. **Wait (ms)**, **Tick Count (ms)**, and **Wait Until Next ms Multiple**, located in the **Time & Dialog** subpalette of the Functions palette, accomplish these tasks.

Wait (ms) causes your VI to wait a specified number of milliseconds before it continues execution.

Figure 6.35
Wait (ms).

Wait Until Next ms Multiple causes LabVIEW to wait until the internal clock equals or has passed a multiple of the millisecond multiple input number before continuing VI execution; it is causing for making loops execute at specified intervals and synchronizing activities. These two functions are similar but not identical. For example, **Wait Until Next ms Multiple** will probably wait less than the specified number of milliseconds in the first loop iteration, depending on the value of the clock when it sorts (that is, how long it takes until the clock is at the next multiple and the VI proceeds). In addition, if the loop is still executing when the clock passes a millisecond multiple, the VI will wait until the clock reaches the *next* multiple, so the VI may become "out of synch" and slow down. Just make sure you take all possibilities into account when you use these functions.

Figure 6.36
Wait Until Next ms Multiple.

Tick Count (ms) returns the value of your operating system's internal clock in milliseconds; it is commonly used to calculate elapsed time, as in the next activity. Be warned that the internal clock doesn't always have great resolution—one tick of the clock can be up to 55 milliseconds (ms) on Windows 95/98, 10 ms on Windows 2000/NT, 17 ms on MacOS 9.x and earlier, and 1 ms on Linux, Solaris, and MacOS X machines; LabVIEW can't work around this operating system limitation.

Figure 6.37
Tick Count (ms).

6.4.3 Activity 6-4: Matching Numbers

Now you'll have the opportunity to work with the Sequence Structure and one of the timing functions. You will build a VI that computes the time it takes to match an input number with a randomly generated number. Remember this algorithm if you ever need to time a LabVIEW operation.

1. Open a new front panel.

2. Build the front panel shown in Figure 6.38.

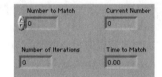

Figure 6.38

3. Change the precision of <u>Number to Match</u>, <u>Current Number</u>, and <u>Number of Iterations</u> to zero by selecting **Format & Precision...** from their pop-up menus. Enter "0" for **Digits of Precision** so that no digits are displayed to the right of the decimal point.

4. Open the diagram window and build the block diagram shown in Figures 6.39 to 6.41.

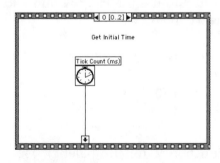

Figure 6.39

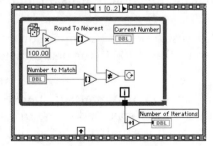

Figure 6.40

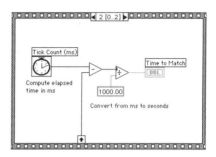

Figure 6.41

5. Place the Sequence Structure (**Structures** palette) in the diagram window. It works like the For Loop and While Loop; click with the structure cursor and drag to define the boundaries you want.

   You will have to build three separate frames of the Sequence Structure. To create a new frame, pop up on the frame border and choose **Add Frame After** from the pop-up menu.

6. Create the sequence local by popping up on the bottom border of Frame 0 and choosing Add Sequence Local from the pop-up menu. The sequence local will appear as an empty square. The arrow inside the square will appear automatically when you wire to the sequence local.

7. Build the rest of the diagram. Some new functions are described here. Make sure to use the Help window to display terminal inputs and outputs when you wire!

Tick Count
Function

Tick Count (ms) function (**Time & Dialog** palette). Returns the value of the internal clock.

Random
Number
Function

Random Number (0–1) function (**Numeric** palette). Returns a random number between 0 and 1.

Multiply
Function

Multiply function (**Numeric** palette). Multiplies the random number by 100 so that the function returns a random number between 0.0 and 100.0.

Round to
Nearest
Function

Round to Nearest function (**Comparison** palette). Rounds the random number between 0 and 100 to the nearest whole number.

Not Equal?
Function

Not Equal? function (**Comparison** palette). Compares the random number to the number specified in the front panel and returns a TRUE if the numbers are not equal; otherwise, this function returns a FALSE.

Increment
Function

Increment function (**Numeric** palette). Adds one to the loop count to produce the <u>Number of Iterations</u> value (to compensate for zero-based indexing).

In Frame 0, the **Tick Count (ms)** function returns the value of the internal clock in milliseconds. This value is wired to the sequence local, so it will be available in subsequent frames. In Frame 1, the VI executes the While Loop as long as the number specified does not match the number returned by the **Random Number (0–1)** function. In Frame 2, the **Tick Count (ms)** function returns a new time in milliseconds. The VI subtracts the old time (passed from Frame 0 through the sequence local) from the new time to compute the time elapsed, and then divides by 1,000 to convert from milliseconds to seconds.

There is a more efficient way to do this exercise, using just a two-frame Sequence Structure (try to build it, if you're feeling adventurous). However, we wanted you to practice using sequence locals.

8. Turn on execution highlighting, which slows the VI enough to see the current generated number on the front panel.

9. Enter a number inside the <u>Number to Match</u> control and run the VI. When you want to speed things up, turn off execution highlighting.

10. Use the **Save** command to save the VI in your MYWORK directory or VI library as **Time to Match.vi**, and then close it. Good job!

6.5 The Formula Node

Now that you know about LabVIEW's four main control flow structures, we'll introduce a structure that doesn't affect program flow. The *Formula Node* is a resizable box that you use to enter algebraic formulas directly into the block diagram. You will find this feature extremely useful when you have a long formula to solve. For example, consider the fairly simple equation $y = x^2 + x + 1$. Even for this simple formula, if you implement this equation using regular LabVIEW arithmetic functions, the block diagram is a little bit harder to follow than the text equations.

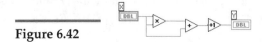

Figure 6.42

You can implement the same equation using a Formula Node, as shown in Figure 6.43.*

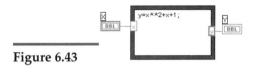

Figure 6.43

With the Formula Node, you can directly enter a formula or formulas, in lieu of creating complex block diagram subsections. Simply enter the formula inside the box. You create the input and output terminals of the Formula Node by popping up on the border of the node and choosing **Add Input** or **Add Output** from the pop-up menu. Then enter variable names into the input and output boxes. Names are case sensitive, and *each formula statement must terminate with a semicolon (;).*

You will find the Formula Node in the **Structures** subpalette of the **Functions** palette.

These operators and functions are available inside the Formula Node (you can get this information from the **Help** window shown in Figure 6.44):

Figure 6.44

```
Context Help

The following operators are allowed in formula statements, arranged from
lowest to highest precedence:
=, op=        assignment, op can be +, -, *, /, >>, <<, &, ^, |, %, **
? :           conditional evaluation
||            logical OR
&&            logical AND
|             bitwise OR
^             bitwise exclusive OR
&             bitwise AND
!=, ==        inequality, equality
>, <, >=, <=  greater, less, greater or equal, less or equal
>>, <<        arithmetic shift right, left
+, -          addition, subtraction
*, /          multiplication, division
**            exponentiation
+,-,!,~,++,-- unary plus, unary negation, logical NOT, bit complement,
pre-increment, post-increment

The following built-in functions are allowed in formula statements:
abs, acos, acosh, asin, asinh, atan, atanh, ceil, cos, cosh, cot, csc, exp,
expm1, floor, getexp, getman, int, intrz, ln, lnp1, log, log2, max, min,
mod, rand, rem, sec, sign, sin, sinc, sinh, sqrt, tan, tanh
```

The following example shows a conditional branching that you could perform inside a Formula Node. Consider the following code fragment, similar

---

* In versions prior to LabVIEW 6.0, the exponentiation operator for the Formula Node was the ^ symbol. In LabVIEW 6.0 and greater, the ^ symbol means something else entirely (bitwise XOR), and the ** symbol is exponentiation.

to Activity 6-3, that computes the square root of **x** if **x** is positive, and assigns the result to **y**. If **x** is negative, the code assigns a value of –99 to **y**.

```
if (x >= 0) then
      y = sqrt(x)
else
      y = -99
end if
```

You can implement the code fragment using a Formula Node, as shown in Figure 6.45.

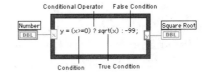

Figure 6.45

6.5.1 Activity 6-5: Formula Fun

You will build a VI that uses the Formula Node to evaluate the equation $y = \sin(x)$ and graph the results.

1. Open a new panel. Select **Waveform Graph** from the **Graph** subpalette of the **Controls** palette. Label it Graph. You'll learn all about graphs in Chapter 8, but this activity would be kind of dull without a nice pictorial representation, so we thought we'd give you a taste of them.

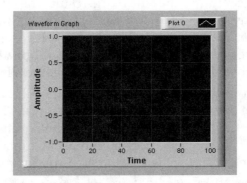

Figure 6.46

2. Build the block diagram shown in Figure 6.47.

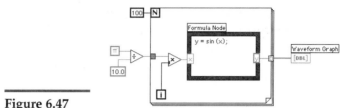

Figure 6.47

With the Formula Node (**Structures** palette), you can directly enter mathematical formulas. Create the input terminal by popping up on the border and choosing **Add Input** from the pop-up menu; then create the output terminal by choosing **Add Output** from the pop-up menu.

When you create an input or output terminal, you must give it a variable name. The variable name must exactly match the one you use in the formula. Remember, variable names are case sensitive.

Notice that a semicolon (;) must terminate the formula statement.

π

Pi Constant

The π constant is located in the **Functions>>Numeric>>Additional Numeric Constants** palette.

During each iteration, the VI multiplies the iteration terminal value by $\pi/10$. The multiplication result is wired to the Formula Node, which computes the sine of the result. The VI then stores the result in an array at the For Loop border. (You will learn all about arrays in Chapter 7. Then you will see why you can wire array data out of a For Loop, whereas scalar data come out of a While Loop by default.) After the For Loop finishes executing, the VI plots the array.

3. Return to the front panel and run the VI. Note that you could also use the existing **Sine** function (**Functions>>Numeric>>Trigonometric** palette) to do the same thing as the Formula Node in this activity, but LabVIEW does not have built-in functions for every formula you'll need and we wanted to give you the practice.

4. Save the VI in your MYWORK directory or VI library and name it **Formula Node Exercise.vi**. Close the VI.

VI Logic

```
for i = 0 to 99
      x = i * (PI/10)
      y = sin(x)
      array[i] = y
next i
Graph (array)
```

6.6 Wrap It Up!

LabVIEW has two structures to repeat execution of a subdiagram—the *While Loop* and the *For Loop*. Both structures are resizable boxes; one places the subdiagram to be repeated inside the border of the loop structure. The While Loop executes as long as the value at the *conditional terminal* is TRUE. The For Loop executes a specified number of times.

Shift registers, available for While Loops and For Loops, transfer values from the end of one loop iteration to the beginning of the next. You can configure shift registers to access values from many previous iterations. For each iteration you want to recall, you must add a new element to the left terminal of the shift register. You can also have multiple shift registers on a loop to store multiple variables.

LabVIEW has two structures to add control to dataflow—the *Case Structure* and the *Sequence Structure*. Only one case or one frame of these structures is visible at a time; you can switch between them using the little arrows at the top of the structure, with the pop-up menu, or by clicking with the Operating tool in the window at the top of the structure.

You use the Case Structure to branch to different subdiagrams depending on the input to its selector terminal, much like an if-then-else statement in conventional languages. Simply place the subdiagrams you want to execute inside the border of each case of the Case Structure and wire an input to the case selector terminal. Case Structures can be Boolean (with two cases), numeric, or string (with up to $2^{15}-1$ cases)—LabVIEW automatically determines which type when you wire a Boolean or numeric control to the selector terminal.

Sometimes the principles of dataflow cause your program to behave differently from the way you want it to, and you need a way to force a certain execution order. The Sequence Structure lets you set a specific order for your diagram functions. The portion of the diagram to be executed first is placed in the first frame (Frame 0) of the Sequence Structure, the subdiagram to be executed second is placed in the second frame, and so on.

You use *sequence locals* to pass values between Sequence Structure frames. The data passed in a sequence local are available only in frames subsequent to the frame in which you created the sequence local, *not* in those frames that precede the frame in which its value is assigned.

With the *Formula Node*, you can directly enter formulas in the block diagram, an extremely useful feature for complex function equations. Remember that variable names are case sensitive and that each formula statement must end with a semicolon (;).

The **Time & Dialog** subpalette of the **Functions** palette provides functions that pop up dialog boxes and control or monitor VI timing. The **One Button Dialog** and **Two Button Dialog** functions pop up a dialog box containing the message of your choice. The **Wait (ms)** function pauses your VI for the specified number of milliseconds. **Wait Until Next ms Multiple** can force loops to execute at a given interval by pausing until the internal clock equals (or has exceeded) a multiple of the millisecond input. These two wait functions are similar but not identical, and you don't really need to worry about the difference right now. **Tick Count (ms)** returns to you the value of the internal clock.

6.7   Additional Activities

Activity 6-6: Equations

Build a VI that uses the Formula Node to calculate the following equations:

```
y1 = x3 + x2 + 5
y2 = (m * x) + b
```

Use only one Formula Node for both equations. (Remember to put a semicolon [;] after each equation in the node.) Name the VI **Equations.vi**.

Activity 6-7: Calculator

Build a VI that functions like a calculator. The front panel should have digital controls to input two numbers and a digital indicator to display the result of the operation (add, subtract, multiply, or divide) that the VI performs on the two numbers. Use a slide control to specify the operation to be performed. Name the VI **Calculator.vi**.

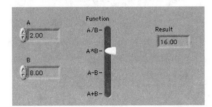

Figure 6.48

You will want to use Text Labels on the slide control, obtained from the pop-up menu of the slide, to specify the function (add, subtract, multiply, or divide). If you don't see this option in the pop-up menu, make sure you're popping up on the slide itself and not the scale. Slides with Text Labels behave very much like text ring controls. When you first select Text Labels, the slide will have two settings, max and min. You can use the Labeling tool to change this text. To add another text marker to the slide, pop up on the text display that appears next to your slide and select Add Item After or Add Item Before and then type in that marker's text.

Activity 6-8: Combination For/While Loop Challenge

Using only a While Loop, build a combination For Loop/While Loop that stops either when it reaches "N" (specified inside a front panel control) or when a user pushes a stop button. Name the VI **Combo For/While Loop.vi**.

*Don't forget that a While Loop only executes while the conditional terminal reads a TRUE value. You might want to use the **And** function (**Boolean** subpalette of the **Functions** palette). Also, remember that while executing a loop, LabVIEW does not update indicators or read controls that are outside of the loop. Your stop button must be inside your loop if you want correct functionality.*

Activity 6-9: Dialog Display

Write a VI that reads the value of a front panel switch, and then pops up a dialog box indicating if the switch is on or off. Name the VI **Dialog Display.vi**. If you've developed the bad habit of using the continuous run button, now is the time to break it, or you will get yourself stuck in an endless loop! If you do get stuck, use the keyboard shortcut to stop your VI: <control-.> under Windows, <command-.> on a Macintosh, <meta-.> on the Sun, and <alt-.> under Linux.

OVERVIEW

In this chapter, you will learn about two new, more complex data types—arrays and clusters. These composite data types allow you great flexibility in data storage and manipulation. You will also see some valuable uses for arrays and clusters and learn how you can use built-in functions to manage them.

GOALS

- Learn about the built-in array manipulation functions
- Grasp the concept of polymorphism
- Learn how to use clusters and how to bundle and unbundle them
- Understand how clusters differ from arrays

KEY TERMS

- Array
- Auto-indexing
- Polymorphism
- Cluster
- Bundle
- Unbundle

LabVIEW's Composite Data: Arrays and Clusters

7

7.1 What Are Arrays?

Until now, we've dealt with scalar numbers only (a scalar is simply a data type that contains a single value, or "nonarray"), and now it's time to move on to something more powerful and compound. A LabVIEW *array* is a collection of data elements that are all the same type, just like in traditional programming languages. An array can have one or more dimensions, and up to 2^{31} elements per dimension (memory permitting, of course). An array data element can have any type except another array, a chart, or a graph.

Array elements are accessed by their indices; each element's *index* is in the range 0 to $N - 1$, where N is the total number of elements in the array. The *one-dimensional* (1D) array shown here illustrates this structure.

Index	0	1	2	3	4	5	6	7	8	9
10-Element array	12	32	82	8.0	4.8	5.1	6.0	1.0	2.5	1.7

Notice that the *first* element has index 0, the *second* element has index 1, and so on. You will find that waveforms (and many other things) are often stored in arrays, with each point in the waveform comprising an element of the array. Arrays are also useful for storing data generated in loops, where each loop iteration generates one element of the array.

7.2 Creating Array Controls and Indicators

It takes two steps to make the controls and indicators for compound data types such as arrays and clusters. First, you create the array control or indicator; second, you combine the *array shell* with a *data object*, which can be numeric, Boolean, path, or string (or cluster, but we'll cover that later). You will find the array shell in the **Array & Cluster** subpalette of the **Controls** palette.

Index display ———

Element display window ———

Figure 7.1

To create an array, drag a data object into the element display window. You can also deposit the object directly by clicking inside the window when you first choose the object from the Controls palette. The element display window resizes to accommodate its new data type, as shown in Figure 7.2, but remains grayed out until you enter data into it. Note that all elements of an array must be either controls or indicators, not a combination.

Figure 7.2

When you first drop an array shell on the front panel, its block diagram terminal is black, characteristic of an undefined data type. The terminal also contains brackets, shown in part (A) of Figure 7.3, which are LabVIEW's way of denoting an array structure. When you assign a data type to the array (by placing a control or indicator in its element display window), then the array's block diagram terminal assumes its new type's color and lettering (although it retains its brackets), as in part (B). You will notice that array wires are thicker than wires carrying a single value.

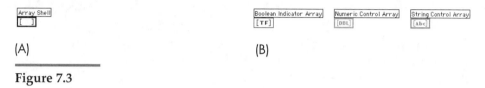

(A)

(B)

Figure 7.3

You can enter data into your array as soon as you assign a data type to it. Use the Labeling or Operating tool to type in a value, or if your data are numeric, click the arrows of the index display to increment or decrement them.

Resizing
Brackets
Cursor
(Positioning
Tool)

If you want to resize the object in the display window, use the Positioning tool and make sure it turns into the standard resizing brackets when you place it on the corner of the window (you will probably have to position it slightly inside the box and you may have to move it around a bit to get the correct brackets). If you want to show more elements at the same time, move the Positioning tool around the window corner until you find the grid cursor, and then stretch either horizontally or vertically (your data are unchanged by the layout of your array). You will then have multiple elements visible. The element closest to the index display always corresponds to the element number displayed there.

Grid Cursor
(Positioning
Tool)

You can create array constants on the block diagram just like you can create numeric, Boolean, or string constants. Choosing **Array Constant** from the **Array** subpalette of the **Functions** palette will give you an array shell; then simply place in an appropriate data type (usually another constant) just like you do on the front panel. This feature is useful when you need to initialize shift registers or provide a data type to a file or network function (which you'll learn about later).

If you want to clear an array control, indicator, or constant of data, pop up on the index display (*not* the element itself or you'll get the wrong menu) and choose **Data Operations>>Empty Array**.

7.3 Using Auto-Indexing

The For Loop and the While Loop can index and accumulate arrays at their boundaries automatically, adding one new element for each loop iteration. This capability is called *auto-indexing.* One important thing to remember is that *auto-indexing is enabled by default on For Loops but disabled by default on While Loops.* Figure 7.4 shows a For Loop auto-indexing an array at its boundary. Each iteration creates the next array element. After the loop completes, the array passes out of the loop to the indicator; none of the array data are available until after the Loop finishes. Notice that the wire becomes thicker as it changes to an array wire type at the Loop border.

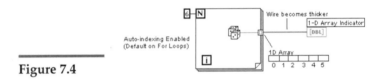

Auto-indexing Enabled
(Default on For Loops)

Figure 7.4

If you need to wire a scalar value out of a For Loop without creating an array, you must disable auto-indexing by popping up on the tunnel (the square with the [ ] symbol) and choosing **Disable Indexing** from the tunnel's pop-up menu.

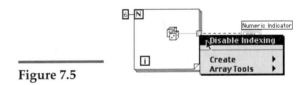

Figure 7.5

Since auto-indexing is disabled by default, whenever you need to wire array data out of a While Loop, you must pop up on the tunnel and select **Enable Indexing**.

In Figure 7.6, auto-indexing is disabled, and only the last value returned from the **Random Number (0–1)** function passes out of the loop. Notice that the wire remains the same size after it leaves the loop. Pay attention to this wire size, because auto-indexing is a common source of problems among beginners. They often create arrays when they don't want to, or don't create them when they need them, and then go crazy trying to figure out why they get a bad wire.

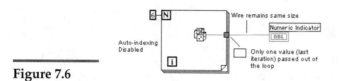

Figure 7.6

Auto-indexing also applies when you are wiring arrays into loops. If indexing is enabled as in loop (A) in Figure 7.7, the loop will index off one element from the array each time it iterates (note how the wire becomes thinner as it enters the loop). If indexing is disabled as in loop (B), the entire array passes into the loop at once.

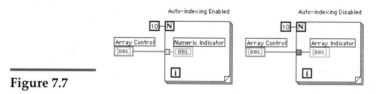

Figure 7.7

 *Because For Loops are often used to process arrays, LabVIEW enables auto-indexing by default when you wire and array into or out of them. By default, LabVIEW does not enable auto-indexing for While Loops. You must pop up on the array tunnel and choose **Enable Indexing** from the pop-up menu if you want your While Loop to auto-index. Pay close attention to the state of your indexing, lest you develop errors that are tricky to spot.*

Using Auto-Indexing to Set the For Loop Count

When you enable auto-indexing on an array *entering* a For Loop, LabVIEW automatically sets the *count* to the array size, thus eliminating the need to wire a value to the count terminal. If you give LabVIEW conflicting counts, for example, by setting the count explicitly and by auto-indexing (or by auto-indexing two different size arrays), LabVIEW sets the count to the smallest of the choices. In Figure 7.8, the array size, and not the value wired to the count terminal, determines the number of For Loop iterations, because the array size is the smaller of the two.

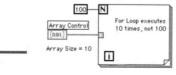

Figure 7.8

7.4 Two-Dimensional Arrays

A two-dimensional, or 2D, array stores elements in a gridlike fashion. It requires two indices to locate an element: a column index and a row index, both of which are zero-based like everything else in LabVIEW. Figure 7.9 shows how a six-column by four-row array that contains six times four elements is stored.

	0	1	2	3	4	5
0						
1						
2						
3						

Figure 7.9
Six-column by four-row array of 24 elements.

Grid Cursor
Positioning
Tool

You can add dimensions to an array control or indicator by popping up on its *index display* (not on the element display) and choosing **Add Dimension** from the pop-up menu. Figure 7.10 shows a 2D array of digital controls. Notice that you now have two indices to specify each element. You can use the grid cursor of the Positioning tool to expand your element display in two dimensions so that you can see more elements.

Figure 7.10

Remove unwanted dimensions by selecting **Remove Dimension** from the index display's pop-up menu.

If you have waveforms from several channels being read from a data ac-quisition (DAQ) board, they will be stored in a 2D array, with each column in the 2D array corresponding to one channel's data.

Creating Two-Dimensional Arrays

You can use two For Loops, one inside the other, to create a 2D array if you don't want to type in values on the front panel. The inner For Loop creates a row, and the outer For Loop "stacks" these rows to fill in the columns of the matrix. Figure 7.11 shows two For Loops creating a 2D array of random numbers using auto-indexing.

Notice that a 2D array wire is even thicker than a 1D array wire.

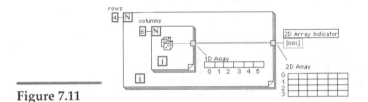

Figure 7.11

7.5 Activity 7-1: Building Arrays with Auto-Indexing

Now we'll give you a chance to better understand arrays and auto-indexing by working with them yourself. In this activity you will open and observe a VI that uses auto-indexing on both a For Loop and a While Loop to create ar-rays of data.

1. Open the **Building Arrays.vi** example, located in EVERYONE\ CH7.LLB. This exercise generates two arrays on the front panel, using a For Loop to create a 2D array and a While Loop to create a 2D array. The For Loop executes a set number of times; you must hit the Stop button to halt the While Loop (or it will stop after 101 iterations).

2. Take a look at the front panel; then switch to the block diagram. No-tice how the nested For Loops create a 2D array's rows and columns

at their borders, respectively, using auto-indexing. Also notice how the auto-indexed wires become thicker as they leave the loop boundaries.

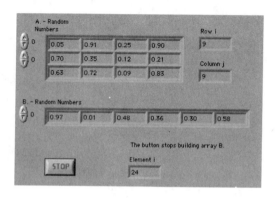

Figure 7.12

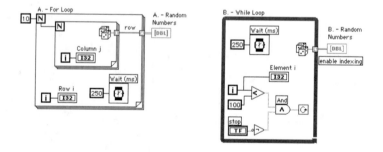

Figure 7.13

3. Before we could get array data out of the While Loop, we had to pop up on the tunnel containing the random number and select **Enable Indexing**. To see how this works, go ahead and pop up on the tunnel, and then select **Disable Indexing**. You will see the wire leaving

the loop break. Pop up on the tunnel again and select **Enable Indexing** to fix it.

This loop uses a little logic algorithm to ensure that if the user does not press the Stop button after a reasonable amount of time (101 iterations), the loop stops anyway. If the user has not pressed the Stop button and the loop has executed fewer than 101 times, the loop continues. If either of those conditions changes, the loop stops.

Why does it execute 101 times, instead of 100? Remember that the While Loop checks the conditional terminal at the end of each iteration. At the end of the 100th iteration, i = 99 (since it starts at zero), which is less than 100, and the loop continues. At the end of the 101st iteration, the count is no longer less than 100 and the loop stops (assuming the Stop button hasn't already stopped it).

4. Now that you understand how it works, run the VI. Remember to hit Stop to halt the While Loop, especially since the front panel indicator does not update until the entire array has been generated (remember, controls and indicators outside of loops are not read or updated while the loop is executing).

5. Close the VI and don't save any changes.

7.6 Functions for Manipulating Arrays

By now, you know a little bit about what arrays are, so let's talk about all the cool stuff you can do with them. LabVIEW has many functions to manipulate arrays in the **Array** subpalette of the **Functions** palette. To avoid a common pitfall, always keep in mind that arrays (and all other LabVIEW structures) are zero indexed—the first element has an index of zero, the second has an index of one, and so on. Some common functions are discussed here, but you might also want to browse through the **Array** subpalette just to see what else is built in for you.

• **Initialize Array** will create and initialize an *n*-dimensional array with the value of your choice. You can configure it for larger dimensions by "growing" it with the Positioning tool to provide more **dimension size** inputs. This function is useful for allocating memory for arrays of a certain size or for initializing shift registers with array-type data.

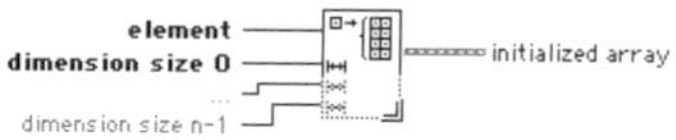

Figure 7.14
Initialize Array.

In Figure 7.15, **Initialize Array** shows how to initialize a ten-element, one-dimensional array, with each element of the array containing a zero.

Figure 7.15

• **Array Size** returns the number of elements in the input array. If the input array is *n* dimensional, **Array Size** returns an *n*-element, one-dimensional array, with each element containing the size of one of the array's dimensions.

Figure 7.16
Array Size.

Figure 7.17
Array Size.

• Depending on how you configure it, **Build Array** concatenates, or combines, two arrays into one, or adds extra elements to an array. The function looks like the icon at left when first placed in the diagram window. You can resize or "grow" this function to increase the number of inputs. **Build Array** has two types of inputs, *array* and *element*, so that it can assemble an array from both array and single-valued inputs.

Figure 7.18
Build Array.

For example, the **Build Array** function shown in Figure 7.19 has been configured to concatenate two arrays and one element into a new array.

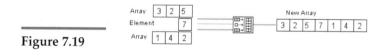

Figure 7.19

*Pay special attention to the input of a **Build Array** function. Array inputs have two "dotted" squares, while element inputs have a single hollow square. Although LabVIEW will adapt the input type depending on what kind of data you wire to it (elements or arrays), they are not interchangeable and can cause lots of confusing bad wires if you're not careful.*

The **Build Array** function input will automatically adapt to an element or an array input, depending on what you wire to it.

As you get more advanced, you'll find that **Build Array** can also build or add elements to multidimensional arrays. To add an element to a multidimensional array, the element must be an array of one size smaller dimension (i.e., you can add a 1D element to a 2D array). You can also build a 2D array by using the **Build Array** function and wiring 1D arrays in as "elements" (each 1D array will become one row of the 2D array). Sometimes you'll want to concatenate several 1D arrays together, instead of building a 2D array. In this case, you'll need to pop up on the **Build Array** function, and choose **Concatenate Inputs**.

• **Array Subset** returns a portion of an array starting at **index** and containing **length** elements. Notice that the third element's index is two because the index starts at zero; that is, *the first element has an index of zero.*

Figure 7.20
Array Subset.

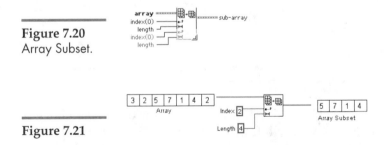

Figure 7.21

• **Index Array** accesses a particular element of an array. An example of an **Index Array** function accessing the third element of an array is shown in Figure 7.22a. Here, the **Index Array** function is extracting a scalar element from an array. So to extract a single scalar element, wire the desired element's row index to the top input and its column index to the bottom input.

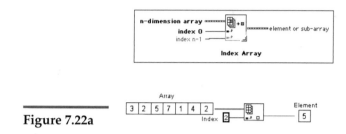

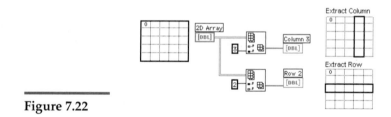

Figure 7.22a

You also can use this function to *slice off* a row, column, or scalar element of a 2D array. To do this, you simply leave one of the **Index Array** function inputs unwired. To slice off a row from the 2D array, wire the row index (the first index input) of the row you want to slice off. To slice off a column from the 2D array, leave the first index input unwired, and wire the column index (the second index input) with the column you want to slice off.

Notice that the index terminal symbol changes from a solid to an empty box when you leave its input unwired. Figure 7.22 shows how to extract a column or a row from a 2D array.

Figure 7.22

7.7 Activity 7-2: Array Acrobatics

Are you confused yet? You'll get a better feel for arrays as you work with them. In this exercise, you will finish building a VI that concatenates two arrays and then indexes out the element in the middle of the new concatenated array.

1. Open **Array Exercise.vi**, located in `EVERYONE\CH7.LLB`.

   The front panel contains two input arrays (each showing three elements), two digital controls, and an output array (showing eight elements). The VI will concatenate the arrays and the control values in the following sequence to create the new array:

Initial Array + Element 1 + Element 2 + Terminal Array

The front panel is already built. You will finish building the diagram. Note that an array control or indicator will appear grayed out until you or the program assigns it some data.

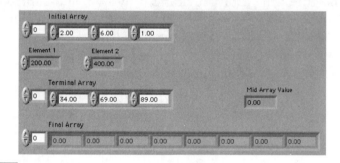

Figure 7.23

2. Build the block diagram shown in Figure 7.24. Use the Help window to find the correct terminals on the functions.

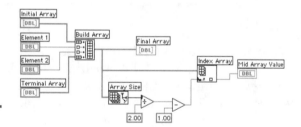

Figure 7.24

Build Array Function

Build Array function (**Array** palette). In this exercise, concatenates the input data to create a new array in the following sequence: Initial Array + Element 1 + Element 2 + Terminal Array.

Build Array

The function looks like the icon at the left when placed in the diagram window. Place the Positioning tool on the lower-right corner and resize the function until it includes four inputs.

Array Input

The inputs will automatically detect whether you are wiring an array or an element. The symbols at the left indicate whether an input is an array or an element.

Element Input

Array Size Function

Array Size function (**Array** palette). Returns the number of elements in the concatenated array.

Index Array
Function

Index Array function (**Array** palette). In this exercise, returns the element in the middle of the array.

LabVIEW builds the array using the **Build Array** function; it then calculates the index for the middle element of the array by taking the length of the array, dividing it by two, and subtracting one (to account for the zero-based array index). Since the array has an even number of elements, the middle element will actually be one of the two middle elements.

3. Return to the front panel and run the VI. Try several different number combinations.

4. Save the finished VI in your MYWORK directory and close the VI.

You might also want to look through the programs in EXAMPLES/ GENERAL/ARRAYS.LLB in the LabVIEW directory to see some other things you can do with arrays.

7.8 Polymorphism

Another handy feature you'll find useful is the *polymorphism* of the LabVIEW arithmetic functions, **Add**, **Multiply**, **Divide**, and so on. *Polymorphism* is just a big word for a simple principle: The inputs to these functions can be of different size and representation. For example, you can add a scalar to an array or add two arrays together using the same function. Figure 7.25 shows some of the polymorphic combinations of the **Add** function.

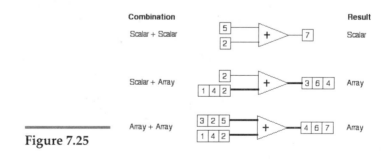

Figure 7.25

In the first combination, the result is a scalar number. In the second combination, the scalar is added to *each* element of the array. In the third combination, each element of one array is added to the corresponding element of the other array. In all instances, the same **Add** function is used, but it performs a different type of operation.

In Figure 7.26, each iteration of the For Loop generates one random number (valued between 0 and 1) that is stored in the array created at the border of the loop. After the loop finishes execution, the **Multiply** function multiplies each element in the array by the scaling factor you set. The front panel array indicator then displays the scaled array.

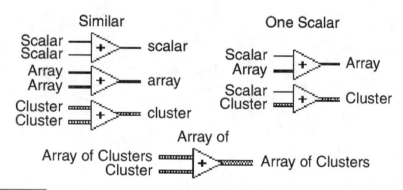

Figure 7.26

Figure 7.27 shows some of the possible polymorphic combinations of the **Add** function. You'll learn about the clusters it depicts in the next section.

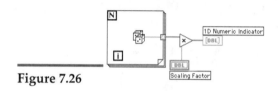

Figure 7.27

If you are doing arithmetic on two arrays with a different number of elements, the resulting array will be the size of the smaller of the two. In other words LabVIEW operates on corresponding elements in the two arrays until one of the arrays runs out of elements. The remaining elements in the longer array are ignored.

7.9 Activity 7-3: Polymorphism

You will build a VI that demonstrates polymorphism on arrays.

1. Open a new panel and recreate the VI shown in Figure 7.28.

Grid Cursor
Positioning
Tool

First, create the arrays. Remember, to create an array, you must first select the **Array** shell from the **Array & Cluster** subpalette of the **Controls** palette. Then put a numeric indicator into the shell's data object window. To see more than one element in the array, you must grab and drag the corner of the filled element display window with the grid cursor of the Positioning tool. You can drag out multiple elements in a 1D array either horizontally or vertically.

All four arrays in this exercise contain indicator elements. Make sure and give them unique labels so that you don't get them confused. If you ever do lose track of which front panel object corresponds to which block diagram terminal, simply pop up on one of them and choose **Find Terminal** or **Find Indicator**; LabVIEW will highlight the corresponding object.

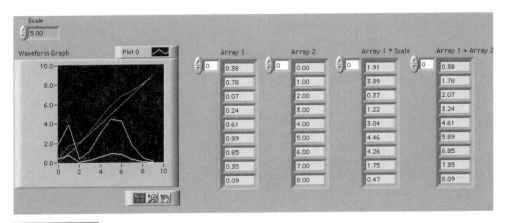

Figure 7.28

2. After you've created the arrays, select a **Waveform Graph** from the **Graph** subpalette of the **Controls** palette. Although you'll learn more about graphs in the next chapter, we wanted to give you a preview and spice up this exercise.

3. Don't forget to create the Scale control.

4. Build the diagram shown in Figure 7.29—be careful because the wiring can get a little tricky!

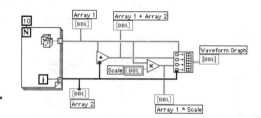

Figure 7.29

Auto-indexing is enabled by default on the For Loop, so your arays will be generated automatically

5. You can find **Add**, **Multiply**, and **Random Number (0–1)** in the **Numeric** palette.

6. Select **Build Array** from the Array palette. You will have to grow it using the Positioning tool so that it has four inputs. By default, the array inputs are not concatenated; the output from **Build Array** is a 2D array. Each input array becomes a row, so the output 2D array is composed of 4 rows and 10 columns.

7. Run the VI. Your graph will plot each array element against its index for all four arrays at once: <u>Array 1</u> data, <u>Array 2</u> data, <u>Array 1* Scale</u>, and <u>Array 1+ Array 2</u>.

   The results demonstrate several applications of polymorphism in LabVIEW. For example, <u>Array 1</u> and <u>Array 2</u> could be incoming waveforms that you wish to scale.

8. Save the VI as **Polymorphism Example.vi** and place it in your MY-WORK directory or VI library. Close the VI.

7.10 Compound Arithmetic

While we're on the subject of arithmetic, we should mention the **Compound Arithmetic** function. Whereas the previous polymorphism example showed you how to add and multiply different-sized data together, this function lets you operate on more than two numbers simultaneously. The **Compound Arithmetic** function eliminates the need for multiple **Add**, **Multiply**, **AND**,

OR, and **XOR** terminals should you need to perform one of these functions on several numbers at once (**AND, OR,** and **XOR** are Boolean arithmetic operations).

Figure 7.30
(Left) Compound **Add** function.
(Right) Compound **AND** function.

The **Compound Arithmetic** function can be found in both the **Numeric** and **Boolean** subpalettes of the **Functions** palette. Like many other functions, you can use the Positioning tool to grow it and provide more input terminals. The **Compound Arithmetic** function has only one form; you configure it to perform the arithmetic function of your choice. To change the function (choices are **Add, Multiply, AND, OR,** or **XOR**), pop up on the output terminal and select **Change Modes.** You can also click on the function with the Operating tool to change the mode.

Select the **Invert** pop-up option to invert the sign of numeric inputs and outputs or the value of Booleans (from FALSE to TRUE or vice versa). A small circle at the input or output symbolizes an inverted value.

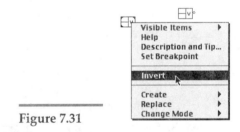

Figure 7.31

A Word about Boolean Arithmetic

LabVIEW's Boolean arithmetic functions, **And, Or, Not, Exclusive Or, Not Exclusive Or, Not And,** and **Not Or,** can be very powerful. You will see occasional Boolean arithmetic throughout this book. If you've never seen it before, we recommend reading about it in a good logic or digital design book. But just as a refresher, we'll mention a few basics. If you can't remember which function does what, use the Help window!

Not is probably the easiest to describe, since it simply inverts the input value. If the input value is TRUE, **Not** will output FALSE; if the input is FALSE, **Not** returns TRUE.

The **And** function outputs a TRUE only if all inputs are TRUE.

The **Or** function outputs a TRUE if at least one input is TRUE.

The **And** and **Or** functions have the following outputs, given the inputs shown:

FALSE **And** FALSE = FALSE	FALSE **Or** FALSE = FALSE
TRUE **And** FALSE = FALSE	TRUE **Or** FALSE = TRUE
FALSE **And** TRUE = FALSE	FALSE **Or** TRUE = TRUE
TRUE **And** TRUE = TRUE	TRUE **Or** TRUE = TRUE

7.11 All about Clusters

Now that you've got the concept of arrays under your belt, understanding clusters should be easy. Like an array, a *cluster* is a data structure that groups data. However, unlike an array, a cluster can group data of different types (i.e., numeric, Boolean, etc.); it is analogous to a *struct* in C or the data members of a *class* in C++ or Java. A cluster may be thought of as a *bundle* of wires, much like a telephone cable. Each wire in the cable represents a different element of the cluster. Because a cluster has only one "wire" in the block diagram (even though it carries multiple values of different data types), clusters reduce wire clutter and the number of connector terminals that sub-VIs need. You will find that the cluster data type appears frequently when you plot your data on graphs and charts.

Figure 7.32
Bundling data.

You can access cluster elements by *unbundling* them all at once or by indexing one at a time, depending on the function you choose; each method has its place. Think of unbundling a cluster as unwrapping a telephone cable and having access to the different-colored wires. Unlike arrays, which can change size dynamically, clusters have a fixed size, or a fixed number of wires in them.

Figure 7.33
Unbundling data.

You can connect cluster terminals with a wire only if they have exactly the same type; in other words, both clusters must have the same number of elements, and corresponding elements must match in both data type and order. The principle of polymorphism applies to clusters as well as arrays, as long as the data types match.

You will often see clusters used in error handling. Figure 7.34 shows the error clusters, **Error In.ctl** and **Error Out.ctl**, used by LabVIEW VIs to pass a record of errors between multiple VIs in a block diagram (for example, many of the data acquisition and file I/O VIs have error clusters built into them). These error clusters are so frequently used that they appear in the **Array & Cluster** subpalette of the Controls palette for easy access.

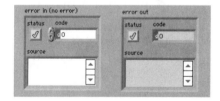

Figure 7.34

7.12 Creating Cluster Controls and Indicators

Create a cluster by placing a **Cluster** shell (**Array & Cluster** subpalette of the **Controls** palette) on the front panel. You can then place any front panel objects inside the cluster. Like arrays, you can deposit objects directly inside when you pull them off of the **Controls** palette, or you can drag an existing object into a cluster. *Objects inside a cluster must be all controls or all indicators.* You cannot combine both controls and indicators inside the same cluster, because the cluster itself must be one or the other. The cluster will be a control or indicator based on the status of the first object you place inside it. Resize the cluster with the Positioning tool if necessary. Figure 7.35 shows a cluster with four controls.

Figure 7.35

You can create block diagram cluster constants in a similar two-step manner.

If you want your cluster to conform exactly to the size of the objects inside it, pop up on the *border* (not inside the cluster) and choose an option in the **Autosizing** menu.

7.13 Cluster Order

Cluster elements have a logical order unrelated to their position within the shell. The first object placed in the cluster is element zero, the second is element one, and so on. If you delete an element, the order adjusts automatically. *You must keep track of your cluster order if you want to connect your cluster to another cluster—the order and data types must be identical.* Also, if you choose to unbundle the cluster all at once, you'll want to know which value corresponds to which output on the cluster function (more about unbundling will be discussed in Section 7.17).

Change the order within the cluster by popping up on the cluster *border* and choosing **Reorder Controls in Cluster...** from the pop-up menu. A new set of buttons appears in the Toolbar, and the cluster appearance changes as shown in Figure 7.36.

The white boxes on the elements show their current places in the cluster order. The black boxes show the new places. Clicking on an element with the cluster order cursor sets the element's place in the cluster order to the number displayed on the Toolbar. You can type a new number into that field before you click on an object.

Revert Button

If you don't like the changes you've made, revert to the old order by clicking on the Revert button. When you have the order the way you want, you

OK Button

can set it and return to your regular front panel by clicking on the OK button and exiting the cluster order edit mode.

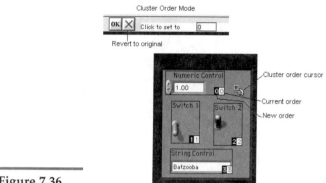

Figure 7.36

7.14 Using Clusters to Pass Data to and from SubVIs

The connector pane of a VI can have a maximum of 28 terminals. You probably don't want to pass information to all 28 terminals when calling a subVI anyway because the wiring can be very tedious and you can easily make a mistake. By bundling a number of controls or indicators into a cluster, you can use a single terminal and wire to pass several values into or out of the subVI. You can use clusters to work around the 28-terminal limit for the connector or just enjoy simplified wiring with fewer (and therefore larger and easier to see) terminals.

7.15 Bundling Your Data

Bundle
Function

The **Bundle** function (**Cluster** palette) assembles individual components into a new cluster or allows you to replace elements in an existing cluster. The function appears as the icon at the left when you place it in the diagram window. You can increase the number of inputs by dragging a corner of the function with the Positioning tool. When you wire to each input terminal, a symbol representing the data type of the wired element appears on the empty terminal. The order of the resultant cluster will be the order of inputs to the **Bundle**.

Figure 7.37
Bundle.

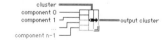

If you just want to create a new cluster, you do not need to wire an input to the center **cluster** input of the **Bundle** function. You do need to wire this input if you are replacing an element in the cluster.

Be careful with this function: If you add an element to your cluster without adding an element to the block diagram **Bundle** to match, your program will break!

7.16 Replacing a Cluster Element

If you want to *replace* an element in a cluster, first size the **Bundle** function to contain the same number of input terminals as there are elements in this cluster (it must be the same size or you will get a bad wire). Then wire the cluster to the middle terminal of the **Bundle** function (symbols for data types inside the cluster will appear in the **Bundle** inputs) and wire the new value(s) of the element(s) you want to replace to the corresponding inputs. You need to wire only to those terminals that have values you want to change.

Figure 7.38

7.17 Unbundling Your Clusters

Unbundle Function

The **Unbundle** function (**Cluster** palette) splits a cluster into each of its individual components. *The output components are arranged from top to bottom in the same order they have in the cluster.* If they have the same data type, the elements' order in the cluster is the only way you can distinguish among them. The function appears as the icon at left when you place it in the diagram window. You can increase the number of outputs by dragging a corner of the function with the Positioning tool. The **Unbundle** function must be resized to contain the same number of outputs as there are elements in the input cluster, or it will produce bad wires. When you wire an input cluster to the correctly sized **Unbundle**, the previously blank output terminals will assume the symbols of the data types in the cluster.

Figure 7.39
Unbundle.

*Knowing the cluster order is essential when accessing cluster data with **Bundle** and **Unbundle**. For example, if you have two Booleans in the same cluster, it would be easy to mistakenly access Switch 2 instead of Switch 1 since they are referenced in the **Unbundle** function by order, not by name. Your VI will wire with no error, but your results will be incorrect.*

Be careful with this function—if you add an element to your cluster without resizing the block diagram **Unbundle** to match, your program will break!

LabVIEW does have a way to bundle and unbundle clusters using element names; we'll talk about it in a minute.

7.18 Activity 7-4: Cluster Practice

In this activity, you will build a VI to teach you how to work with clusters. You will create a cluster, unbundle it, and then rebundle it and display the values in another cluster.

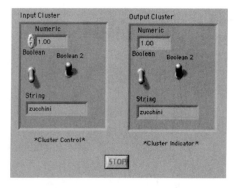

Figure 7.40

1. Open a new panel and place a **Cluster** shell (**Array & Cluster** palette) on it. Label it <u>Input Cluster</u>. Enlarge the shell (make sure to grab the cluster border or nothing will happen).

2. Place a digital control, two Boolean switches, and a string control inside the <u>Input Cluster</u> shell.

3. Now create <u>Output Cluster</u> by cloning <u>Input Cluster</u>. Then pop up on an object in the cluster (or on the cluster border) and select **Change to Indicator**. Also change the new cluster's label. (Remember, <control>-drag under Windows, <option>-drag on the Mac, <meta>-drag on Sun, and <alt>-drag on Linux to clone an object. This technique ensures correct cluster order and is very efficient.)

You could also create <u>Output Cluster</u> the same way you created <u>Input Cluster,</u> using indicators instead of controls (make sure to put elements into the cluster in the same order).

4. Verify that <u>Input Cluster</u> and <u>Output Cluster</u> have the same cluster order by popping up on each cluster border and choosing **Reorder Controls in Cluster....** If the cluster orders are different, change one of them so that both are the same.

5. Finally, place a **Rectangular Stop Button** (**Boolean** palette) on the front panel. Note that this button is FALSE by default. Do not change its state.

Stop if TRUE

6. Build the block diagram shown in Figure 7.41. Notice that even though each cluster contains four objects, you see only one terminal per cluster on the block diagram. Make sure you change the conditional terminal of the While Loop so that it is set to **Stop if True** (you can do this by popping up on the conditional terminal).

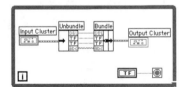

Figure 7.41

Unbundle
Function

Unbundle function (**Cluster** palette). This function will break out the cluster so that you can access individual elements. Resize it so that it has four outputs. The data-type labels will appear on the **Unbundle** function after you wire <u>Input Cluster</u> to it.

Bundle
Function

Bundle function (**Cluster** palette). This function reassembles your cluster. Resize it so that it contains four inputs.

*You can also access **Bundle** and **Unbundle** by popping up the cluster terminal you plan to wire to and choosing the function you want from the **Cluster Tools** menu. The chosen function will appear containing the correct number of input or output terminals.*

7. Return to the front panel and run the VI. Enter different values for the control cluster and watch how the indicator cluster echoes the values. Press the Stop button to halt execution.

You might have noticed that you could really just wire from the <u>Input Cluster</u> terminal to the <u>Output Cluster</u> terminal and the VI would do the same thing, but we wanted you to practice with **Bundle** and **Unbundle**.

8. Close and save the VI as **Cluster Exercise.vi** in your MYWORK directory or VI library.

7.19 Bundling and Unbundling by Name

Sometimes you don't need to assemble or disassemble an entire cluster—you just need to operate on an element or two. You can use **Bundle By Name** and **Unbundle By Name** to do just that!

Bundle By Name, found in the **Cluster** palette, references elements by name instead of by position (as **Bundle** does). Unlike **Bundle**, you can access only the elements you need. However, **Bundle By Name** cannot create new clusters; it can only replace an element in an existing cluster. Unlike **Bundle**, you must always wire to **Bundle By Name**'s middle input terminal to tell the function in which cluster to replace the element.

Figure 7.42
Bundle By Name.

Unbundle By Name, also located in the **Cluster** palette, returns elements whose name(s) you specify. You don't have to worry about cluster order or correct **Unbundle** function size.

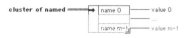

Figure 7.43
Unbundle By Name.

*Make sure all cluster elements have owned labels when you are using the **By Name** functions. Obviously, you can't access by name if you don't have names—LabVIEW won't know what you want!*

For example, if you wanted to replace the value of <u>Boolean 2</u> in the last exercise, you could use the **Bundle By Name** function without having to worry about cluster order or size.

Figure 7.44

Similarly, if you only needed to access the value of <u>String</u>, you would want to use the **Unbundle By Name** function.

Figure 7.45

As soon as you wire the cluster input of **Bundle By Name** or **Unbundle by Name**, the name of the first element in the cluster appears in the **name** input or output. To access another element, click on the name input or output with the Operating or Labeling tool. You should see a list of the names of all labeled elements in the cluster. Select your element from this list, and the name will appear in the **name** terminal. You can also access this list by popping up on name and choosing **Select Item>>**.

Figure 7.46

Both functions can be resized to accommodate as many elements as you need; select each component you want to access individually. As an added plus, you no longer have to worry about your program breaking if you resize your cluster. The **By Name** functions do not break unless you remove an element they reference.

7.20 Activity 7-5: More Fun with Clusters

In this activity, you will build a VI that checks whether the value of the Numeric 1 digital control in the input cluster is greater than or equal to zero. If it is less than zero, the VI will take the absolute value of all controls. If Numeric 1 is greater than or equal to zero, the VI does not take the absolute value of any controls. Regardless of the value of Numeric 1, the VI multiplies all values by 0.5 and displays the results in Output Cluster, demonstrating how you can use polymorphism with clusters.

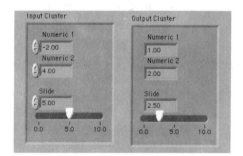

Figure 7.47

1. Open a new panel and place a **Cluster** shell (**Array & Cluster** palette) on it. Label it Input Cluster.

2. Create the Numeric 1, Numeric 2, and Slide controls from the **Numeric** palette. As you select them from the palette, click to place them inside the cluster shell. Make sure to create them in the order specified (since you will be wiring the Input Cluster to the Output Cluster) and give them labels.

3. Now create Output Cluster the same way using indicators (and make sure to put elements into the cluster in the same order). Or you might create Output Cluster by cloning Input Cluster and then changing its label.

4. Build the block diagram shown in Figure 7.48. Make sure to build both the TRUE and FALSE cases of the Case Structure.

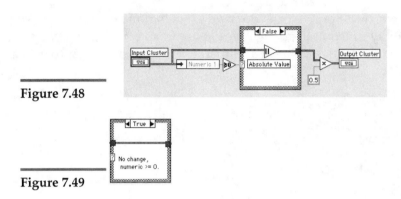

Figure 7.48

Figure 7.49

Unbundle by
Name Function

Unbundle By Name function (**Cluster** palette) extracts <u>Numeric 1</u> from <u>Input Cluster</u> so that you can compare it to zero. If "Numeric 1" isn't showing in the output name area, click with the Operating tool to choose it from the list of cluster elements.

Greater Or
Equal to 0?
Function

Greater Or Equal to 0? function (**Comparison** palette). Function returns true if the numeric input is zero or a positive number.

Absolute Value

Absolute Value (**Numeric** palette) returns the input number if that number is greater than or equal to zero; returns the inverse of the input number if the input number is less than zero. In this activity, it takes the absolute value of the entire cluster.

5. Run the VI. Try both positive and negative values of <u>Numeric 1</u>. Note the use of polymorphism to multiply all values in the cluster by 0.5 at the same time and to find the absolute value of the entire cluster.

6. Save the VI as **Cluster Comparison.vi** in your MYWORK directory or VI library.

7.21 Interchangeable Arrays and Clusters

Sometimes you will find it convenient to change arrays to clusters, and vice versa. This trick can be extremely useful, especially since LabVIEW includes many more functions that operate on arrays than clusters. For example, maybe you have a cluster of buttons on your front panel and you want to reverse the order of the buttons' values. Well, **Reverse 1D** Array would be perfect, but it only works on arrays. Have no fear—you can use the **Cluster to Array** function to change the cluster to an array, use **Reverse 1D** Array to switch the values around, and then use **Array to Cluster** to change back to a cluster.

array ——⬛— cluster   cluster ══════⬛········· array

Array to Cluster   Cluster to Array

Figure 7.50

Cluster To Array converts a cluster of N elements of the same data type into an array of N elements of that type. Array index corresponds to cluster order (i.e., cluster element 0 becomes the value at array index 0). You cannot use this function on a cluster containing arrays as elements, because Lab-VIEW won't let you create an array of arrays. Note that, to use this function, all elements in the cluster must have the same data type.

Array To Cluster converts an N-element, 1D array into a cluster of N elements of the same data type; you must pop up on the **Array To Cluster** terminal and choose **Cluster Size. . .** to specify the size of the output cluster since clusters don't size automatically like arrays do. The cluster size defaults to nine; if your array has fewer than the number of elements specified in the cluster size, LabVIEW will automatically fill in the extra cluster values with the default value for the data type of the cluster. However, if the input array has a greater number of elements than the value specified in the cluster size window, the block diagram wire going to the output cluster will break until you adjust the size.

Both functions are very handy if you want to display elements in a front panel cluster control or indicator but need to manipulate the elements by index value on the block diagram. They can be found in both the **Array** and **Cluster** subpalettes of the Functions palette.

7.22 Wrap It Up!

An *array* is a collection of ordered data elements of the same type. In Lab-VIEW, arrays can be of any data type, except chart, graph, or another array. You must create an array using a two-step process: First, place an array shell

(**Array & Cluster** subpalette of the **Controls** palette) in the window, and then add the desired control or indicator to the shell.

LabVIEW offers many functions to help you manipulate arrays, such as **Build Array** and **Index Array**, in the **Array** subpalette of the **Functions** palette. Most often, you will use these array functions to work with only 1D arrays; however, these functions are smart and will work similarly with multidimensional arrays (although sometimes you need to resize them first).

Both the For Loop and the While Loop can accumulate arrays at their borders using *auto-indexing,* a useful feature for creating and processing arrays. Remember that by default LabVIEW enables indexing in For Loops and disables indexing in While Loops.

Polymorphism is a fancy name for the ability of a function to adjust to inputs of different-sized data. We talked about polymorphic capabilities of arithmetic functions; however, many other functions are also polymorphic.

Clusters also group data, but unlike arrays, they will accept data of different types. You must create them on the front panel in a two-step process: First, place a cluster shell (**Array & Cluster** subpalette of the **Controls** palette) on the front panel, and then add the desired controls or indicators to the shell. Keep in mind that objects inside a cluster must be either all controls or all indicators. You cannot combine both controls and indicators within one cluster.

Clusters are useful for reducing the number of wires or terminals associated with a VI. For example, if a VI has many front panel controls and indicators that you need to associate with terminals, it is easier to group them as a cluster and have only one terminal.

The **Unbundle** function (**Cluster** palette) splits a cluster into each of its individual components. **Unbundle By Name** works similarly to **Unbundle** but accesses elements by their label. You can access as many or as few elements as you like using **Unbundle By Name**, whereas you have access to the whole cluster using **Unbundle** (and consequently have to worry about correctly sized terminals and cluster order).

The **Bundle** function (**Cluster** palette) assembles individual components into a single cluster or replaces an element in a cluster. **Bundle By Name** can't assemble clusters, but it can replace individual elements in a cluster without accessing the entire cluster In addition, with **Bundle By Name**, you don't have to worry about cluster order or correct **Bundle** function size. Just make sure all cluster elements have names when using **Bundle By Name** and **Unbundle By Name**!

7.23 Additional Activities

Activity 7-6: Reversing the Order Challenge

Build a VI that reverses the order of an array containing 100 random numbers. For example, `array[0]` becomes `array[99]`, `array[1]` becomes `array[98]`, and so on. Name the VI **Reverse Random Array.vi.**

Activity 7-7: Taking a Subset

Build a VI that generates an array containing 100 random numbers and displays a portion of the array, for example, from index 10 to index 50. Name the VI **Subset Random Array.vi.**

Use the **Array Subset** function (**Array** palette) to extract the portion of the array.

Activity 7-8: Dice! Challenge

Build a VI that simulates the roll of a die (possible values 1–6) and keeps track of the number of times that the die rolls each value. Your input is the number of times to roll the die, and the outputs include (for each possible value) the number of times the die fell on that value. Name the VI **Die Roller.vi.**

You will need to use a shift register to keep track of values from one iteration of a loop to the next.

Activity 7-9: Multiplying Array Elements

Build a VI that takes an input 1D array, and then multiplies pairs of elements together (starting with elements 0 and 1) and outputs the resulting array. For example, the input array with values 1, 23, 10, 5, 7, 11 will result in the output array 23, 50, 77. Name the VI **Array Pair Multiplier.vi.**

OVERVIEW

LabVIEW's charts and graphs let you display plots of data in a graphical form. Charts interactively plot data, appending new data to old so that you can see the current value in the context of previous data, as the new data become available. Graphs plot pregenerated arrays of values in a more traditional fashion, without retaining previously generated data. In this chapter, you will learn about charts and graphs, several ways to use them, and some of their special features. You will also learn about LabVIEW's special intensity charts and graphs, 3D graphs, and digital waveform graphs. Finally, you will take a look at the waveform data type, a useful LabVIEW representation for time-based data.

GOALS

- Understand the uses of charts and graphs
- Be able to recognize a chart's three modes: strip, scope, and sweep
- Understand mechanical action of Boolean switches
- Recognize the difference in functionalities of charts and graphs
- Customize the appearance of charts and graphs by changing the scales and using the palette, legend, and cursors
- Be familiar with the intensity and 3D charts and graphs to plot three dimensions of data
- Know about the digital waveform graph for displaying digital signals
- Understand the waveform data type, what its components are, and when to use it

KEY TERMS

- Plot
- Waveform chart
- Plot legend
- Scale legend
- Graph palette
- Cursor
- Waveform graph
- *XY* graph
- Intensity charts and graphs
- 3D graphs
- Digital waveform graph
- Waveform data type
- *t0, dt, Y*

LabVIEW's Exciting Visual Displays: Charts and Graphs

8

8.1 Waveform Charts

A plot is simply a graphical display of X versus Y values. Often, Y values in a plot represent the data value, while X values represent time. The *waveform chart*, located in the **Graph** subpalette of the **Controls** palette, is a special numeric indicator that can display one or more plots of data. Most often used inside loops, charts retain and display previously acquired data, appending new data as they become available in a continuously updating display. In a chart, the Y values represent the new data, and X values represent time (often, each Y value is generated in a loop iteration, and so the X value represents the time for one loop). LabVIEW has only one kind of chart, but the chart has three different update modes for interactive data display. Figure 8.1 shows an example of a multiple-plot waveform chart.

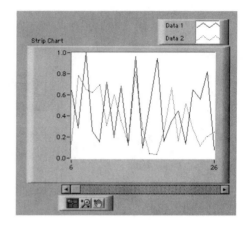

Figure 8.1

8.1.1 Chart Update Modes

The waveform chart has three update modes—*strip chart mode, scope chart mode,* and *sweep chart mode,* shown in Figure 8.2. The update mode can be changed by popping up on the waveform chart and choosing one of the options from the **Advanced>>Update Mode>>** menu. If you want to change modes while the VI is running (and is subsequently in run mode, where the menus are slightly different), select **Update Mode** from the chart's runtime pop-up menu.

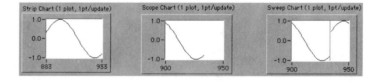

Figure 8.2

The strip chart has a scrolling display similar to a paper strip chart. The scope chart and the sweep chart have retracing displays similar to that of an oscilloscope. On the scope chart, when the plot reaches the right border of the plotting area, the plot erases, and plotting begins again from the left border. The sweep chart acts much like the scope chart, but the display does not go blank when the data reaches the right border. Instead, a moving vertical line marks the beginning of new data and moves across the display as new data are added. These distinctions are much easier to understand when you actually see the different modes in action, so don't worry if it sounds confusing now. You'll get to experiment with them in the next activity.

Because there is less overhead in retracing a plot, the scope chart and the sweep chart operate significantly faster than the strip chart.

8.1.2 Single-Plot Charts

The simplest way to use a chart is to wire a scalar value to the chart's block diagram terminal, as shown in Figure 8.3. Each loop iteration draws one more point on the displayed waveform.

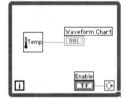

Figure 8.3

8.1.3 Wiring a Multiple-Plot Chart

Waveform charts can also accommodate more than one plot. However, because you can't wire from multiple block diagram sources to a single chart terminal, you must first bundle the data together using the **Bundle** function (**Cluster** palette). In Figure 8.4, the **Bundle** function "bundles" or groups the outputs of the three different VIs that acquire temperature into a cluster so that they can be plotted on the waveform chart. Notice the change in the waveform chart terminal's appearance when it's wired to the **Bundle** function. To add more plots, simply increase the number of **Bundle** input terminals by resizing using the Positioning tool.

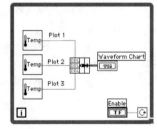

Figure 8.4

*For an online example of charts, their modes, and their expected data types, open and run **Charts.vi** in CH8.LLB of the EVERYONE directory.*

8.1.4 Show the Digital Display?

Like many other numeric indicators, charts have the option to show or hide the digital display (pop up on the chart to get the **Visible Items>>** option). The digital display shows the most recent value displayed by the chart.

8.1.5 The Scrollbar

Charts also have scrollbars that you can show or hide. You can use the scrollbar to display older data that have scrolled off the chart.

8.1.6 Clearing the Chart

Sometimes you will find it useful to remove all previous data from the chart display. Select **Data Operations>>Clear Chart** from the chart's pop-up menu to clear a chart from edit mode. (Remember, you are usually in edit mode if your VI is not running. To switch between modes when the VI is not running, choose **Change to Run/Edit Mode** from the **Operate** menu.) If you are in run mode, **Clear Chart** is a pop-up menu option instead of being hidden under **Data Operations**.

8.1.7 Stacked and Overlaid Plots

If you have a multiple-plot chart, you can choose whether to display all plots on the same *Y* axis, called an overlaid plot, or you can give each plot its own *Y* scale, called a stacked plot. You can select **Stack Plots** or **Overlay Plots** from the chart's pop-up menu to toggle the type of display. Figure 8.5 illustrates the difference between stacked and overlaid plots.

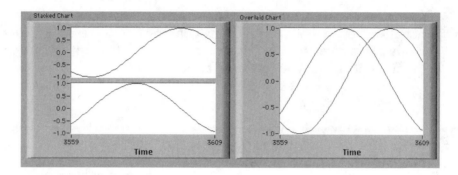

Figure 8.5

8.1.8 Multiple *X* and *Y* scales

If you have a multi-plot chart, sometimes you will want an overlaid plot that has different scales for each plot; for example, if one plot's *Y* range is from 0.0 to 1.0 and the other is –1,000 to +1,000, it would be difficult to see both of them overlaid without a separate scale for each. You can create separate scales for both the *X* and *Y* axes by popping up on the corresponding axis and selecting **Y/X Scale>>Duplicate Scale.** Figure 8.6 illustrates two *Y* scales for a multi-plot chart, with each scale on one side of the chart.

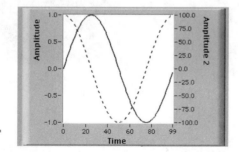

Figure 8.6

8.1.9 Chart History Length

By default, a chart can store up to 1,024 data points. If you want to store more or less data, select **Chart History Length...** from the pop-up menu and specify a new value of up to 100,000 points. Changing the buffer size does not affect how much data are shown on the screen—resize the chart to show more or less data at a time. However, increasing the buffer size does increase the amount of data you can scroll back through.

8.2 Activity 8-1: Temperature Monitor

You will build a VI to measure temperature and display it on the waveform chart. This VI will measure the temperature using the **Thermometer** VI you built as a subVI in a previous lesson.

1. Open a new front panel. You will recreate the panel shown in Figure 8.7 (but feel free not to type in the comments—they're for your benefit).

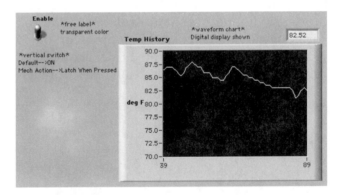

Figure 8.7

2. Place a vertical switch (**Boolean** palette) in the front panel window. Label the switch <u>Enable</u>. You will use the switch to stop the temperature acquisition.

3. Place a waveform chart (**Graph** palette) in the panel window. Label the waveform chart <u>Temp History</u>. The waveform chart will display the temperature in real time.

4. The waveform chart has a digital display that shows the latest value. Pop up on the waveform chart and choose **Visible Items>>Digital Display** from the pop-up menu.

5. Because the temperature sensor measures room temperature, rescale the waveform chart so that you can see the temperature (otherwise it will be "off the chart"). Using the Labeling tool, double-click on "10.0" in the waveform chart scale, type 90, and click outside the text area. The click enters the value. You also can press <enter> to input your change to the scale. Change "0.0" to 70 in the same way.

6. Open the block diagram window and build the diagram shown in Figure 8.8.

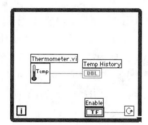

Figure 8.8

7. Place the **While Loop** (**Structures** palette) in the block diagram window and size it appropriately.

8. Place the two terminals inside the While Loop if they aren't already there.

9. Import the **Thermometer** subVI.

Thermometer.vi

Thermometer.vi. This VI returns one temperature measurement from a temperature sensor (or a simulation, depending on your setup). You should have written it in Chapter 4 and modified it in Chapter 5. Load it using the **Select A VI...** button on the **Functions** palette. It should probably be in your MYWORK directory. If you don't have it, you can use **Thermometer.vi** in CH5.LLB or **Digital Thermometer.vi**, located in the **Tutorial** subpalette of the **Functions** palette.

10. Wire the block diagram as shown in Figure 8.8.

11. Return to the front panel and turn on the vertical switch by clicking on it with the Operating tool. Run the VI.

Remember, the While Loop is an indefinite looping structure. The subdiagram within its border will execute as long as the specified condition is TRUE. In this example, as long as the switch is on (TRUE), the **Thermometer** subVI will return a new measurement and display it on the waveform chart.

12. To stop the acquisition, click on the vertical switch. This action gives the loop conditional terminal a FALSE value and the loop ends.

13. The waveform chart has a display buffer that retains a number of points after they have scrolled off the display You can show this scrollbar by popping up on the waveform chart and selecting **Visible Items>>Scrollbar** from the pop-up menu. You can use the Positioning tool to adjust the scrollbar's size and position.

    To scroll through the waveform chart, click on either arrow in the scrollbar.

    To clear the display buffer and reset the waveform chart, pop up on the waveform chart and choose **Data Operations>>Clear Chart** from the pop-up menu. If you want to clear the chart while you're in run mode, select **Clear Chart** from the runtime pop-up menu.

14. Make sure the switch is TRUE and run the VI again. This time, try changing the update mode of the chart. Pop up and choose **Update Mode>>Scope Chart** from the chart's runtime menu. Notice the difference in chart display behavior. Now choose **Sweep Chart** and see what happens.

Using Mechanical Action of Boolean Switches

Take a step out of this activity for a second. You've certainly noticed by now that each time you run this VI, you first must turn on the vertical <u>Enable</u> switch before clicking the run button, or the loop will only execute once. You can modify the *mechanical action* of a Boolean control to change its behavior and circumvent this inconvenience. LabVIEW offers six possible choices for the mechanical action of a Boolean control.

Switch When
Pressed
   Switch When Pressed action changes the control's value each time you click on the control with the Operating tool. This action is the default for Booleans and is similar to that of a ceiling light switch. It is not affected by how often the VI reads the control.

Switch When
Released
   Switch When Released action changes the control's value only when you release the mouse button during a mouse click within the graphical boundary of the control. The action is not affected by how often the VI reads the control. This mode is similar to what happens when you click on a check mark in a dialog box; it becomes highlighted but does not change until you release the mouse button.

Switch Until
Released

Switch Until Released action changes the control's value when you click on the control. It retains the new value until you release the mouse button, at which time the control reverts to its original value. The action is similar to that of a door buzzer and is not affected by how often the VI reads the control.

Latch When
Pressed

Latch When Pressed action changes the control's value when you click on the control. It retains the new value until the VI reads it once, at which point the control reverts to its default value. This action happens whether or not you continue to press the mouse button. **Latch When Pressed** is similar in functionality to a circuit breaker and is useful when you want the VI to do something only once for each time you set the control, such as to stop a While Loop when you press a <u>STOP</u> button.

Latch When
Released

Latch When Released action changes the control's value only after you release the mouse button. When your VI reads the value once, the control reverts to the old value. This action guarantees at least one new value. As with **Switch When Released**, this mode is similar to the behavior of buttons in a dialog box; the button becomes highlighted when you click on it and latches a reading when you release the mouse button.

Latch Until
Released

Latch Until Released changes the control's value when you click on the control. It retains the value until your VI reads the value once or until you release the mouse button, whichever occurs last.

For example, consider a vertical switch—its default value is off (FALSE).

15. Modify the vertical switch in your VI so that you do not need to turn the switch to TRUE each time you run the VI.

   a) Turn the vertical switch to *on* (TRUE).

   b) Pop up on the switch and choose **Data Operations>>Make Current Value Default** from the pop-up menu to make the *on* position the default value.

   c) Pop up on the switch and choose **Mechanical Action>>Latch When Pressed** from the pop-up menu.

16. Run the VI. Click on the vertical switch to stop the acquisition. The switch will move to the *off* position briefly and then automatically change back to *on* after the While Loop conditional terminal reads one FALSE value.

You cannot use mechanical action on a Boolean object if you will be modifying the object's value using local variables. We'll tell you why when we talk about locals in Chapter 12.

Adding Timing

When you run the VI in this activity, the While Loop executes as quickly as possible. You may want to take data at certain intervals, however, such as once per second or once per minute. You can control loop timing using the **Wait Until Next ms Multiple** function (**Time & Dialog** menu).

17. Modify the VI to take a temperature measurement about once every half-second by placing the code segment shown in Figure 8.9 into the While Loop.

Figure 8.9

Wait Until Next
ms Multiple

Wait Until Next ms Multiple function (**Time & Dialog** menu) ensures that each iteration waits the specified amount of time (in this case, a half-second or 500 milliseconds) before continuing.

18. Run the VI. Run it several more times, trying different values for the number of milliseconds.

19. Save and close the VI. Name it **Temperature Monitor.vi** and place it in your MYWORK directory or VI library. Excellent job!

8.3 Graphs

Unlike charts, which plot data interactively, graphs plot pregenerated arrays of data all at once; they do not have the ability to append new values to previously generated data. LabVIEW provides several types of graph for greater flexibility: *waveform graphs, XY graphs, intensity graphs, 3D graphs, digital waveform graphs*, and some specialized graphs (*Smith plots, Polar charts, Min–Max*, and *distribution* plots). We'll talk about waveform and *XY* graphs now and cover intensity and 3D graphs in Section 8.8. Both waveform graphs and *XY* graphs look identical on the front panel of your VI but have very different functionality.

An example of a graph with several graph options enabled is shown in Figure 8.10.

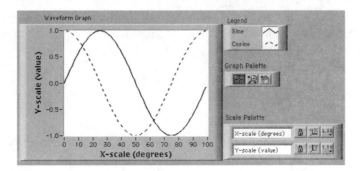

Figure 8.10

You can obtain both types of graph indicators from the **Graph** subpalette of the **Controls** palette. The waveform graph plots only single-valued functions (only one Y value for every X) with uniformly spaced points, such as acquired time-sampled, amplitude-varying waveforms. The waveform graph is ideal for plotting arrays of data in which the points are evenly distributed.

The *XY graph* is a general-purpose, Cartesian graph, ideal for plotting data with varying time bases or data with several Y values for every X value, such as circular shapes. The two types of graph look the same but take different types of input, so you must be careful not to confuse them.

8.3.1 Single-Plot Waveform Graphs

For basic single-plot graphs, you can wire an array of Y values directly to a waveform graph terminal, as shown in Figure 8.11. This method assumes that the initial X value is zero and that the delta X value (i.e., the increment between X values) is one. Notice that the graph terminal in the block diagram appears as an array indicator here.

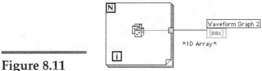

Figure 8.11

Sometimes you will want the flexibility to change the time base for the graph. For example, you start sampling at a time other than "initial $X = 0$" (or $X_0 = 0$), or your samples are spaced more than one unit apart, or "delta $X = 1$" (also written $\Delta X = 1$). To change the time base, bundle the X_0 value, ΔX value, and the data array into a cluster; then wire the cluster to the graph. Notice in Figure 8.12 that the graph terminal now appears as a cluster indication.

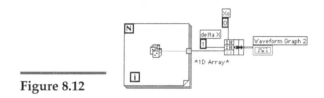

Figure 8.12

8.3.2 Multiple-Plot Waveform Graphs

You can show more than one plot on a waveform graph by creating an array (or a 2D array) of the data types used in the single-plot examples. Notice how graph terminals change appearance depending on the structure of data wired to them (array, cluster, array of clusters, etc.) and the data type (I16, DBL, etc.).

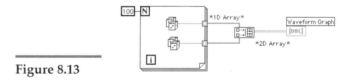

Figure 8.13

In Figure 8.13, it is assumed the initial X value is 0 and the delta X value is 1 for both arrays. The **Build Array** function creates a 2D array out of two 1D arrays. Notice that this 2D array has two rows with 100 columns per row—a 2×100 array. By default, graphs plot each *row* of a 2D array as a separate waveform. If your data are organized by column, you must make sure to transpose your array when you plot it! Transposing means simply switching row values with column values; for example, if you transpose an array with 3 rows and 10 columns, you end up with an array with 10 rows and 3 columns. LabVIEW makes it easy to do this—simply pop up on the graph and select **Transpose Array** (this menu option is grayed out if the graph does not have a 2D array wired to it). You can also use the **Transpose 2D Array** function found in the **Array** subpalette of the **Functions** menu.

In Figure 8.14, the X_0 value and ΔX (or delta X) value for each array are specified. These X parameters do not need to be the same for both sets of data.

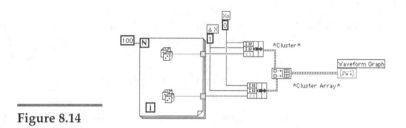

Figure 8.14

In this diagram, the **Build Array** function (**Array** palette) creates an array out of its two cluster inputs. Each input cluster consists of an array and two scalar numbers. The final result is an array of clusters, which the graph can accept and plot. In this case, the specified X_0 and ΔX are the same as the default values, but you can always use other values as well. You're now starting to learn about complex data structures—as long as you take them bit by bit, they're not too confusing!

8.4   Activity 8-2: Graphing a Sine on a Waveform Graph

You will build a VI that generates a sine wave array and plots it in a waveform graph. You will also modify the VI to graph multiple plots.

1. Open a new VI, and build the front panel shown in Figure 8.15.

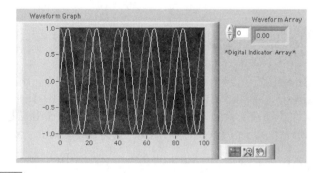

Figure 8.15

2. Place an array shell (**Array & Cluster** palette) in the front panel window. Label the array shell <u>Waveform Array</u>. Place a digital indicator (**Numeric** palette) inside the Data Object window of the array shell to display the array contents.

3. Place a waveform graph (**Graph** palette) in the front panel window. Label the graph <u>Waveform Graph</u> and enlarge it by dragging a corner with the Positioning tool.

   Hide the legend by popping up on the graph and selecting **Visible Items>>Plot Legend**.

   Disable autoscaling by popping up and choosing **Y Scale>>AutoScale Y**. Modify the Y axis limits by selecting the scale limits with the Labeling tool and entering new numbers; change the Y axis minimum to -1.0 and the maximum to 1.0. We'll talk more about autoscaling in the next section.

4. Build the block diagram shown in Figure 8.16.

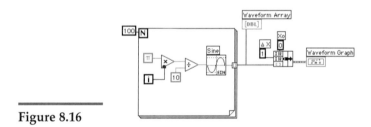

Figure 8.16

Sine Function

Sine function (**Numeric>>Trigonometric** palette) computes $\sin(x)$ and returns one point of a sine wave. The VI requires a scalar index input that it expects in radians. In this exercise, the x input changes with each loop iteration and the plotted result is a sine wave.

Pi Constant

Pi constant (**Numeric>>Additional Numeric Constants** palette) is the value of π to Pi.

Bundle
Function

Bundle function (**Cluster** palette) assembles the plot components into a single cluster. The components include the initial X value (0), the delta X value (1), and the Y array (waveform data). Use the Positioning tool to resize the function by dragging one of the corners.

Each iteration of the For Loop generates one point in a waveform and stores it in the waveform array created at the loop border. After the loop finishes execution, the **Bundle** function bundles the initial value of X, the delta value for X, and the array for plotting on the graph.

5. Return to the front panel and run the VI. Right now the graph should show only one plot.

6. Now change the delta X value to 0.5 and the initial X value to 20 and run the VI again. Notice that the graph now displays the same 100 points of data with a starting value of 20 and a delta X of 0.5 for each point (shown on the X axis).

7. Place the Positioning tool on the lower-right corner of the array until the tool becomes a grid, and drag. The indicator now displays several elements with indices ascending as you go from left to right (or top to bottom), beginning with the element that corresponds to the specified index, as illustrated in Figure 8.17. Don't forget that you can view any element in the array simply by entering the index of that element in the index display. If you enter a number greater than the array size, the display will dim.

Figure 8.17

In the block diagram in Figure 8.16, you specified an initial X and a delta X value for the waveform. Frequently, the initial X value will be zero and the delta X value will be 1. In these instances, you can wire the waveform array directly to the waveform graph terminal, as shown in Figure 8.18, and take advantage of the default delta X and initial X values for a graph.

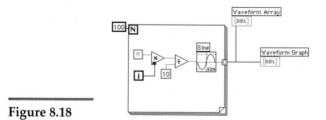

Figure 8.18

8. Return to the block diagram window. Delete the **Bundle** function and the **Numeric Constants** wired to it; then select **Remove Broken**

Wires from the **Edit** menu or use the keyboard shortcut. Finish wiring the block diagram as shown in Figure 8.18.

9. Run the VI. Notice that the VI plots the waveform with an initial X value of 0 and a delta X value of 1, just as it did before, with a lot less effort on your part to build the VI.

Multiple-Plot Graphs

You can also create multiple-plot waveform graphs by wiring the data types normally passed to a single-plot graph first to a **Build Array** function and then to the graph. Although you can't build an array of arrays, you can create a 2D array with each input array as a row.

1. Create the block diagram shown in Figure 8.19.

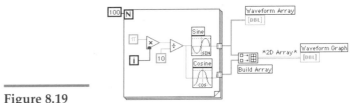

Figure 8.19

Build Array
Function

Build Array function (**Array** palette) creates the proper data structure to plot two arrays on a waveform graph. Enlarge the **Build Array** function to include two inputs by dragging a corner with the Positioning tool. You will want to make sure you *don't* select "concatenate inputs" (which is the default value selected in the pop-up menu) to the **Build Array** so that the output will be a 2D array.

Cosine
Function

Cosine function (**Numeric>>Trigonometric** palette) computes $\cos(x)$ and returns one point of a cosine wave. The VI requires a scalar index input that it expects in radians. In this exercise, the x input changes with each loop iteration and the plotted result is a cosine wave.

2. Switch to the front panel. Run the VI. Notice that the two waveforms both plot on the same waveform graph. The initial X value defaults to 0 and the delta X value defaults to 1 for both data sets.

3. Just to see what happens, pop up on the graph and choose **Transpose Array**. The plot changes drastically when you swap the rows

with the columns, doesn't it? Choose **Transpose Array** again to return things to normal.

4. Save and close the VI. Name it **Graph Sine Array.vi** in your MYWORK directory or VI library.

8.5 XY Graphs

The waveform graphs you have been using are designed for plotting evenly sampled waveforms. However, if you sample at irregular intervals or are plotting a mathematical function that has multiple Y values for every X value, you will need to specify points using their (X,Y) coordinates. *XY graphs* plot this different type of data; they require input of a different data type than waveform graphs. A single-plot XY graph and its corresponding block diagram are shown in Figures 8.20 and 8.21.

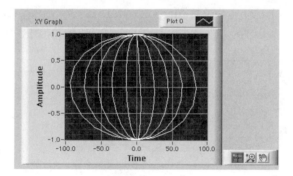

Figure 8.20

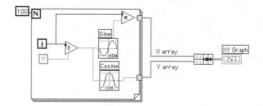

Figure 8.21

The XY graph expects an input of a bundled X array (the top input) and a Y array (the bottom input). The **Bundle** function (**Cluster** palette) combines the X and Y arrays into a cluster wired to the XY graph. The XY graph terminal now appears as a cluster indicator.

For a multiple-plot *XY* graph, simply build an array of the clusters of *X* and *Y* values used for single plots, as shown in Figure 8.21.

 *It is very easy to confuse the **Bundle** and **Build Array** functions when assembling data for graphs! Pay close attention to which one you need to use!*

 *For some online graph examples, look at **Waveform Graph.vi** and **XY Graph.vi** in* CH8.LLB *in the* EVERYONE *directory.*

8.6 Chart and Graph Components

Graphs and charts have many powerful features that you can use to customize your plots. This section covers how to configure these options.

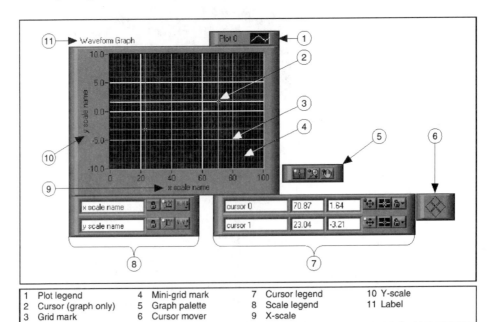

1 Plot legend	4 Mini-grid mark	7 Cursor legend	10 Y-scale
2 Cursor (graph only)	5 Graph palette	8 Scale legend	11 Label
3 Grid mark	6 Cursor mover	9 X-scale	

Figure 8.22

8.6.1 Playing with the Scales

Charts and graphs can automatically adjust their horizontal and vertical scales to reflect the points plotted on them; that is, the scales adjust themselves to show all points on the graph at the greatest resolution. You can turn this *autoscaling* feature on or off using the **AutoScale X** and **AutoScale Y** options from the **X Scale** menu, or the **Y Scale** menu of the object's pop-up menu. You can also control these autoscaling features from the Scale Legend (which we'll get to in a minute). LabVIEW defaults to autoscaling on for graphs and off for charts. However, using autoscaling may cause the chart or graph to update more slowly, depending upon the computer and video system you use, because new scales must be calculated with each plot.

If you don't want to autoscale, you can change the horizontal or vertical scale directly using the Operating or Labeling tool to type in a new number, just as you can with any other LabVIEW control or indicator, and turn autoscaling off.

X and *Y* Scale Menus

The *X* and *Y* scales each have a submenu of options as shown in Figure 8.23.

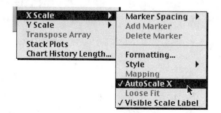

Figure 8.23

Use **AutoScale** to turn the autoscaling option on or off.

Normally, the scales are set to the exact range of the data when you perform an autoscale operation. You can use the **Loose Fit** option if you want LabVIEW to round the scale to "nicer" numbers. With a loose fit, the numbers are rounded to a multiple of the increment used for the scale. For example, if the markers increment by five, then the minimum and maximum values are set to a multiple of five instead of the exact range of the data.

The **Formatting...** option brings up a dialog box, shown in Figure 8.24, that allows you to configure the following things:

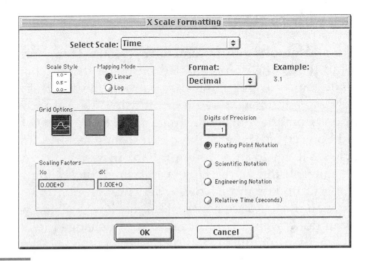

Figure 8.24

- The **Select Scale** is a pull-down menu allowing you to select the scale by name that you are formatting. In the previous figure, the *X* scale is labeled "Time."

- The **Scale Style** menu lets you select major and minor tick marks for the scale, or none at all. Click on this icon to see your choices. A major tick mark corresponds to a scale label, while a minor tick mark denotes an interior point between labels. This menu also lets you select the markers for a given axis as either visible or invisible.

- The **Mapping Mode** lets you select either a linear or logarithmic scale for the data display.

- The **Grid Options** lets you choose between no gridlines, gridlines only at major tick mark locations, or gridlines at both major and minor tick marks. You can also change the color of the gridlines here. Click on the grid buttons to see a pullout menu of your options.

You can set **X₀**, the *X* value you want to start at, and **dX**, the increment between *X* values (same as delta *X*), in the **Scaling Factors** section.

The **Format and Precision** section lets you choose **Numeric** format or **Time & Date** format. If you choose **Numeric**, you can then pick the number of digits of precision as well as the notation (**Floating Point**, **Scientific**, **Engineering**, or **Relative Time**) of the scale display. You can also specify your numbers in **Decimal**, **Unsigned Decimal**, **Hexadecimal**, **Octal**, or **Binary** format. If you choose **Time & Date** format, you can choose how you would like the time and date displayed.

The Scale Legend

The *scale legend* lets you create labels for the X and Y scales (or for multiple XY scales, if you have more than one) and have easy pop-up access to their configuration. The scale legend gives you buttons for scaling the X or Y axis, changing the display format, and autoscaling.

Popping up on the graph or chart and choosing **Visible Items>>Scale Legend** will give you a box like the one shown in Figure 8.25.

Figure 8.25

In the scale legend, you can type the names of your scales in the text box. This text will show up on the graph or chart's box X or Y axis.

You can click on the buttons to configure the same options as in the **X I Y Scale Formatting** pop-up window just described. It's just a more convenient place to access this information for some people.

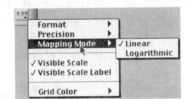

Figure 8.26

Scale Lock
Button

Use the Operating tool to click on the Scale Lock button to toggle autoscaling for each scale, the visibility of scales, etc.

You can resize the Scale Legend with the Positioning tool, just like an array, to create, show, or hide additional scales on your chart or graph.

8.6.2 The Plot Legend

Charts and graphs use a default style for each plot unless you have created a custom plot style for it. The *plot legend* lets you label, color, select line style, and choose point style for each plot. Show or hide the legend using the **Visible Items>>Plot Legend** option of the chart or graph pop-up menu. You can also specify a name for each plot in the legend. An example of a legend is shown in Figure 8.27.

Figure 8.27

When you select **Plot Legend**, only one plot shows in the box that appears. You can show more plots by dragging down a corner of the legend with the Positioning tool. After you set plot characteristics in the **Plot Legend**, the plot retains those settings, regardless of whether the legend is visible. If the chart or graph receives more plots than are defined in the legend, LabVIEW draws the extra plots in default style.

When you move the chart or graph body, the legend moves with it. You can change the position of the legend relative to the graph by dragging only the legend to a new location. *Resize the legend on the left to give labels more room in the window or on the right to give plot samples more room.*

By default, each plot is labeled with a number, beginning with zero. You can modify this label the same way you modify other LabVIEW labels—just start typing with the Labeling tool. Each plot sample has its own menu to change the plot, line, color, and point styles of the plot. You can also access this menu by clicking on the legend with the Operating tool.

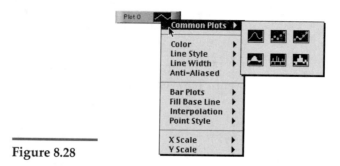

Figure 8.28

- The **Common Plots** option lets you easily configure a plot to use any of six popular plot styles, including a scatter plot, a bar plot, and a fill to zero plot. Options in this subpalette configure the point, line, and fill styles of the graph in one step (instead of setting these options individually, as listed next).

- The **Color** option displays the Color palette so that you can select the plot color. You can also color the plots on the legend with the Color tool. You can change the plot colors while running the VI.

- The **Line Style** and **Line Width** offer different types of patterns and widths for the lines to plot.

- The **Anti-Aliased** option makes line plots appear smoother and nicer overall, but be aware that anti-aliased line drawing can be computationally intensive and slow performance.

- The **Bar Plots** option lets you create bar plots of 100%, 75%, or 1% width, either horizontally or vertically. The **Fill Baseline** option controls the baseline of the bars; plots can have no fill or they can fill to zero, negative infinity, or infinity.

- The **Interpolation** option determines how LabVIEW draws lines between data points. The first option does not draw any lines, making it suitable for a scatter plot (in other words, you get points only). The option at the bottom left draws a straight line between points. The four stepped options link points with a right-angled elbow, useful for histograms.

Figure 8.29

- The **Point Style** options display different point styles (none, round, square, hollow, filled, etc.) from which you can choose.

- The **X Scale** and **Y Scale** are useful for specifying which plot corresponds to which *X* and *Y* scale, when you have multiple scales on your chart or graph.

8.6.3 Activity 8-3: Using an *XY* Graph to Plot a Circle

You will build a VI that plots a circle on dependent *X* and *Y* arrays.

1. Open a new front panel. You will recreate the panel shown in Figure 8.30.

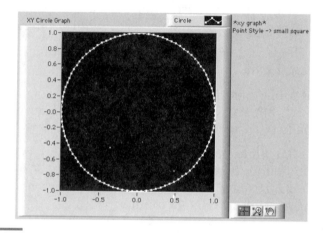

Figure 8.30

2. Place an **XY Graph** (Graph palette) in the panel window. Label the graph <u>XY Circle Graph</u>.

3. Enlarge the graph by dragging a corner with the Positioning tool. Try to make the plot region approximately square.

 You can maintain the aspect ratio while dragging by holding down the <shift> key and dragging diagonally.

4. Pop up on the graph and select **Visible Items>>Plot Legend**. Resize the legend from the left side and enter the `Circle` label using the Labeling tool. Pop up on the line in the legend and select the small square from the **Point Style** palette. Then select a new plot color from the **Color** palette.

5. Build the block diagram shown in Figure 8.31.

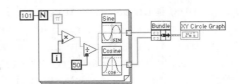

Figure 8.31

Sine Function

Cosine
Function

Bundle
Function

Pi Constant

Sine and **Cosine** functions (**Numeric>>Trigonometric** palette) calculate the sine and cosine, respectively, of the input number. In this exercise, you use the function in a For Loop to build an array of points that represents one cycle of a sine wave and one cycle of a cosine wave.

Bundle function (**Cluster** palette) assembles the sine array (X values) and the cosine array (Y values) to plot the sine array against the cosine array.

Pi constant (**Numeric>>Additional Numeric Constants** palette) is used to provide radian input to sine and cosine functions.

Using a **Bundle** function, you can graph the one-cycle sine array versus the one-cycle cosine array on an XY graph, which produces a circle.

6. Return to the front panel and run the VI. Save it as **Graph Circle.vi** in your MYWORK directory or VI library. Congratulations! You're getting the hang of this stuff!

8.6.4 Using the Graph Palette

The *graph palette* is the little box that shows up at the right bottom of the graph or chart when you first drop it on the front panel. With the graph palette, you can pan (i.e., scroll the display area) and focus in on a specific area using palette buttons (called zooming). You can also move cursors around (we'll talk about cursors shortly). The palette, which you access from the **Visible Items** menu of the chart or graph pop-up menu, is shown in Figure 8.32.

Figure 8.32

Standard Operate Mode   The remaining three buttons let you control the operation mode for the graph. Normally, you are in standard mode, meaning that you can click on the graph cursors to move them around. If you press the pan button, then you switch to a mode where you can scroll the visible data by dragging sections of the graph with the pan cursor. If you click on the zoom button, you get a pop-up menu that lets you choose from several methods of zooming (focusing on specific sections of the graph by magnifying a particular area).

Zoom Button

Pan Button

Here's how these zoom options work:

Zoom by Rectangle. Drag the cursor to draw a rectangle around the area you want to zoom in on. When you release the mouse button, your axes will rescale to show only the selected area.

Zoom by Rectangle

Zoom by Rectangle in X provides zooming restricted to X data (the Y scale remains unchanged).

Zoom by Rectangle in X

Zoom by Rectangle in Y provides zooming restricted to Y data (the X scale remains unchanged).

Zoom by Rectangle in Y

Undo Last Zoom is pretty obvious. It also resets the scales to their previous setting.

Undo Last Zoom

Zoom In about a Point. If you hold down the mouse on a specific point, the graph will continuously zoom in until you release the mouse button.

Zoom In about a Point

Zoom Out about a Point. If you hold down the mouse on a specific point, the graph will continuously zoom out until you release the mouse button.

Zoom Out about a Point

> *Note:* For the last two modes, ***Zoom In*** and ***Zoom Out about a Point***, shift-clicking will zoom in the other direction.

8.6.5 Graph Cursors

LabVIEW graphs have *cursors* to mark data points on your plot to further animate your work. Figure 8.33 shows a picture of a graph with cursors and the **Cursor Legend** visible. You won't find cursors on charts, so don't bother to look.

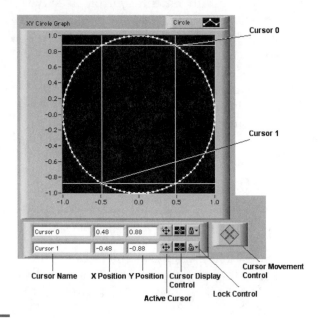

Figure 8.33

You can view the Cursor palette by selecting **Visible Items>>Cursor Legend** from the graph's pop-up menu. When the palette first appears, it is grayed out. Click on the Active Cursor button to make a cursor active (or type text in one of the Cursor Name fields with the Labeling tool). You can move cursors around manually, or programmatically using property nodes, which you'll learn about in Chapter 12.

To manually move a cursor, drag it around the graph with the Operating tool. If you drag the intersection point, you can move in all directions. If you drag the horizontal or vertical lines of the cursor, you can drag only horizontally or vertically, respectively.

You can also use the Cursor Movement Control to move a cursor up, down, left, and right.

A graph can have as many cursors as you want. The **Cursor** legend helps you keep track of them, and you can stretch it to display multiple cursors.

What does this **Cursor** legend do, you ask? You can label the cursor in the first box at the left of the Cursor legend. The next box shows X position, and the next after that shows Y position. Click on the little crosshair Active

Cursor button to make the cursor on that line active and select it for movement (a little green light denotes an active, selected cursor). When you click a button on the Cursor Movement Control, all active cursors move.

Click with the Operating tool (do not pop up!) on the Cursor Display Control (which looks like a crosshair) to specify things like cursor style, point style, and color.

The little lock at the far right will lock the cursor to a plot. Locking restricts cursor movement so it can reside only on plot points, but nowhere else in the graph. Click on the lock with the Operating tool to access its menu (don't pop up or you'll get the wrong one!).

To delete a cursor, pop up on the **Cursor** legend, and choose **Data Operations>>Delete Element**.

8.7   Activity 8-4: Temperature Analysis

You will build a VI that measures temperature approximately every 0.25 seconds for 10 seconds. During the acquisition, the VI displays the measurements in real time on a waveform chart. After the acquisition is complete, the VI plots the data on a graph and calculates the minimum, maximum, and average temperatures.

1. Open a new front panel and build the VI shown in Figure 8.34.

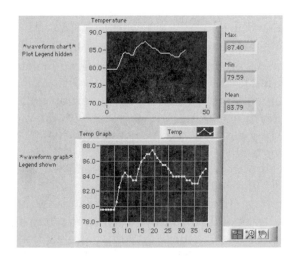

Figure 8.34

2. Rescale the chart so that it ranges from 70.0 to 90.0. Also, make sure autoscaling is on for both axes of the graph (it should be, because that is the default).

3. In the Plot Legend, using the Labeling tool, type in Temp as shown. Now pop up (or click with the Operating tool) on the <u>Temp</u> plot representation in the legend and change the **Point Style** to small squares. Feel free to color your plots as well.

   The <u>Temperature</u> chart displays the temperature as it is acquired. After acquisition is finished, the VI plots the data in <u>Temp Graph</u>. The <u>Mean</u>, <u>Max</u> and <u>Min</u> digital indicators will display the average, maximum, and minimum temperatures, respectively.

4. Build the block diagram shown in Figure 8.35. Make sure to use the Help window to display the inputs and outputs of these functions, and pay attention to the wire whiskers and tip strips, or you will almost certainly wire to the wrong terminal!

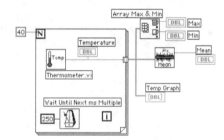

Figure 8.35

Thermometer.vi

Thermometer.vi. Use the **Select A VI...** palette to access the one you built. It should probably be in your MYWORK directory. If you don't have it, you can use **Thermometer.vi** in CH5.LLB or **Digital Thermometer.vi**, located in the **Tutorial** subpalette of the **Functions** palette. **Thermometer** returns one temperature measurement each time it is called.

Wait Until Next
ms Multiple

Wait Until Next ms Multiple function (**Time & Dialog** palette) causes the For Loop to execute every 0.25 seconds (250 ms).

Array Max &
Min

Array Max & Min function (**Array** palette) returns the maximum and minimum values in the array. In this case it returns the maximum and minimum temperatures measured during the acquisition.

Mean.vi

Mean.vi (**Mathematics>>Probability and Statistics** palette) returns the average of the temperature measurements. Make sure you wire to the right output terminal to get the mean.

5. The For Loop executes 40 times. The **Wait Until Next ms Multiple** function causes each iteration to take place approximately every 250 ms. The VI stores the temperature measurements in an array created at the For Loop border using auto-indexing. After the For Loop completes, the array passes to various nodes. The **Array Max & Min** function returns the maximum and minimum temperature. **Mean.vi** returns the average of the temperature measurements.

6. Return to the front panel and run the VI.

7. Using the Scale Legend (show it by choosing **Visible Items>>Scale Legend** from the pop-up menu on the graph), change the precision so that the graph shows three decimal places on the *Y* scale.

Zoom Button

8. Using the Graph Palette, click on the Zoom button, select a zooming mode, and zoom in on the graph.

9. Pop up on the graph and select **Visible Items>>Cursor Legend**. It will first appear grayed out, so click on the Active Cursor button on the top line to make the first cursor active.

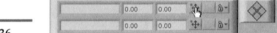

Figure 8.36

Use the Operating tool to drag the cursor around on the graph; pay attention to how the *X* and *Y* values in the cursor display change. These values can help you determine the value of a particular point on the graph. Now use the Cursor Movement Control buttons to move the cursor around. Make the second cursor active and use the Cursor Movement Control to move both cursors at the same time. Click on the Cursor Display Control (the crosshairs) with the Operating tool (if you pop up, you will get a different menu) and change the color of one of your cursors.

Are you starting to see what cursors can do? If you want, select the Cursor 0 text and replace it with a cursor name of your own. Finally click on the lock button with the Operating tool and deselect **Allow Drag** from the menu that appears. Then choose **Lock to Plot** from the same menu. You can no longer drag the cursor around on the graph with the mouse. However, try using the Cursor Movement Control and you'll find that the cursor will track the plot it's assigned to.

10. Close and save the VI. Name it **Temperature Analysis.vi** and place it in your MYWORK directory or VI library

8.8   Intensity Charts and Graphs — Color as a Third Dimension

So what do you do if you want to plot three variables against each other, instead of just two? If you're using Windows, you can use the ActiveX 3D Graph Controls, which we'll discuss next.

On all versions of LabVIEW, you can also make an intensity plot. *Intensity charts* and *graphs* display three dimensions of data on a 2D plot by using color to display values of the third dimension of data (the "Z" values). Like the waveform chart, the intensity chart features a scrolling display, whereas the intensity graph display is fixed. Intensity plots are extremely useful for displaying patterned data such as terrain, where color represents altitude over a two-dimensional area, or temperature patterns, where color represents temperature distribution.

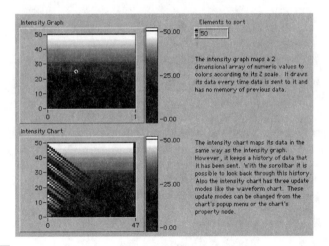

Figure 8.37

Intensity plots function much like two-dimensional charts and graphs in most ways, with the addition of color to represent the third variable. A Color Scale is available so that you can set and display the color mapping scheme. Intensity graph cursor displays also include a Z value.

Intensity charts and graphs accept 2D arrays of numbers, where each number in the array is a color value. The indices of each array element represent the plot location for that color. Not only can you define the mapping of numbers to colors using the color scale (which works like a color ramp, if you've played with that), but you can also do it programmatically using property nodes (Chapter 12 will tell you all about them). The simple example in Figure 8.38 shows a 3 × 4 array plotted on an intensity graph. The colors mapped are red (1.0), brown (2.0), and green (3.0).

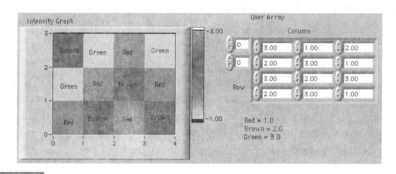

Figure 8.38

To assign a color to a value on the Color Scale, pop up on the corresponding marker and select **Marker Color>>**. The Z scale of intensity charts and graphs has Arbitrary Marker Spacing by default, so you can alter the "gradient" of colors by dragging a marker with the Operating tool. Create new markers by selecting **Add Marker** from the Color Scale pop-up menu; then drag them to a desired location and assign a new color. If you want to learn more about intensity charts and graphs, we suggest playing around with them. You can also look in LabVIEW's online documentation for more exact instructions.

8.8.1 Activity 8-5: The Intensity Graph

In this activity, you will look at a VI that displays wave interference patterns. This VI will also show you how the intensity graph maps its input 2D array to the display.

1. To get an idea of how intensity graphs and charts work, open and run the **Intensity Graph Example VI**, located in EVERYONE\CH8.LLB. Run the VI. You will see an intricate interference waveform plotted on the graph. The color range is defined in the block diagram using the intensity graph attribute node. Modify the color range by clicking on the color boxes in the first frame of the Sequence Structure with the Operating tool and then choosing a new color from the palette that appears. Run the VI again.

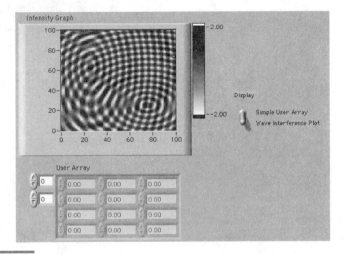

Figure 8.39

2. Switch the <u>Display</u> switch to Simple User Array and enter values between 0.0 and 10.0 in the <u>User Array</u> control (the color range for the graph has been defined in the block diagram using the intensity graph property node—blue (0.0) to red (10.0). After you enter all of the values, run the VI. Notice how the magnitude of each element is mapped to the graph. Now change your values and run it again.

3. Take a look at the block diagram to see how the VI works.

4. Close the VI and don't save any changes.

For more examples of intensity plots, open some of the VIs that ship with LabVIEW in example\general\graphs\intgraph.llb. Also check out **Simulation of Tomography** and **Heat Equation Example** in examples\analysis\mathxmpl.llb.

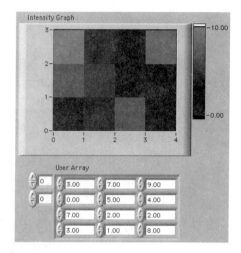

Figure 8.40

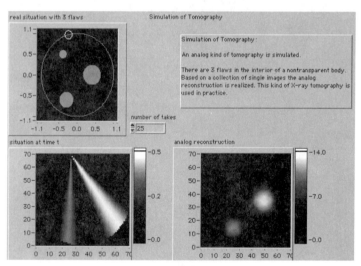

Figure 8.41

8.8.2 3D Graphs

For true three-dimensional graphs, LabVIEW for Windows* Professional Version offers a **3D Surface Graph**, **3D Parametric Graph**, and **3D Curve Graph** from the **Graph** palette.

---

* If you're a Mac or Linux user and feel left out, let National Instruments know you'd like this feature across all platforms.

These 3D graphs are only for Windows and are not present in the Base package of LabVIEW.

Figures 8.42 and 8.43 show a VI's front panel and block diagram that presents an example of a "doughnut"—called a torus—plotted on the **3D Parametric Graph** control. Unlike the previous charts and graphs you've seen, 3D graphs don't have a simple block diagram terminal; they actually use special subVI functions that are automatically created when you drop a 3D graph on the front panel.

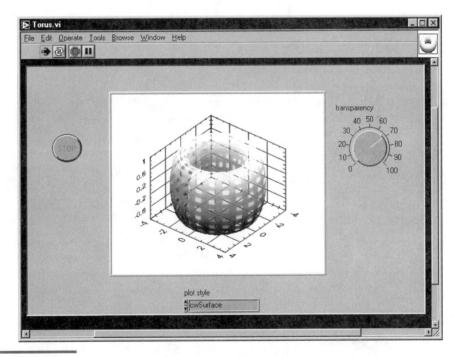

Figure 8.42

The three types of 3D graph can all plot in three dimensions, but each one works slightly differently:

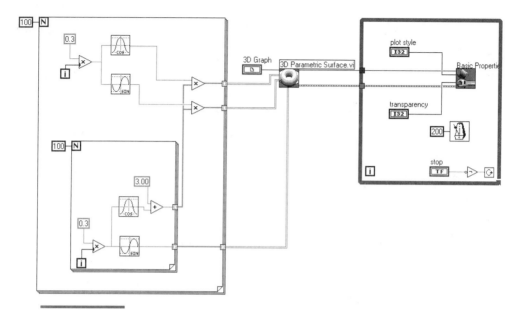

Figure 8.43

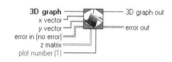

Figure 8.44
3D Surface Graph.

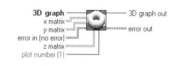

Figure 8.45
Parametric Surface.

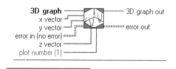

Figure 8.46
3D Curve.

A **3D Surface Graph** plots a simple surface from the z matrix. The surface is modified by the x and y vectors that cause the surface to shift in relation to the x and y planes. It accepts one 2D array and the two optional 1D arrays.

3D Parametric Surface plots a surface in terms of x, y, and z surfaces. It takes in three 2D arrays or matrices that specify each of the x, y, and z planes.

3D Curve describes a line in terms of x, y, and z points. This VI has three 1D array or vector inputs that specify each point in the plot.

Three-dimensional graphs can be more complicated than the simple charts and graphs with which you've worked. They are really an advanced topic, and so we won't say more about them here; we just wanted to let you know they are there if you need them. You can browse LabVIEW's 3D Graph examples to get an idea of how to use them.

8.8.3 Digital Waveform Graphs

There's one more advanced type of graph we'll mention briefly—the **Digital Waveform Graph** (from the **Graph**) palette. All the graphs we've seen so far work very well for *analog* data. If you want to use LabVIEW to display and analyze *digital* signals (signals that usually only have two states, such as true or false, 0 or 5 V, on or off, etc.), you may find this special type of graph useful, particulary if you work with timing diagrams or logic analyzers.

Figure 8.47 shows an example of a Digital Waveform Graph that is plotting eight digital signals over time.

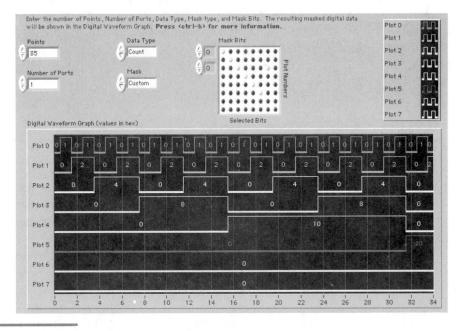

Figure 8.47

On the block diagram (Figure 8.48), the digital waveform graph has cluster terminal whose inputs are

- X_0 (initial X value)
- ΔX (delta X)
- Data (array of unsigned bytes)
- Number of ports
- Masks (array of integers)—optional input

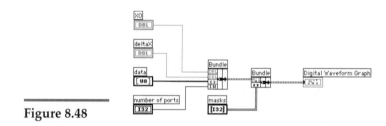

Figure 8.48

If you've worked with digital logic before, the digital waveform graph isn't hard to use; otherwise, don't even worry about learning it! You can find some good examples on using digital waveform graphs in LabVIEW's built-in examples (`examples\general\graphs\DigitalWaveformGraph.llb`).

8.9 Waveforms

In many engineering and scientific applications, much of the data you handle are a collection of values that vary over time. For example, audio signals are pressure values versus time; an EKG is voltage versus time; the variations in the surface of a liquid as a pebble drops in are (x,y,z) coordinates versus time. LabVIEW provides you with a convenient way to organize and work with this kind of time-varying data: the *waveform* data type. A waveform data type allows you to store not only the main values of your data but also the timestamp of when the first point was collected, the time delay between each data point, and notes about the data. It is similar to other LabVIEW data types like arrays and clusters; you can add, subtract, and perform many other operations directly on the waveforms. You can create a

Waveform control on the front panel from the **I/O** palette (Figure 8.49). The corresponding block diagram representation is a brown terminal (Figure 8.50).

Figure 8.49
Waveform control.

Figure 8.50
Waveform terminal.

Examining the waveform data type a little more closely, we see it is really just a special type of cluster that consists of four components, called **Y, t0, dt,** and **Attributes**:

Y: This component is a 1D array of numeric data points, which can be either a single point or another waveform, depending on the operation. The representation of the 1D array is DBL.

t0: This component is a scalar value that represents the time (according to the system clock) when the first point in the Y array was acquired. It is also referred to as the initial time or the timestamp.

dt (Δt, or delta-t): This is a scalar value that represents the time between data points in the Y array.

Attribute: By default, this component is hidden (you can see it by popping up and selecting **Visible Items>> attribute**). It is a string data type that allows you to bundle other information along with your waveform, such as the device number or channel number of your data acquisition system.

In Figure 8.49, the waveform control shows that the first point of the waveform starts at 7:21:19 PM, on Nov. 10, 2002, and that the time between each point is 0.5 seconds.

8.9.1 Waveforms versus Arrays

In many ways, you can think of waveforms as just 1D arrays that hold your data—with some extra information attached about the time and timing of the data points. Waveforms are most often useful in analog data acquisition, which we'll discuss in Chapters 10 and 11.

Of course, you do not necessarily need to use waveform data types; you can always just use arrays to hold your data. However, waveform data types offer several advantages over arrays:

- **The presence of t0**. Before the waveform data type existed, you could not determine when your data were taken. The waveform data type automatically returns the time of day and the date in the **t0** component, which gives you a real-world acquisition time for your data.

- **Easier graphing**. The waveform data type also simplifies graphing your data. In previous versions of LabVIEW, you had to bundle the value of the initial point (X_0) and the time between points (delta X) with your data (Y array). Because waveform data type already contains these elements, all you have to do is wire it to the graph.

- **Easier multiple-plot graphing**. The waveform data type also simplifies multiple plot graphs. In previous versions of LabVIEW, you had to bundle your X_0, delta X, and Y array for each plot and then send them to a build array to get a multiple-plot graph. Using a waveform data type, you just wire a 1D array of waveforms to the graph for a multiple plot. If, for example, you are acquiring data on multiple channels with an analog input VI, the VI automatically returns a 1D array, so all you do is wire it directly to the graph.

8.9.2 Waveform Functions

In the **Functions** palette, you'll find an entire subpalette dedicated to the waveform manipulation, aptly named **Waveform** (Figure 8.51).

Since waveforms are really a special type of cluster, you'll find functions similar to **Unbundle** and **Bundle**: **Get Waveform Components** and **Build Waveform**.

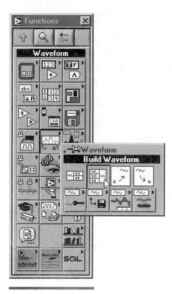

Figure 8.51
The Waveform palette.

Figure 8.52
Get Waveform Components.

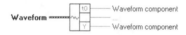

Figure 8.53
Build Waveform.

Returns the waveform components you specify. You specify components by right-clicking and selecting **Add Element** and creating an indicator. This function is expandable.

Builds a waveform or modifies an existing waveform. If you do not wire an input to waveform, **Build Waveform** creates a new waveform based on the components you enter. If you do wire an input in waveform, the waveform is modified based on the components you specify. This function is expandable.

The **Waveform** palette also has subpalettes with many useful functions and operations you can perform on waveforms.

Figure 8.54

The functions in **Waveform Operations** allow you to perform arithmetic and comparison functions on waveforms, such as adding, subtracting, multiplying, finding the max and min points, concatenating, etc. Note that in most waveform operations that involve two or more waveforms, the waveforms involved must all have the same **dt** values.

Figure 8.55

The functions in **Waveform File I/O** allow you to write waveform data to and read waveform data from files.

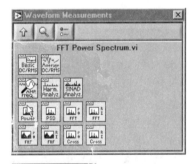

Figure 8.56

The **Waveform Measurements** palette allows you to perform common time and frequency domain measurements such as DC, RMS, Tone Frequency/Amplitude/Phase, Harmonic Distortion, SINAD, and Averaged FFT measurements.

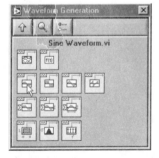

Figure 8.57

The **Waveform Generation** palette allows you to generate different types of single and multitone signals, function generator signals, and noise signals. For example, you can generate a sine wave, specifying the amplitude, frequency, etc.

One last thing you should know about waveforms: When you need to plot a waveform, you can wire it directly to a **Waveform Chart** or a **Waveform Graph.** The terminal automatically adapts to accept a waveform data type and will reflect the timing information on the X axis.

Let's look at a simple example of how to use and plot a waveform in the next activity.

8.9.3 Activity 8-6: Generate and Plot a Waveform

In this activity, you will generate a sine waveform, set its initial timestamp to the current time, and plot it on a chart.

1. Open a new front panel.
2. On the front panel, place dial (from the **Numeric** palette), a chart, and a waveform indicator. Label the dial <u>Frequency</u>, so that it looks like Figure 8.58.

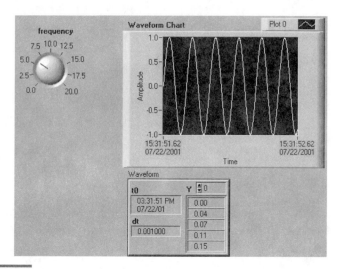

Figure 8.58

3. Build a block diagram that will generate a sine waveform and plot it. Because the waveform generation functions do not provide a value for the **t0** component of the waveform (they return the default value of 7:00:00 PM 12/31/1903), use the **Build Waveform** to set the **t0** to the current time. Here are the functions you will use:

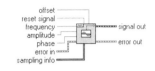

Figure 8.59
Sine Waveform.vi.

The **Sine Waveform** function, from the **Waveform>>Waveform Generation** palette, allows you to create a sine wave with the parameters you specify.

Figure 8.60
Get Date/Time in Seconds.

The **Get Date/Time in Seconds** returns the current system date and time.

Figure 8.61
Build Waveform.

The **Build Waveform** function allows you to modify components of the waveform; in this activity, it is just **t0**.

Your block diagram should look like Figure 8.62.

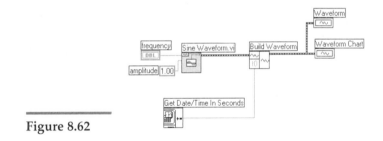

Figure 8.62

4. Run and test your VI at different frequencies. You may want to pop up on the chart and select **X Scale>>Autoscale X** and **Y Scale>>Autoscale Y**.

5. Save your VI as `Waveform Exercise.vi`.

We'll take another look at waveforms again in Chapter 11 where waveforms are the data type we use in analog I/O functions.

8.10 Wrap It Up!

You can create exciting visual displays of data using LabVIEW's charts and graphs. *Charts* append new data to old data, interactively plotting one point (or one set of points) at a time, so you can see a current value in context with previous values. *Graphs,* in contrast, display a full block of data after it has been generated. *Waveforms*, a new data type we learned about, can be used with both charts and graphs.

LabVIEW provides several kinds of graphs: *waveform* graphs, *XY* graphs, *intensity* graphs, *3D* graphs, and *digital waveform* graphs.

The *waveform graph* plots only single-valued points that are evenly distributed with respect to X as time-varying waveforms. In other words, graph plots a Y array against a set time base.

The *XY graph* is a general-purpose, Cartesian graph that lets you plot multi-valued functions such as circular shapes. It plots a Y array against an X array.

Intensity plots are excellent for displaying patterned data since they can plot three variables of data against each other on a 2D display. Intensity charts and graphs use color to represent the third variable. They accept a 2D array of numbers, where each number is mapped to a color and the number's indices in the array specify location for the color on the graph or chart. In most other ways, intensity plots function like standard two-variable charts and graphs.

Three-dimensional graphs (Windows only) are more sophisticated and perspective-oriented 3D graphs that allow you to plot (x,y,z) coordinates in 3D space. A variety of functions are provided in LabVIEW for manipulating and configuring the 3D graphs.

The *digital waveform graph* is a special type of graph used for plotting digital time-domain data; it is particularly useful for showing true/false states changing over time.

You can configure the appearance of charts and graphs using the plot legend, the scale legend, and the graph palette. You can also change the scales to suit your data and bring up cursors to mark your plots.

Both charts and graphs can draw multiple plots at a time. Data types can get tricky, so you may want to refer to the examples in this chapter or those that ship in the `examples` directory as a template while writing your own graphing VIs.

Mechanical action of Boolean switches allows you to control how they behave when you click on them. You can set a switch to return to its default value after its new value has been read once—that way it's all ready to be used again. This type of action is called latch action. You can also specify if you want the mouse click to register when you press the mouse button or release it.

Waveforms are a special LabVIEW data type that store information about the initial timestamp and time interval between a series of data points. There is an entire **Waveforms** palette on the **Functions** palette that provides you with all sorts of functions to perform on waveforms. You can wire waveform data directly to a chart or graph to plot it.

8.11 Additional Activities

Activity 8-7: Temperature Limit

Build a VI that continuously measures the temperature once per second and displays the temperature on a chart in scope mode. If the temperature goes above or below the preset limits, the VI turns on a front panel LED. The chart should plot the temperature as well as the upper and lower temperature limits. You should be able to set the limits from the front panel. Take a look at the front panel shown in Figure 8.63 for a start. Name the VI **Temperature Limit.vi**.

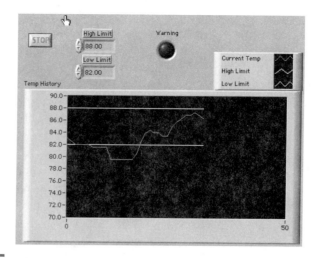

Figure 8.63

Activity 8-8: Max/Min Temperature Limit

Modify the VI you created in Activity 8-6 to display the maximum and minimum values of the temperature trace. Name the VI **Temp Limit (max/min).vi**.

*You must use shift registers and the **Array Max & Min** function (**Array** palette).*

Activity 8-9: Plotting Random Arrays

Build a VI that generates a 2D array (3 rows by 10 columns) containing random numbers. After generating the array, index off each row and plot each row on its own graph. Your front panel should contain three graphs. Name the VI **Extract 2D Array.vi**.

OVERVIEW

This chapter introduces some of the powerful things you can do with strings. Lab-VIEW has many built-in string functions, similar to its array functions, that let you manipulate string data for screen display, instrument control, or any number of reasons. You will also learn how to save data to and retrieve data from a disk file.

GOALS

- Learn more about options for string controls and indicators
- Understand how to use LabVIEW's string functions
- Convert numeric data to string data, and vice versa
- Use the file input and output (I/O) VIs to save data to a disk file and then read it back into LabVIEW

KEY TERMS

- Scrollbar
- Table
- Spreadsheet file
- Formatting string
- Regular expression

Exploring Strings and File I/O

9

9.1 More about Strings

We introduced strings in Chapter 4: A string is simply a collection of ASCII characters.* Often, you may use strings for more than simple text messages. For example, in instrument control, you pass numeric data as character strings. You then convert these strings to numbers to process the data. Storing numeric data to disk can also use strings; in many of the file I/O VIs, LabVIEW first converts numeric values to string data before it saves them to a file.

9.1.1 Choose Your Own Display Type

String controls and indicators have several options you might find useful. For example, they can display and accept characters that are normally nondisplayable, such as backspaces, carriage returns, and tabs. If you choose

---

* At time of this writing, the current release of LabVIEW, 6.0, does not support Unicode, although it's a feature that may appear soon.

'\' **Codes Display** (instead of **Normal Display**) from a string's pop-up menu, nondisplayable characters appear as a backslash (\) followed by the appropriate code. Table 9.1 shows what these codes mean.

Table 9.1 *LabVIEW '\' Codes.*

Code	LabVIEW Implementation
\00 – \FF	Hexadecimal value of an 8-bit character; alphabetical characters must be uppercase
\b	Backspace (ASCII BS, equivalent to \08)
\f	Formfeed (ASCII FF, equivalent to \0C)
\n	New Line (ASCII LF, equivalent to \0A)
\r	Return (ASCII CR, equivalent to \0D)
\t	Tab (ASCII HT, equivalent to \09)
\s	Space (equivalent to \20)
\\	Backslash (ASCII \, equivalent to \5C)

You must use uppercase letters for hexadecimal characters and lowercase letters for the special characters, such as formfeed and backspace. LabVIEW interprets the sequence \BFare as hex BF followed by the word "are," whereas LabVIEW interprets \bFare and \bfare as a backspace followed by the words "Fare" and "fare." In the sequence \Bfare, \B is not the back-space code, and \Bf is not a valid hex code. In a case like this, when a back-slash is followed by only part of a valid hex character, LabVIEW assumes a zero follows the backslash, and so LabVIEW interprets \B as hex 0B. Any time a backslash is not followed by a valid character code, LabVIEW ignores the backslash character.

Don't worry, the data in the string do not change when the display mode is toggled; only the display of certain characters changes. '\'**Codes Display** mode is very useful for debugging programs and for specifying nondisplayable characters required by instruments, the serial port, and other devices.

Strings also have a **Password Display** option, which sets the string control or indicator to display a "*" for every character entered into it, so that no one

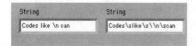

This string shows the Backslash code for the New Line character in Normal Display Mode

After switching to "\" Codes Display Mode, the string shows the spaces entered by the user.

Figure 9.1

can see what you type. While the front panel shows only a stream of "****," the block diagram reads the actual data in the string. Obviously, this display can be useful if you need to programmatically password-protect all or part of your VIs.

Figure 9.2

Figure 9.3

If you want to see your string as hexadecimal characters instead of alphanumeric characters, use the **Hex Display** option.

9.1.2 Single-Line Strings

If you choose **Limit to Single Life**, the string will never get married. If you choose **Limit to Single Line** from a string's pop-up menu, your string cannot exceed one line of text; that is, no carriage returns are allowed in the string. If you hit <enter> or <return>, text entry will be automatically terminated. If strings are not limited to a single line, hitting <return> causes the cursor to jump to a new line to let you type more.

9.1.3 Updating While You Type

Normally, string controls don't change their value on their terminal in the block diagram until you finish typing and hit <enter>, click outside the string box, or click on the "√" button on the VI toolbar to indicate the string entry is complete. Most of the time you will want this behavior, because you don't want your block diagram code to evaluate an incomplete string before a user is finished typing.

If you do want the value to be udpated as you type (just like a knob, for example), pop up on the string control and select **Update Value While Typing**.

9.1.4 The Scrollbar

If you choose the **Visible Items>>Scrollbar** option from the string pop-up **Visible Items** submenu, a vertical scrollbar appears on the string control or indicator. You can use this option to minimize the space taken up on the front panel by string controls that contain a large amount of text. Note that this option will be grayed out unless you've increased the size of your string enough for a scrollbar to fit.

9.1.5 Tables

A table is a flat piece of furniture . . . no—wait—wrong context! In Lab-VIEW, a table is a special structure that displays a two-dimensional array of strings. You can find it in the **Lists & Table** subpalette of the **Controls** palette. A table in all its glory is shown in Figure 9.4.

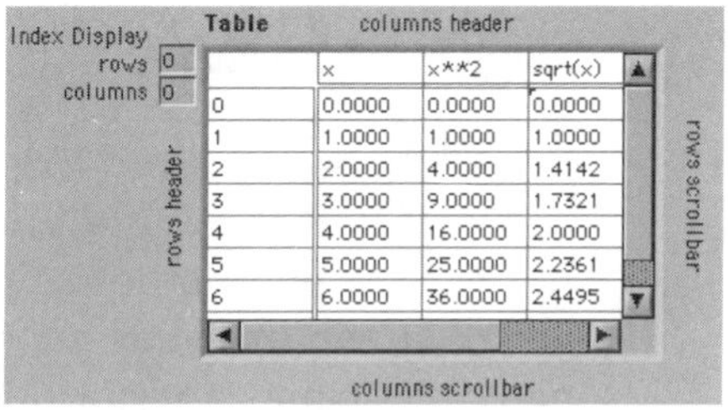

Figure 9.4

Tables have row and column headings that you can show or hide; the headings are separated from the data space by a thin open border space. You can enter text headings using the Labeling tool or Operating tool (like everything else). You can also update or read headings using property nodes, which you've heard so much about and will learn how to use soon enough.

Like an array index display, a table index display indicates which cell is visible at the upper-left corner of the table.

For a good example of a table and how to use it, open and run **Building Tables.vi**, located in EVERYONE\CH9.LLB.

9.1.6 Listboxes

There are two kinds of listboxes in LabVIEW, **Listbox** and **Multi-column Listbox**, from the **List & Table** palette.

A listbox in LabVIEW is similar to a table, but it behaves very differently at runtime. During edit mode, you can type any text into a listbox just like you would a table. When you run the VI, the listbox acts as a "multiple choice" menu where you can click on any row to highlight it and select it.

Figure 9.5 shows a multicolumn listbox.

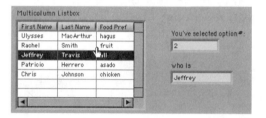

Figure 9.5

Listboxes are useful when you want to present data (either in single columns or in multiple columns) to a user but you just want them to choose an option, not type in text.

For a simple example of how a listbox works, see **Listbox Example.vi** in EVERYONE\CH9.LLB.

9.2 Using String Functions

Like arrays, strings can be much more useful when you take advantage of the many built-in functions provided by LabVIEW. This section examines a few of the functions from the **String** subpalette of the **Functions** palette. You might also browse through the rest of this palette to see what other functions are built in.

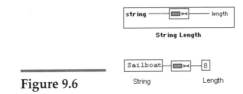

Figure 9.6

String Length returns the number of characters in a given string.
Concatenate Strings concatenates all input strings into a single output string.

Figure 9.7
Concatenate Strings.

The function appears as the icon at the left when you place it on the block diagram. You can resize the function with the Positioning tool to increase the number of inputs.

Concentrate
Strings
Function

Figure 9.8

In addition to simple strings, you can also wire a one-dimensional array of strings as input; the output will be a single string containing a concatenation of strings in the array.

Figure 9.9

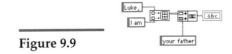

In many instances, you must convert strings to numbers or numbers to strings. The **Format Into String** and **Scan From String** functions have these capabilities (as do various other functions, but we'll concentrate on these). We'll talk about **Format Into String** now and **Scan From String** in a little while.

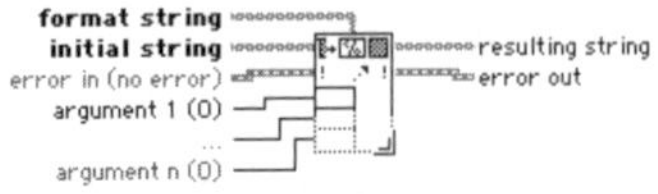

Figure 9.10
Format Into String.

format string
initial string
error in (no error)
argument 1 (0)

argument n (0)

resulting string
error out

Simply put, **Format Into String** converts numeric data into string data.

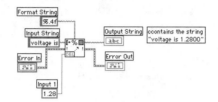

Figure 9.11

In the example shown in Figure 9.11, the function converts the floating-point number 1.28 to the 6-byte string "1.2800."

Format Into String formats the input argument (which is in numeric format) as a string, according to the format specifications in **format string**. These specifications are listed in detail in LabVIEW manuals and Online Reference (search on "String Function Overview"). The function appends the newly converted string to the input wired to **initial string**, if there is one, and outputs the results in **resulting string**. Table 9.2 gives some examples of **Format Into String**'s behavior.

Table 9.2

Initial String	Format String	Number	Resulting String
(empty)	score=%2d%%	87	score=87%
score=	%2d%%	87	score=87%
(empty)	level=%7.2eV	0.03642	level=3.64E-2V
(empty)	%5.3f	5.67 N	5.670 N

The "%" character begins the formatting specification. Given "%*number1.number2*," *number 1* specifies field width of the resulting string and *number 2* specifies the precision (i.e., number of digits after the decimal point). An "f" formats the input number as a floating-point number with fractional format, "d" formats it as a decimal integer, and "e" formats it as a floating-point number with scientific notation.

Format Into String can be resized to convert multiple values to a single string simultaneously.

Get Date/Time String (found in the **Time & Dialog** palette) outputs the date string, which contains the current date, and the time string, which contains the current time. This function is useful for timestamping your data. Note that you don't have to wire any inputs to **Get Date/Time String**; it can use the default values.

Figure 9.12
Get Date/Time String.

9.3 Activity 9-1: String Construction

It's time for you to practice and give a new meaning to the phrase "strings attached." You will build a VI that converts a number to a string and concatenates that string with other strings to form a single output string. The VI also determines the length of the output string.

1. Build the front panel shown in Figure 9.13.

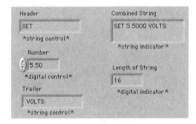

Figure 9.13

The VI will concatenate the input from the two string controls and the digital control into a single output string, which is displayed in the string indicator. The digital indicator will display the string's length.

2. Build the block diagram pictured in Figure 9.14.

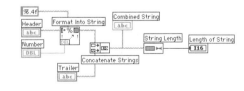

Figure 9.14

Format Into
String Function

Format Into String (**String** palette) converts the number you specify in the <u>Number</u> digital control to a string with fractional format and four digits of precision.

Concatenate
Strings function

Concatenate Strings function (**String** palette) combines all input strings into a single output string. To increase the number of inputs, stretch the icon using the Positioning tool.

String Length
function

String Length function (**String** palette) returns the number of characters in the concatenated string.

3. Return to the front panel and type text inside the two string controls and a number inside the digital control. Make sure to add spaces at the end of the header and the beginning of the trailer strings, or your output string will run together. Run the VI.

4. Save and close the VI. Name it **Build String.vi** and place it in your MYWORK directory or VI library. Do you feel like a LabVIEW expert yet? You're getting there!

9.4 Parsing Functions

Sometimes you will find it useful to take strings apart or convert them into numbers, and these parsing functions can help you accomplish these tasks.

String Subset accesses a particular section of a string. It returns the substring beginning at **offset** and containing **length** number of characters. Remember, the first character's offset is zero.

Figure 9.15
String Subset.

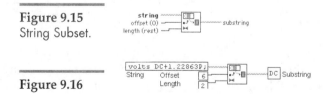

Figure 9.16

Match Pattern is used to look for a given pattern of characters in a string. It searches for and returns a **matched substring**. **Match Pattern** looks for the **regular expression** in a string, beginning at **offset**; if it finds a match, it splits the string into three substrings. If no match is found, the match substring is empty and **offset past match** is set to –1.

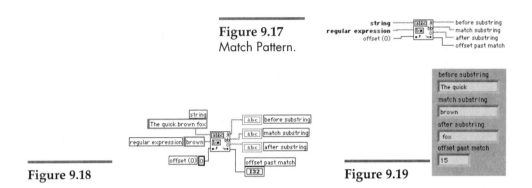

Figure 9.17
Match Pattern.

Figure 9.18

Figure 9.19

Scan From String, the "opposite" of **Format Into String**, converts a string containing valid numeric characters (0 to 9, +, −, e, E, and period) to numeric data. This function starts scanning the **input string** at **initial search location** and converts the data according to the specifications in format string. (To learn more about the specifications, see the LabVIEW manuals or "String Function Overview" in the Online Reference.) **Scan From String** can be re-sized to convert multiple values simultaneously.

Figure 9.20
Scan From String.

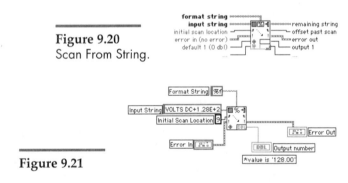

Figure 9.21

In this example, **Scan From String** converts the string "VOLTS DC+1.28E+2" to the number 128.00. It starts scanning at the eighth character of the string (which is the + in this case—remember that the first character offset is zero).

Both **Format Into String** and **Scan From String** have an **Edit Scan String** interface that you can use to create the format string. In this dialog box, you can specify format, precision, data type, and width of the converted value. Double-click on the function or pop up on it and select **Edit Format String** to access the **Edit Scan String** or **Edit Format String** dialog box.

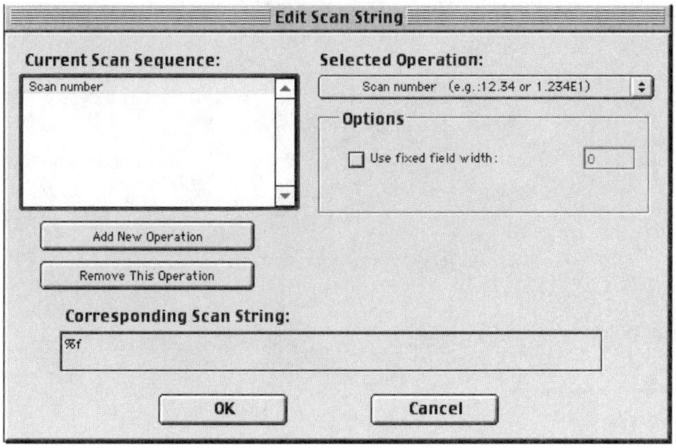

Edit Scan String
Current Scan Sequence:
Scan number
Selected Operation:
Scan number (e.g.:12.34 or 1.234E1)
Options
Use fixed field width:
0
Add New Operation
Remove This Operation
Corresponding Scan String:
%f
OK
Cancel

Figure 9.22

After you create the format string and click the Create String button, the dialog box creates the string constant and wires it to the **format string** input for you.

9.5 Activity 9-2: More String Parsing

You will create a VI that parses information out of a longer string by taking a subset of a string and converting the numeric characters in that subset into a numeric value.

1. Build the front panel shown in Figure 9.23.

Input String
VOLTS DC +1.345E+02
Subset Offset
6
Subset Length
2
Number Offset
9
String Subset
DC
Number
134.50
Password
*****
Password Display enabled
Matched Pattern
No Match

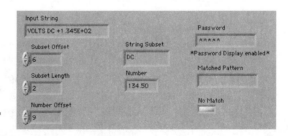

Figure 9.23

2. Set the Password string to display only asterisks by selecting **Password Display** from its pop-up menu.

3. Create the block diagram pictured in Figure 9.24.

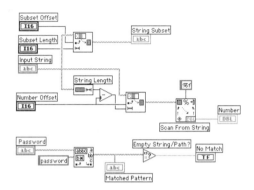

Figure 9.24

String Subset
Function

The **String Subset** function, found in the **String** palette, returns a subset of given length from the input string, according to the offset specified.

Scan From
String Function

The **Scan From String** function, located in the **String** palette, converts a string containing valid numeric characters (0 to 9, +, −, E, and period) to a number.

Match Pattern
Function

Match Pattern, also found in the **String** palette, compares the user's input password string to a given password string. If there is a match, it is displayed; if not, the string indicator shows an empty string.

Empty
String/Path?
Function

Empty String/Path?, from the **Comparison** palette, returns a Boolean TRUE if it detects an empty string from the match substring output of **Match Pattern**.

String Length
Function

String Length function (**String** palette) returns the number of characters in the string.

4. Run the VI with the inputs shown. Notice that the string subset of "DC" is picked out of the input string. Also notice that the numeric part of the string was parsed out and converted to a number. You can try different control values if you want, just remember that strings, like arrays, are indexed starting at zero.

Also note how the Password string shows only "*****." **Match Pattern** checks the input password against a password string (which in this case contains the characters, "password") and then returns a match if it finds one. If it finds no match, it returns an empty string.

5. Close the VI by selecting **Close** from the **File** menu. Save the VI in your MYWORK directory or VI library as **Parse String.vi**.

9.6 File Input/Output

File input and output (I/O) operations retrieve information from and store information in a disk file. LabVIEW has a number of very versatile file I/O functions, as well as some simple functions that take care of almost all aspects of file I/O in one shot. We'll talk about the simple file functions in this chapter. All are located in the **File I/O** subpalette of the **Functions** palette.

9.6.1 How They Work

The File functions expect a file path input, which looks kind of like a string. A path is a specific data type that provides a platform-specific way to enter a path to a file. We talked about them briefly in Chapter 4, and they'll come up again in Chapter 12. If you don't wire a file path, the File functions will pop up a dialog box asking you to select or enter a filename. When called, the File functions open or create a file, read or write the data, and then close the file. The files created with the VIs we'll talk about now are just ordinary text files. Once you have written data to a file, you can open the file using any word processing program to see your data.

One very common application for saving data to file is to format the text file so that you can open it in a spreadsheet program. In most spreadsheets, tabs separate columns and EOL (End of Line) characters separate rows. **Write To Spreadsheet File** and **Read From Spreadsheet File** deal with files in spreadsheet format.

Write Characters To File writes a character string to a new file or appends the string to an existing file.

Figure 9.25
Write Characters to File.vi.

Read Characters From File reads a specified number of characters from a file beginning at a specified character offset.

Figure 9.26
Read Characters From File.vi.

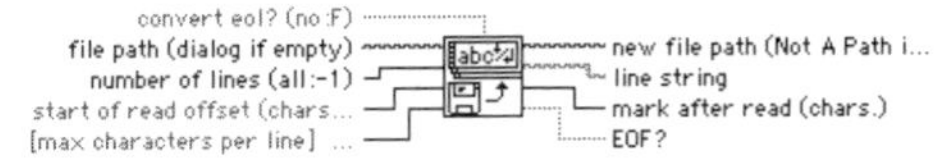

Figure 9.27
Read Lines From File.vi.

Read Lines From File reads a specified number of lines from a file beginning at a specified character offset.

Write To Spreadsheet File converts a 2D or 1D array of single-precision numbers to a text string and then writes the string to a new file or appends

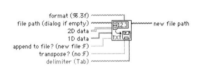

Figure 9.28
Write To Spreadsheet File.vi.

the string to an existing file. You can optionally transpose the data. Do not wire inputs for both 1D and 2D data (or one will be ignored). The text files created by this VI are readable by most spreadsheet applications.

Read From Spreadsheet File reads a specified number of lines or rows from a numeric text file, beginning at a specified character offset, and con-

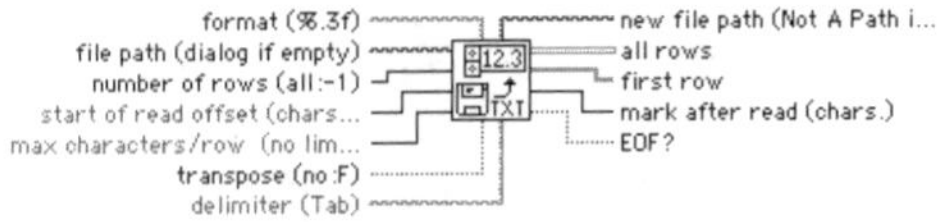

Figure 9.29
Read From Spreadsheet File.vi.

verts the data to a 2D single-precision array of numbers. You can optionally transpose the array. This VI will read spreadsheet files saved in text format.

These file functions are very high level and easy to use. All are found in the **File I/O** palette. LabVIEW contains other file functions that are much more versatile but more complicated, and we'll tell you about them in Chapters 12 and 15.

9.7   Activity 9-3: Writing to a Spreadsheet File

You will modify an existing VI to save data to a new file in ASCII format. Later you can access this file from a spreadsheet application.

1. Open **Graph Sine Array.vi**, which you built in Chapter 8. If you didn't finish building that VI, you can find a finished version in EVERYONE\CH8.LLB. As you recall, this VI generates two data arrays and plots them on a graph. You will modify this VI to write the two arrays to a file in which each column contains a data array.

2. Open the diagram of **Graph Sine Array.vi** and modify the VI by adding the diagram code shown inside the oval.

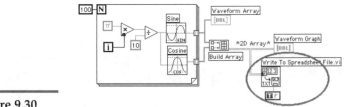

Figure 9.30

Write to
Spreadsheet
File VI

The **Write To Spreadsheet File** VI (**File I/O palette**) converts the 2D array to a spreadsheet string and writes it to a file. If no path name is specified (as in this activity), then a file dialog box will pop up and prompt you for a filename.

Boolean
Constant

The **Boolean Constant** (**Boolean** palette) controls whether or not the 2D array is transposed before it is written to file. To change it to TRUE, click on the constant with the Operating tool. In this case, you do want the data transposed because the data arrays are row specific (each row of the 2D array is a data array). Since you want each column of the spreadsheet file to contain data for one waveform, the 2D array must first be transposed.

3. Return to the front panel and run the VI. After the data arrays have been generated, a file dialog box will prompt you for the filename of

the new file you are creating. Type in a filename (or if you don't see this option, click on the "New..." button from the dialog box, and choose "File") and click the OK button. Remember the name and location of the file, as you will read in the data in the next exercise.

Do not attempt to write data files in VI libraries with the File I/O VIs. Doing so may overwrite your library and destroy your previous work.

4. Save the VI in your MYWORK directory or VI library, name it **Graph Sine Array to File.vi**, and close the VI.

5. Use spreadsheet software if you have it, or a simple text editor, to open and view the file you just created. You should see two columns of 100 elements each.

9.8   Activity 9-4: Reading from the Spreadsheet File

You will write a VI to read in the data from the file written in the last exercise and plot it on a graph.

1. Open a new VI and place a waveform graph on its front panel. Make sure autoscaling is on.

2. Create the little block diagram shown in the following illustration. Use the **Read From Spreadsheet File** function to bring in data and display it on the graph.

3. Using the TRUE **Boolean Constant** you must transpose the array when you read it in, because graphs plot data by row and it has been stored in the file by column. Note that if you hadn't transposed the data in the last exercise to store it in columns in the file, you wouldn't have to transpose it back now.

4. Run the VI. Since you are not providing a file path, a dialog box will prompt you to enter a filename. Select the file you created in Activity 9-3.

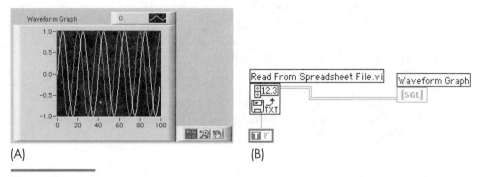

(A)   (B)

Figure 9.31

The VI will read the data from the file and plot both waveforms on the graph.

5. Save the VI in your MYWORK directory or VI library as **Read File.vi**.

9.9   Wrap it Up!

LabVIEW contains many functions for manipulating strings. These functions can be found in the **String** subpalette of the **Functions** palette. With them, you can determine string length, combine two strings, peel off a string subset, convert a string to a number (or vice versa), and perform many other useful things.

Using the functions in the **File I/O** subpalette of the **Functions** palette, you can write data to or read data from a disk file. **Write Characters To File** will save a text string to a file. **Read Characters From File** and **Read Lines From File** can then read that file back into LabVIEW. If you want to save an array of numbers, you must use the **Write To Spreadsheet File** function. You can read that data back in and convert it to numeric format using the **Read From Spreadsheet File** function.

Congratulations! Go out and celebrate today! You've mastered the fundamentals of LabVIEW! You have a strong foundation now and should have the background to investigate almost any LabVIEW topic that interests you. The Advanced section of this book, coming up next, will teach you about many of the very cool, more complex features LabVIEW contains that make your programming job easier, so stay tuned for more exciting LabVIEW adventures!

9.10 Additional Activities

Activity 9-5: Temperatures and Timestamps

Build a VI that takes 50 temperature readings inside a loop, once every 0.25 seconds, and plots each on a chart. It also converts each reading to a string, and then concatenates that string with a Tab character, a timestamp, and an End of Line character. The VI writes all these data to a file. Save the VI as **Temperature Log.vi**.

- *Use the **Tab** and **End of Line** constants in the **String** palette.*
- *Use **Concatenate Strings** to put all of the strings together.*
- *Use **Write Characters To File** to save the data.*
- *You can write data to file one line at a time, but it is much faster and more efficient to collect all of the data in one big string using shift registers and **Concatenate Strings**, and then write it all to file at one time.*

You can look at your file using any word processing program, but it should look something like this:

```
78.9    11:34:38
79.0    11:34:39
79.0    11:34:50
```

Activity 9-6: Spreadsheet Exercise

Build a VI that generates a 2D array (3 rows × 100 columns) of random numbers and writes the transposed data to a spreadsheet file. The file should contain a header for each column as shown in Figure 9.32. Use the VIs from the **String** and **File I/O palettes** for this activity. Save the VI as **Spreadsheet Exercise.vi**.

*Use the **Write Characters To File** VI to write the header and then the **Write to Spreadsheet File VI** (with the append to file input set to TRUE) to write the numerical data to the same file.*

	A	B	C	
1	Waveform 1	Waveform 2	Waveform 3	——Header
2	0.668	0.601	0.04	
3	0.164	0.884	0.695	
4	0.799	0.827	0.685	
5	0.775	0.1	0.008	
6	0.723	0.264	0.464	
7	0.253	0.179	0.145	
8	0.749	0.227	0.036	
9	0.745	0.133	0.347	
10	0.063	0.02	0.358	
11	0.308	0.566	0.392	

Figure 9.32

Advanced Topics

Introduction to the Advanced Section

If you've just completed the first section of this book, congratulations! You've learned the basics of creating a virtual instrument in LabVIEW—choosing front panel controls and indicators, wiring the block diagram, using structures such as the While Loop or Case Structure, and making simple, usable VIs. *Are we having fun yet*? Now that you're getting a grasp on graphical dataflow programming, it's time to delve further into some of the more powerful functions and features LabVIEW offers.

The Advanced section of this book, Chapters 10–16, expands on many of the remaining palette functions and focuses on techniques and tools for writing better programs. It also teaches you the basics of data acquisition from LabVIEW using plug-in data acquisition boards. You won't find as many walk-through exercises in this section—we've oriented the material more toward *informing* ("Wow, I didn't know you could do that in LabVIEW!") and *writing applications* ("How do I acquire data into my PC with LabVIEW?"). Because LabVIEW's functions and features are so extensive, you should find this section to be a valuable guide to getting an overview of what you can do with LabVIEW while leaving many of the gory details to the manuals and online reference files.

Here's a final tip before we start: Feel free to skip sections that you won't need. Unlike the basic section of the book, this material doesn't build on itself in most areas. For example, some of the exercises and examples require using data acquisition hardware. If you don't have this hardware, don't worry; you can safely skim through those sections and go on to other fun LabVIEW stuff.

OVERVIEW

This chapter will give you a deeper look at what we touched on in Chapter 2: data acquisition and instrument control. LabVIEW enables users to turn their computers into virtual instruments by gathering data from the real world, and this is one of the main reasons people use it. We'll take a look at the various options you have for taking or making data, including the use of existing instruments: serial communications, GPIB interfaces, and plug-in DAQ cards. You'll also learn some signal theory and about the kind of hardware used for these systems.

GOALS

- Finally find the meaning of all those acronyms that everyone thinks you know
- Become familiar with the hardware options you have for acquiring or sending data
- Learn some signal theory, including the classification of signals, measurement types, signal conditioning, and sampling
- Get some hints on picking and installing a DAQ board that suits your needs
- Discover the GPIB interface
- Investigate serial communications

KEY TERMS

- DAQ
- GPIB
- Serial
- Signals
- Analog
- Digital
- Frequency

- Grounded signal
- Floating signal
- Ground reference
- Sampling rate
- Nyquist frequency
- Signal conditioning
- MAX

- NI-DAQ
- Virtual channel
- Differential measurement
- Single-ended measurement
- Instrument driver

Getting Data into and out of Your Computer: Data Acquisition and Instrument Control

10

10.1 Acronyms Unlimited

"Let's go ahead and apply CASE tools and UML to designing the PCI interface using that new XML standard."

Admit it: How many times has someone mentioned an acronym in a technical discussion and everyone pretends to understand because nobody wants to ask what it stands for and look ignorant? Well, here's your chance to see what all the acronyms in this chapter stand for and where they came from. Use this list to put your colleagues to the test!

AC: Alternating Current. This acronym originally referred to how a device was powered, with AC being the plug in the wall and DC (direct current) being batteries. Now it's used more generally to refer to any kind of signal (not just current) that varies "rapidly" (whatever you want that to mean) with time.

ADC or **A/D**: Analog-to-Digital Conversion. This conversion takes a real-world analog signal and converts it to a digital form (as a series of bits) that the computer can understand. Many times the chip used to perform this operation is called "the ADC."

DAQ: Data AcQuisition. This little phrase just refers to collecting data in general, usually by performing an A/D conversion. Its meaning is sometimes expanded to include, as in this book, data generation. Don't confuse DAQ with DAC, which sound the same when pronounced in English. (**DAC**, or **D/A**, stands for Digital-to-Analog Conversion, usually referring to the chip that does this.)

DC: Direct Current. This is the opposite of AC. DC longer refers to current specifically. Sometimes people use DC to mean a constant signal of zero frequency. In other cases, such as in DAQ terminology, DC also refers to a very low frequency signal, such as something that varies less than once a second. Obviously the border between an AC and DC signal is subjective.

DMA: Direct Memory Access. You can use plug-in DAQ boards that have built-in DMA, or buy a separate DMA board. DMA lets you throw the data you're acquiring directly into the computer's RAM (there we go, another acronym), thus increasing data transfer speed. Without DMA you still acquire data into memory, but it takes more steps and more time because the software has to direct it there.

GPIB: General Purpose Interface Bus. GPIB is less commonly known as HP-IB (Hewlett-Packard Interface Bus) and IEEE 488.2 bus (Institute of Electrical and Electronic Engineers standard 488.2). It has become a world standard for almost any instrument to communicate with a computer. Originally developed by Hewlett-Packard in the 1960s to allow their instruments to be programmed in BASIC with a PC, now IEEE has helped define this bus with strict hardware protocols that ensure uniformity across instruments.

IVI: Interchangeable Virtual Instruments. This is a standard for instrument drivers (software that you can use to control external instruments), which can work with a wide variety of different instruments.

MXI: Multisystem eXtension Interface. MXI is a standard for connecting the VXI mainframe chassis and conventional computers in a fashion similar to GPIB.

PXI: PCI eXtensions for Instrumentations. **PCI** (Peripheral Component Interconnect) is a standard bus, used on most computers for plugging in dedicated device cards. PXI refers to an open hardware architecture, embraced by National Instruments, for integrating high-performance, modular components for data acquisition, instrument control, image processing, and more.

RS-232: Recommended Standard #232. RS-232 is a standard proposed by the Instrument Society of America for serial communications. It's used interchangeably with the term "serial communication," although serial communication more generally refers to communicating one bit at a time. A few other standards you might see are RS-485, RS-422, and RS-423.

SCXI: Signal Conditioning eXtensions for Instrumentation. This is a high-performance signal conditioning system devised by National Instruments, using an external *chassis* that contains I/O modules for signal conditioning, multiplexing, etc. The chassis is wired into a DAQ board in the PC.

SISTA: Sometimes I'm Sick of These Acronyms. Just kidding!

USB: Universal Serial Bus. USB is a standard bus on most PCs for connecting external peripherals.

VISA: Virtual Instrument Standard Architecture. This is a driver software architecture developed by National Instruments. Its purpose is to try to unify instrumentation software standards, whether the instrument uses GPIB, DAQ, VXI, or RS-232.

VXI: Talk about acronym abuse; this is an acronym for an acronym: VME eXtensions for Instrumentation. VME stands for Versa-Modular Eurocard. VXI is a very high performance system for instrumentation. You can buy VXI instruments that are small modules (instead of the regular big instrument with a front panel) that plug into a VXI chassis. The VXI chassis often has an embedded computer motherboard, so you don't have to use an external PC. VXI is an open industry standard, which means that companies besides National Instruments support it.

10.2 How to Connect Your Computer to the Real World

You've got a good PC, you're excited about using LabVIEW, and you're ready to build an application that will do something *outside* your computer Maybe you need to monitor the electroencephalogram (EEG, brain waves) of some research subjects, or plot force-versus-displacement curves to test the strength of some new plastic. Or perhaps you need something more elaborate, like a whole process control system for a semiconductor manufacturing facility, and you need to provide control signals to the plant.

Whatever your application, you need a way to gather data into your computer. Several solutions are usually possible, but the best solution will decidedly depend on what you can afford and the trade-offs you must make. Before you run out and buy hardware, you need to analyze and understand what kind of signals you're trying to measure (in some cases such as serial communication, you may not even need any additional hardware). Figure 10.1 shows you the most common ways you can get data into your computer.

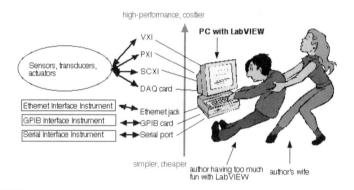

Figure 10.1

One of the first things you should decide when designing your data acquisition system is whether you're going to use traditional external instruments, such as a multimeter. Do you want a "physical" instrument to do some of the data acquisition and processing, or do you want to write a LabVIEW virtual instrument that will do everything via a plug-in DAQ board?

Issues such as cost, scheduling, and flexibility will play a part in this decision. For example, if you wanted to read some low-voltage data, you might use a simple plug-in DAQ card. You could write a VI that is very specific to your application and creates the exact virtual instrument you need. And generally speaking, it's much cheaper to buy a plug-in DAQ card than a standalone instrument. However, if you already have an existing instrument you want to use (a multimeter for example), it might be cheaper to use your existing meter to acquire the voltages and send the data via the GPIB bus to the computer, if your meter has a GPIB interface.

Sometimes there's no question about using an external instrument. If you need to perform mass spectroscopy, for example, we don't know of any mass spectrometer available as a card that plugs into your PC!

One more factor that might influence your decision to use an external instrument is the availability of an *instrument driver*. Contrary to a popular perception, an instrument driver is not a "virtual" copy of the instrument's front panel in LabVIEW; in fact, most instrument drivers look pretty boring. An instrument driver usually consists of a collection of subVIs, each of which sends a specific command or group of commands to the instrument. These subVIs have all the low-level code written in them so that you can quickly put together an application just by using these subVIs. National Instruments has a database of instrument drivers for hundreds of popular instruments that use the GPIB, serial, or VXI interface. You can get these for free from NI! Other instrument drivers are often available from third-party vendors, such as software consultants, or the instrument manufacturer itself.

Finally, if you are planning to buy or already have bought a plug-in DAQ card, make good use of it! Most people don't fully realize the potential their computer has when a DAQ card is running and humming inside it. Need to view an AC signal? Wait, don't go borrow that oscilloscope; just look at your signal right on your screen with one of the DAQ example VIs that comes with LabVIEW! With one DAQ card, you can create as many virtual instruments as you need. And, when it's time to upgrade your plug-in card or move to another platform, you may not even need to change a thing in your block diagram. That's right, LabVIEW's DAQ VIs work (with a few exceptions) *independently of whatever board you have in your computer.*

The rest of this chapter is essentially divided into two parts: DAQ and instrument control. The DAQ sections will cover hardware considerations such as signal theory, types of hardware, and configuration instructions. The instrument control section will give you some more details about serial and GPIB protocols.

10.3 Signals 101

Before we delve completely into data acquisition, we want to talk a little about *what* you'll be acquiring. A signal is simply a representation of any physical quantity whose magnitude and variation with time (or occasionally some other variable) contain information.

10.3.1 Timing Is Everything

Although it may not be obvious at first, *time* is usually the most critical aspect of almost any measurement. Whether we want to observe how an engine's temperature changes over time, see what a filtered audio signal looks like, or close some valves when a gas mixture reaches its optimum ratio, time is the deciding factor in data acquisition and control. We want to know not just *what* happens, but *when*. Even so-called DC signals are not really steady state; if they never changed over time, we would always know their constant value, so why would we want to measure them?

Timing is important in designing your data acquisition software for a couple of reasons. First, you need to figure out how to set the *sampling rate*, or how often your computer takes a measurement. Second, you need to be able to allocate processor time to other tasks such as file I/O.

If you only need to read data once or twice a second or less and you don't need much timing accuracy between the sample points, you can probably use LabVIEW's timing functions to control the sampling rate "directly" from your program, by putting a **Wait** function in a VI that acquires one point, for example, in a loop. For more precise applications or AC signals, you'll let the hardware and low-level software set the sampling rate by configuring your measurement on the DAQ board accordingly. We'll discuss these details along with some examples in Chapter 11.

10.3.2 Signal Classification

Let's say you want to take a measurement. For signal conditioning hardware to condition a signal, or for the DAQ board to measure it directly, you must first convert it to an electrical signal such as voltage or current. A *transducer* performs this conversion. For example, if you wish to measure temperature, you must somehow represent temperature as a voltage that the DAQ board can read. A variety of temperature transducers exist that use some principles of heat and physical properties of materials to convert the temperature to an electrical signal.

Once the physical quantity is in an electrical signal form, you can then measure the signals to extract some type of useful information conveyed through one or more of the following parameters: state, rate, level, shape, and frequency content.

Figure 10.2

Strictly speaking, all signals are analog time-varying signals. However, to discuss signal measurement methods, you should classify a given signal as one of five signal types. Classify the signal by the way it conveys the needed information. First, you can classify any signal as *analog* or *digital*. A digital, or binary, signal has only two possible discrete levels—a high (on) level or low (off) level. An analog signal, in contrast, contains information in the continuous variation of the signal with respect to time.

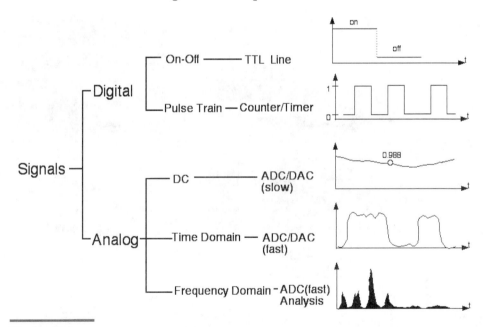

Figure 10.3

Engineers often classify digital signals into two more types and analog signals into three more types. The two digital signal types are the on–off signal and the *pulse train* signal. The three analog signal types are the *DC* signal, the *time domain* (or *AC*) signal, and the *frequency domain* signal. The two digital and three analog signal types are unique in the information each conveys.

You will see that the five signal types closely parallel the five basic types of signal information: state, rate, level, shape, and frequency content.

Digital Signals

The first type of digital signal is the on–off, or *state*, signal. A state signal conveys information concerning the digital state of the signal. Therefore, the instrument needed to measure this signal type is a simple digital state detector. The output of a transistor–transistor logic (TTL) switch is an example of a digital on–off signal. Another example is the state of an LED, as show in Figure 10.4.

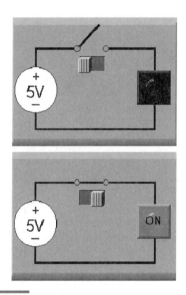
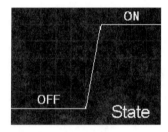

Position of
the switch
determines the
state of the signal

Figure 10.4
State example.

The second type of digital signal is the pulse train, or **rate**, signal. This signal consists of a series of state transitions. Information is contained in the number of state transitions occurring, the rate at which the transitions occur, or the time between one or more state transitions. The output signal of an optical encoder mounted on the shaft of a motor is an example of a digital pulse train signal. In some instances, devices require a digital input for operation. For example, a stepper motor requires a series of digital pulses as an input to control the motor position and speed.

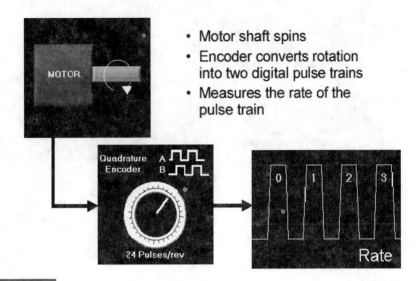

- Motor shaft spins
- Encoder converts rotation into two digital pulse trains
- Measures the rate of the pulse train

Figure 10.5
Rule example.

Analog Level Signals

Analog DC, or *level*, signals are static or slowly varying analog signals. The most important characteristic of the level signal is that the level, or amplitude, of the signal at a given instant conveys information of interest. Because the analog DC signal varies slowly, the accuracy of the measured level is of more concern than the time or rate at which you take the measurement. The instrument or plug-in DAQ board that measures DC signals operates as an analog-to-digital converter (ADC), which converts the analog electrical signal into a digital value for the computer to interpret.

As shown in Figure 10.6, common examples of DC signals include temperature, battery voltage, pressure, and static loads. In each case, the DAQ system monitors the signal and returns a single value indicating the magnitude of the signal at that time. Therefore, these signals are often displayed through LabVIEW indicators such as meters, gauges, strip charts, and numerical readouts.

Your DAQ system should meet the following specifications when acquiring analog DC signals:

- High accuracy/resolution—accurately measure the signal level
- Low bandwidth—sample the signal at low rates (software timing should be sufficient)

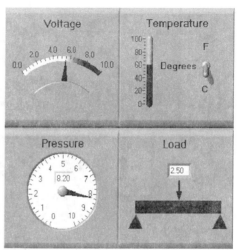

Common examples
of level measurements

signal = slow
accuracy required = high

Figure 10.6
Level examples.

Analog Time Domain Signals

Analog time domain, or *shape*, signals differ from other signals in that they convey useful information not only in the signal level, but also in how this level varies with time. When measuring a shape signal, often referred to as a waveform, you are interested in some characteristics of the waveform shape, such as slope, locations and shapes of peaks, and so on.

To measure the shape of a time domain signal, you must take a precisely timed sequence of individual amplitude measurements, or points. These measurements must be taken at a rate that will adequately reproduce the shape of the waveform. Also, the series of measurements should start at the proper time, to guarantee that the useful part of the signal is acquired. Therefore, the instrument or plug-in DAQ board that measures time domain signals consists of an ADC, a sample clock, and a trigger. A sample clock accurately times the occurrence of each A/D conversion. To ensure that the desired portion of the signal is acquired, the trigger starts the measurement at the proper time according to some external condition.

There are an unlimited number of different time domain signals, a few of which are shown in Figure 10.7. What they all have in common is that the shape of the waveform (level versus time) is the main feature of interest.

Your DAQ system should meet the following specifications when acquiring analog time domain signals:

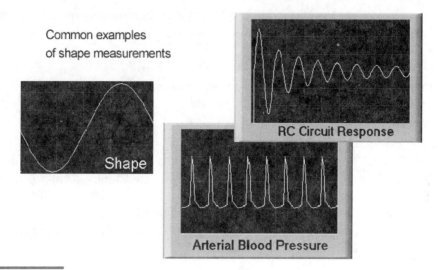

Figure 10.7
Shape examples.

- Higher bandwidths—sample the signal at high rates

- Accurate sample clock—sample the signal at precise intervals (hardware timing needed)

- Triggering—start taking the measurements at a precise time

Analog Frequency Domain Signals

Analog frequency domain signals are similar to time domain signals because they also convey information on how the signals vary with time. However, the information extracted from a frequency domain signal is based on the signal frequency content, as opposed to the shape or time-varying characteristics of the waveform.

Like the time domain signal, the instrument used to measure a frequency domain signal must include an ADC, a sample clock, and a trigger to accurately capture the waveform. Additionally, the instrument must include the necessary analysis capability to extract frequency information from the signal. You can perform this type of digital signal processing (DSP) using application software or special DSP hardware designed to analyze the signal quickly and efficiently.

Your DAQ system should meet the following specifications when acquiring analog frequency domain signals:

- Higher bandwidths—sample the signal at high rates
- Accurate sample clock—sample the signal at precise intervals (hardware timing needed)
- Triggering—start taking the measurements at a precise time
- Analysis functions—convert time information to frequency information

Figure 10.8 shows a few examples of frequency domain signals. While you can analyze any signal in the frequency domain, certain signals and application areas lend themselves especially to this type of analysis. Among these areas are speech, acoustics, geophysical signals, vibration, and system transfer functions.

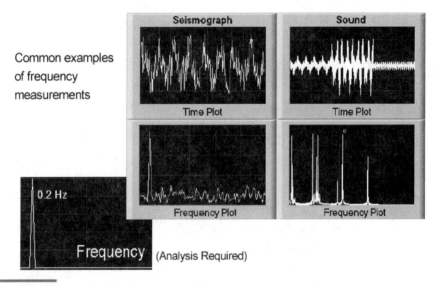

Figure 10.8
Frequency examples.

One Signal—Five Measurement Perspectives

The five classifications of signals presented in this section are not mutually exclusive. A particular signal may convey more than one type of information. Therefore, a signal can be classified as more than one type of signal, and thus you can measure it in more than one way. You can use simpler measurement techniques with the digital on–off, pulse train, and DC signals because they are just simpler cases of the analog time domain signals.

You can measure the same signal with different types of systems, ranging from a simple digital input board to a sophisticated frequency analysis system. The measurement technique you choose depends on the information you want to extract from the signal. Look at Figure 10.9. It demonstrates how one signal—a series of voltage pulses—can provide information for all five signal classes.

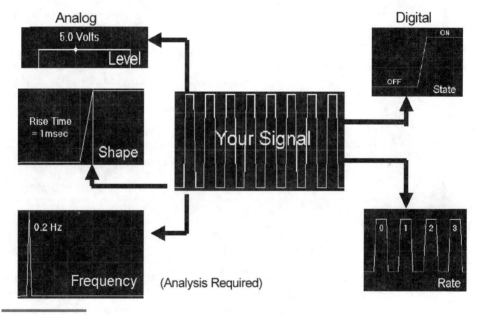

Figure 10.9
Five ways to measure the same signal.

Activity 10-1: Classifying Signals

Classify the following signals into one of the five signal types described earlier (from the perspective of data acquisition). In some cases, a signal can have more than one classification. Choose the type that you think best matches the signal. Circle the number on the left margin as follows:

1: Analog DC
2: Analog AC

3: *Digital on–off*
4: *Digital pulse/counter*
5: *Frequency*

1	2	3	4	5	Voltage level of a battery
1	2	3	4	5	State of a solid-state relay
1	2	3	4	5	Data at your PC's parallel port during printing
1	2	3	4	5	Glitch or spike noise in power source
1	2	3	4	5	Transfer function of a digital filter
1	2	3	4	5	Data flowing over the Internet
1	2	3	4	5	Relative humidity outside
1	2	3	4	5	Car engine's RPM while driving around town
1	2	3	4	5	EEG (brain waves)
1	2	3	4	5	Speech through a microphone
1	2	3	4	5	Absolute pressure in an engine cylinder

Transducers

When you're setting up your DAQ system, remember that ultimately every-thing you're going to measure will have to become an electrical voltage or current. The way you convert measurable phenomena such as temperature, force, sound, light, stupidity, etc., to an electrical signal is by using a *trans-ducer*. Table 10.1 lists some common transducers used to convert physical phenomena into a measurable quantity.

10.3.3 Signal Conditioning

Now that you've figured out what kind of signals you need to acquire, you can just plug the output of your transducers directly into the DAQ board, right? Wrong! In perhaps 50% of the cases or more, the answer is no. We may not always be aware of it, but we live in a very electrically noisy world.

Table 10.1

Phenomena	Transducer
Temperature	Thermocouples Resistance temperature detectors (RTDs) Thermistors Integrated circuit sensor
Light	Vacuum tube photosensors Photoconductive cells
Sound	Microphone
Force and pressure	Strain gauges Piezoelectric transducers Load cells
Position (displacement)	Potentiometers Linear voltage differential transformer (LVDT) Optical encoder
Fluid flow	Head meters Rotational flowmeters Ultrasonic flowmeters
pH	pH electrodes

By the time your signal makes it to the DAQ board, it may have picked up so much noise or have so many other problems that it renders your measurement useless.

You usually need to perform some type of signal conditioning on analog signals that represent physical phenomena. What is signal conditioning, anyway? Simply put, it is a manipulation of your signal to prepare it for digitizing at the DAQ board. Your signal has to arrive as clean as possible, within the voltage (usually ±5 V or 0 to 10 V) and current (usually 20 mA) limits of your DAQ board, with enough precision for your application. It's hard to be more specific unless you can specify what kind of transducers you're going to use. For example, signal conditioning for audio data from a microphone may involve nothing more than grounding the system properly and perhaps using a low-pass filter. However, if you want to measure ionization levels in a plasma chamber sitting at 800 V and you don't want to fry your computer, you'd need to provide some more complex circuitry that includes isolation amplifiers with a step-down gain.

For signals that need special conditioning, or for systems that have very many signals, National Instruments devised *SCXI* (Signal Conditioning eXtensions for Instrumentation). An SCXI system provides a chassis where modular units can be inserted to build a custom system. These modular units include analog input multiplexers, analog output boards, "blank" breadboards, signal conditioning modules for thermocouples, etc. For more information on SCXI, see the National Instruments catalog. Remember that you don't necessarily need SCXI to do signal conditioning.

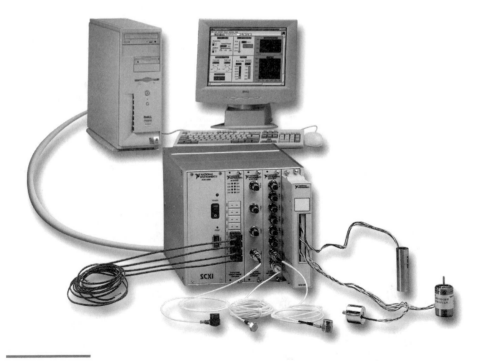

Figure 10.10
An SCXI chassis with several signal conditioning modules connected to a PC with LabVIEW.

Figure 10.11
SCXI systems are often used in industrial rack-mount enclosures for high-channel count applications, as shown in this picture.

Some common types of signal conditioning are:

- Amplification
- Transducer excitation

- Linearization
- Isolation
- Filtering

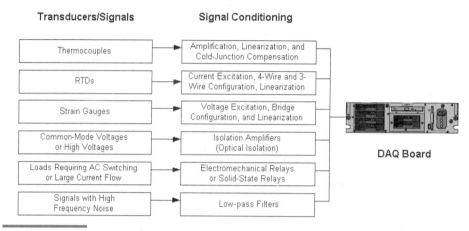

Figure 10.12

10.3.4 Finding a Common Ground

Some physical properties are absolute: luminosity or mass, for example (I know, I know, the physicists won't agree—things act weird when you approach the speed of light, and then almost nothing is absolute, except the speed of light itself). But in any case, voltage is decidedly *not* absolute; it always requires a reference to be meaningful. Voltage is always the measure of a potential *difference* between two bodies. One of these bodies is usually picked to be the reference and is assigned "0 V." So to talk about a 3.47 V signal really means nothing unless we know with respect to what reference. If you've noticed, though, often a reference isn't specified. That's because the 0 V reference is usually the famous *ground*. Herein lies the source of much confusion, because "ground" is used in different contexts to refer to different reference potentials.

Earth ground refers to the potential of the earth below your feet. Most electrical outlets have a prong that connects to the earth ground, which is also usually wired into the building electrical system for safety. Many instruments also are "grounded" to this earth ground, so often you'll hear the term *system ground*. This is the ground that is usually tied to the third ground prong on electrical outlets. The main reason for this type of grounding is safety, and not because it is used as a reference potential. In fact, you can bet that no two sources that are connected to the earth ground are at the same reference level; the difference between them can easily be several hundred millivolts. Thus, we're usually not talking about earth, or safety ground, when we need to specify a reference voltage.

Reference ground, sometimes called a return path or signal common, is usually the reference potential of interest. The common ground may or may not be wired to earth ground. The point is that many instruments, devices, and signal sources provide a reference (the negative terminal, common terminal, etc.) which gives meaning to the voltages we are measuring.

The ground symbols in Figure 10.13 are used in this book when you see wiring diagrams. Be aware, however, that you will find these same symbols used inconsistently among engineers.

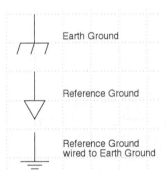

Figure 10.13

The DAQ boards in your computer are also expecting to measure voltage with respect to some reference. What reference should the DAQ board use? You have your choice, which will depend on the kind of signal source you're connecting. Signals can be classified into two broad categories:

- Grounded
- Floating

Let's examine these categories a bit further:

Grounded Signal Source

A grounded source is one in which the voltage signals are referenced to a system ground, such as earth or building ground. Because they use the system ground, they share a common ground with the DAQ board. The most common examples of grounded sources are devices that plug into the building ground through wall outlets, such as signal generators and power supplies.

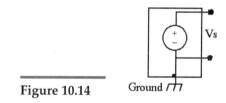

Figure 10.14   Ground

Floating Signal Source

A floating source is a source in which the voltage signal is not referenced to any common ground, such as earth or building ground. Some common examples of floating signal sources are batteries, thermocouples, transformers, and isolation amplifiers. Notice, as shown in Figure 10.15, that neither terminal of the source is connected to the electrical outlet ground. Thus, each terminal is independent of the system ground.

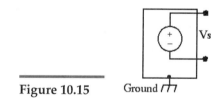

Figure 10.15   Ground

10.3.5 Measuring Differences

To measure your signal, you can almost always configure your DAQ board to make measurements that fall into one of these three categories:

• Differential
• Referenced single-ended
• Nonreferenced single-ended

Differential Measurement System

In a differential measurement system, neither input is connected to a fixed reference such as earth or building ground. Most DAQ boards with instrumentation amplifiers* can be configured as differential measurement systems. Figure 10.16 depicts the eight-channel differential measurement system used in the E-series boards. Analog multiplexers increase the number of measurement channels while still using a single instrumentation amplifier. For this board, the pin labeled AIGND (the analog input ground) is the measurement system ground.

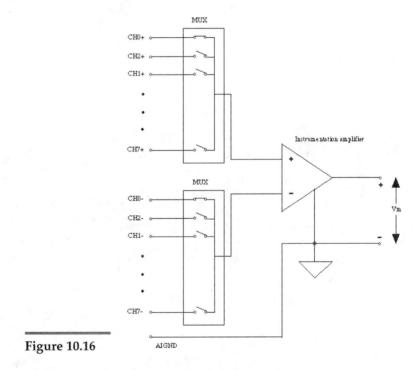

Figure 10.16

Before we discuss single-ended systems, it is worth noting that SCXI systems always use the popular differential measurement system, whereas most plug-in DAQ boards give you a choice.

---

* An *instrumentation amplifier* is a special kind of circuit (usually embedded in a chip) whose output voltage with respect to ground is proportional to the difference between the voltages at its two inputs.

For True Geeks Only

An ideal differential measurement system reads only the potential *difference* between its two terminals—the (+) and (–) inputs. Any voltage present at the instrumentation amplifier inputs with respect to the amplifier ground is referred to as a common-mode voltage. An ideal differential measurement system completely rejects (does not measure) common-mode voltage. Practical devices, however, limit this ability to reject the common-mode voltage. The common-mode voltage range limits the allowable voltage swing on each input with respect to the measurement system ground. Violating this constraint results not only in measurement error but also in possible damage to components on the board. The common-mode voltage range specification quantifies the ability of a DAQ board, operating in differential mode, to reject the common-mode voltage signal.

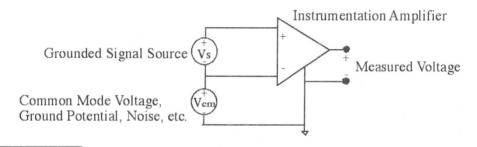

Figure 10.17

You measure the common-mode voltage, V_{cm}, with respect to the DAQ board ground, and you calculate it using the following formula:

$$V_{cm} = \frac{V^+ + V^-}{2}$$

where

V^+ = Voltage at the *noninverting* terminal of the measurement system with respect to the instrumentation amplifier ground.

V^- = Voltage at the *inverting* terminal of the measurement system with respect to the instrumentation amplifier ground.

You cannot have an arbitrarily high common-mode voltage when measuring with a DAQ board. All plug-in DAQ boards specify a maximum working voltage (MWV) that is the maximum common-mode voltage the board can tolerate and still make accurate measurements.

Referenced Single-Ended Measurement System

A referenced single-ended (RSE) measurement system, also called a grounded measurement system, is similar to a grounded signal source, in that the measurement is made with respect to earth ground. Figure 10.18 depicts a 16-channel RSE measurement system.

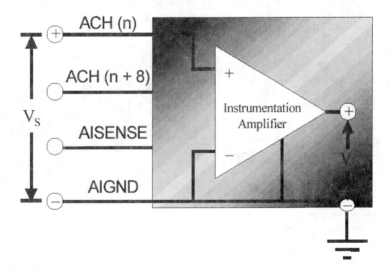

Figure 10.18

Be careful when connecting a voltage or current source to your DAQ board. Make sure the signal source will not exceed the maximum voltage or current that the DAQ board can handle. Otherwise, you could damage both the board and your computer.

NRSE Measurement System

DAQ boards often use a variant of the RSE measurement technique, known as the *nonreferenced single-ended* (NRSE) measurement system. In an NRSE measurement system, all measurements are made with respect to a common reference ground, but the voltage at this reference can vary with respect to the measurement system ground. Figure 10.19 depicts an NRSE measurement system where AISENSE is the common reference for taking measurements and AIGND is the system ground.

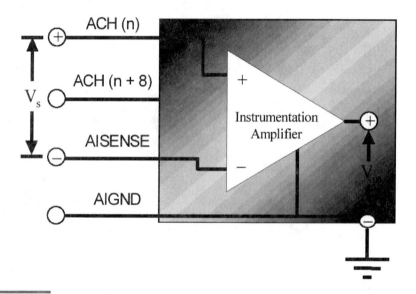

Figure 10.19

Incidentally, your measurement system is determined by how you configure your DAQ board. Most boards from National Instruments can be configured for differential, RSE, or NRSE from a software utility called NI-MAX. Some of their older boards also have to be configured at the board by placing jumpers in certain position. When you configure a particular DAQ board for a particular measurement system type, all your input channels will follow that measurement type. You should note that you can't change this from LabVIEW—you have to decide ahead of time what kind of measurement you're making.

The general guideline for deciding which measurement system to pick is to measure grounded signal sources with a differential or NRSE system and floating sources with an RSE system. The hazard of using an RSE system with a grounded signal source is the introduction of *ground loops*, a possible source of measurement error. Similarly, using a differential or NRSE system to measure a floating source will very likely be plagued by *bias currents,* which cause the input voltage to drift out of the range of the DAQ board (although you can correct this problem by placing bias resistors from the inputs to ground).

Figure 10.20 summarizes the measurement configurations for each signal type.

Grounded Signal Sources
Examples: instruments with non-isolated inputs

Floating Signal Sources
Examples: thermocouples, battery devices, signal conditioning with isolated outputs

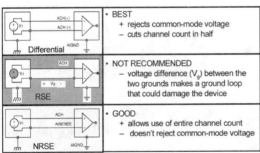
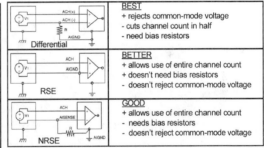

Figure 10.20

10.3.6 Sampling, Aliasing, and Mr. Nyquist

The last, and perhaps most important, part of signal theory we're going to cover here is sampling.

Zeno's arrow aside, real-world signals are continuous things. To represent these signals in your computer, your DAQ board has to check the level of the signal every so often and assign that level a discrete number that your computer will accept; this is called an analog-to-digital conversion. The computer then sort of "connects the dots" and, hopefully, gives you something that looks similar to the real-world signal (that's why we say it *represents* the signal).

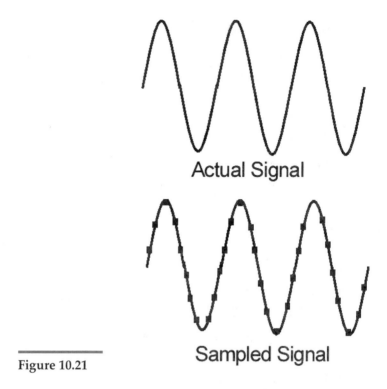

Actual Signal

Sampled Signal

Figure 10.21

The *sampling rate* of a system simply reflects how often an analog-to-digital conversion (ADC) takes place. Each vertical line in Figure 10.21 represents one ADC. If the DAQ system is making one ADC every half a second, we say the sampling rate is 2 samples/second, or 2 Hz. Alternatively, we can specify the sampling period, which is the inverse of the sampling rate (in this example, 0.5 seconds). It turns out that the sampling rate has a terribly important effect on whether your digitized signal looks anything like the real-world signal.

When the sampling rate isn't high enough, a scary thing happens: aliasing. Although not intuitive, aliasing is easy to observe (see Figure 10.22).

Aliasing has the effect of introducing high-frequency components into your data that didn't exist in the real-world signal, thereby severely distorting your signal. Once you have aliased data, you can never go back: There is no way to remove the "aliases." That's why it's so important to sample at a high enough rate.

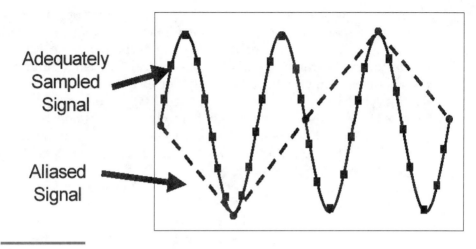

Adequately
Sampled
Signal

Aliased
Signal

Figure 10.22

How do you determine what your sampling rate should be? A guy named Nyquist figured it out, and his principle, called **Nyquist's theorem**, is simple to state:

To avoid aliasing, the sampling rate must be greater than twice the maximum frequency component in the signal to be acquired.

So, for example, if you know that the signal you are measuring is capable of varying as much as 1,000 times per second (1,000 Hz), you'd need to choose a sampling rate higher than 2 kHz. Notice that the Nyquist sampling theorem implies that you know what the highest frequency component will be. It is imperative that you find out if you don't know already; if you can't know ahead of time what the highest frequency component will be, you'll need to filter the signal to remove potential high-frequency components, as we describe next.

The Nyquist theorem only deals with accurately representing the frequency of the signal. It doesn't say anything about accurately representing the shape of your signal. If you need to preserve the shape of your signal, you should sample at a much higher rate than the Nyquist frequency, generally between 5 and 10 times the maximum frequency component of your signal.

Another reason for knowing the frequency range of your signal is to choose proper *anti-aliasing filters* (low-pass filters). In many real-world applications, signals pick up a great deal of high-frequency noise, glitches, or spikes that will greatly exceed the theoretical frequency limit of the frequency measurement you are making. For example, a common biomedical signal is the electrocardiogram (ECG or EKG), a voltage that is related to heart activity. Although these signals rarely have components beyond 250 Hz, the electrode leads easily pick up RF (radio-frequency) noise in the 100 kHz and MHz range! Rather than sample at extremely high frequencies, these DAQ systems implement some low-pass filters that cut out waveforms above 250 Hz. The DAQ board can then breathe easier and sample only at, say, 600 Hz.

The only case where sampling rate is not important is in the so-called DC signals, such as temperature or pressure. The physical nature of these signals is such that they cannot vary by much more than perhaps once or twice a second. In these cases, a low sampling rate such as 10 Hz should do.

10.3.7 In Conclusion . . .

We've covered a lot of issues involving the path from the physical phenomena to the DAQ board. If you didn't grasp many or even most of these concepts at first, don't worry. DAQ theory is a complex subject, and unless you've had some experience in this area of electrical engineering, it can take some practice before you understand it all.

You've seen a summary of how signals are classified, what kind of transducers are often used, the importance of signal conditioning, the different measurement configurations for digitizing grounded or floating signal sources, and the necessity of using Nyquist's sampling theorem. You'd need to take a couple of electrical engineering courses to thoroughly cover the whole topic of data acquisition and instrumentation; we've just skimmed the surface in this section. Nonetheless, it should be enough to get you started with your measurements.

10.4 Selecting and Configuring DAQ Measurement Hardware

10.4.1 Choosing Your Hardware

Figure 10.23

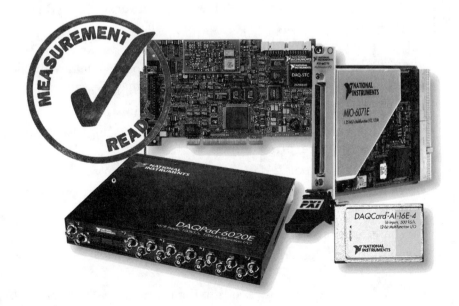

Figure 10.24
Many different type of DAQ boards that work with LabVIEW are available from National Instruments.

Once you know what kind of signals you want to measure and/or generate, it's time to choose a plug-in DAQ board (assuming a DAQ board will meet your requirements). Generally speaking, we recommend always using a National Instruments board if you're going to use LabVIEW. Although it's possible to use boards from other vendors with LabVIEW, you could be in for a long, low-level, device-driver programming marathon. National Instruments offers a huge selection of all types of boards with a good selection of platforms, performance, functionality, and price range. You can browse their catalog online (http://ni.com), where you will find a "DAQ Designer" utility that will help you determine what hardware suits your needs.

To pick the best hardware for your system, you need to understand well what your system requirements are, most noticeably the "I/O count" (how many inputs and how many outputs). The following checklist should be useful in determining if you have all the information you need to select a board:

- What type of system am I using (Windows, Linux, etc.)?
- What type of bus or connector is available (PCI, PC-card for laptops, etc.)?
- How many analog inputs will I need? (Multiply by 2 if you need differential inputs.)
- How many analog outputs will I need?
- Are the analog inputs voltage, current, or both? What are the ranges?
- How many digital input and output lines will I need?
- Do I need any counting or timing signals? How many?
- Do any of the analog I/O signals require special signal conditioning (e.g., temperature, pressure, strain)?
- Will any of the analog I/O signals exceed I/O V or 20 mA?
- What is the minimum sampling rate required on any one channel?
- What is the minimum scan rate for all the channels?
- What precision, or resolution (12-bit or 16-bit), will I need?
- Is portability, ruggedness, isolation, or cost an issue? If so, what are the trade-offs?
- Have I accounted for my needs in the future (expansion, new DAQ systems, etc.)?

Now you're ready to pick your DAQ board(s). The most popular type of National Instruments boards are the *E-series* boards, such as the PCI-MIO-16E-4 board, which provides you with 16 analog inputs, 2 analog outputs channels, 8 digital I/O channels, and 2 counters. As the name implies, this board typically contains various combinations of ADCs, DACs, digital I/O lines, and counter/timer circuitry, making it useful for a wide range of applications. Interboard connections, such as the Real-Time System Integration (RTSI) bus, transfer timing and trigger signals among boards, permitting synchronization of the operations on multiple boards. This board fits well with most applications, which usually require several analog inputs and occasionally an output or digital signal. If you need more analog outputs, you can get a board such as the NI-6704, which provides 32 analog outputs. You can also obtain timing boards, high-count digital I/O boards, PC-cards for laptops, external USB-connected DAQ boards, and just about any combination or specialization of features you can imagine.

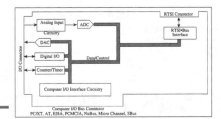

Figure 10.25

The analog input section of the MIO board consists of the ADC and the *analog input circuitry*. The analog input circuitry contains the analog multiplexer, the instrumentation amplifier, and the sample-and-hold circuitry. For more detailed information on the ADC, refer to the manual that accompanies your DAQ board.

In addition to the circuitry shown in Figure 10.25, older DAQ boards can also contain several components, such as jumpers and dip switches, that configure the DAQ hardware. These components, which configure parameters such as base address, direct memory access (DMA) channels, and interrupt levels, are discussed in the next section. The *Plug&Play* standard is making this obsolete.

What if you're concerned about cost? Most plug-in boards set you back about $1,000 to $2,000. In general, two factors are directly proportional to and most influential on a board's price: the sampling rate and the number of analog I/O channels. So if you are making DC measurements, there's no need to shell out extra dough for a board with a one-MHz sampling rate. Also, digital I/O is usually cheap. The board prices shouldn't really seem expensive once you pause to consider what you're getting, however. How much would an oscilloscope, spectrum analyzer, strain gauge meter; and hundreds of other instruments all together cost you if you had to buy them as "nonvirtual" instruments?

One of the best resources to help you figure out what hardware you need is a free utility from National Instruments called **DAQ Designer.** You can run it online at http://ni.com/catalog or request it as a CD. Of course, it will only recommend NI's own hardware. If you want to look at a broader range of options, you should consult with an experienced system integration company (such as Rayodyne, http://www.rayodyne.com) who can advise you on what works best for you.

Finally, be aware that boards not made by National Instruments will only work with LabVIEW if the manufacturer of that board provides a LabVIEW driver (or if you write one yourself).

10.5 Activity 10-2: Measurement System Analysis

Here are a couple of more challenging signal measurement problems. Your objective is to specify the needed information.

Answer the following for each scenario:

1. What kind of signals need to be measured?
2. What signal measurement type do you recommend?
3. What should the sampling rate be for tire signals? What is the Nyquist frequency?
4. Is any signal conditioning needed? If so, what sort?
5. What hardware would you pick for this system?
A. Professor Harry Phace, of the biomedical engineering lab, wants to acquire heart signals from human subjects—hopefully without electrocuting them. He wants to measure electrocardiograms from two subjects at a time. Each subject has four electrodes connected to his body at different places. The objective is to measure in real time the

potential between each of three electrodes, with the fourth electrode designated as a reference. The leads from the electrodes to the DAQ system have no isolation or ground shields. The maximum amount of detail expected in the waveforms are segments 2 ms wide. The signals are within a 0.024 mV range.

B. Ms. I. M. Aynurd needs to measure how the resistance of a flexible material changes under stress and high temperatures. To do so, she has a special chamber with a machine that twists and stretches the material over and over. The chamber also functions as an oven with a variable temperature control. She wants to observe in real time how the resistance of each of 48 strands of this material changes inside the chamber. The resistance is measured by applying a known voltage to each strand and measuring the current. The temperature of the chamber is monitored through a thermocouple. The stress machine is turned on and off through a solid-state relay. Finally, the number of cycles from the stress machine needs to be measured.

Answers are at the end of the chapter.

10.6 Installing the Boards

All plug-in boards use *drivers*, nasty and painful pieces of low-level code that convince your computer that the boards really are inside it and can be used. The good news is that, assuming your drivers are installed properly, you should not have to really mess with them to use your DAQ board. All National Instruments boards come with driver software, collectively referred to as *NI-DAQ*. In fact, NI-DAQ is installed by default when you install LabVIEW, so if you have the full version of LabVIEW, chances are it's already on your machine.

Between NI-DAQ and LabVIEW, there is a utility called *MAX* (Measurement and Automation Explorer). MAX is a Windows software interface that gives you access to all your National Instruments boards (whether they are DAQ, GPIB, VXI, etc.). MAX is mainly useful for configuring and testing your hardware. This is very useful to do before you try to access the hardware in LabVIEW. Again, MAX is installed by default when you install LabVIEW; you should see the shortcut icon on your Windows desktop.

Figure 10.26

We'll talk more about MAX briefly.

> *MAX is a utility available on Windows only. For configuring DAQ boards on other operating systems, such as the MacOS or Linux, use the NI-DAQ configuration utility provided for that OS.*

Figure 10.27 shows the relationships among NI-DAQ, MAX, and LabVIEW.

Figure 10.27

10.6.1 Analog I/O Settings

A DAQ board has several analog I/O parameters that control the operation of the ADC (analog-to-digital converter) and DAC (digital-to-analog converter). On almost all boards, you will configure these settings in the MAX software utility (some older boards require you to set jumpers on the board). The type of settings you can configure include parameters like range, input mode, reference, and polarity; for example, on most E-series boards, you can set the following:

- ADC Input Range   Unipolar 0 to +10 V
                      Bipolar ±5 V
                      Bipoloar ±10 V (default)

- ADC Input Mode   Ground-referenced single-ended
                      Nonreferenced single-ended
                      Differential (default)

- DAC Reference   Internal (default)
                      External

- DAC Polarity   Unipolar—straight binary mode
                      Bipolar—Two's complement mode (default)

10.6.2 Measurement and Automation Explorer (MAX)

MAX is mainly used to configure and test your National Instruments hardware, but it does offer other functionality such as checking to see if you have the latest version of NI-DAQ installed.

The functionality of MAX is divided into four categories:

- Data Neighborhood
- Devices and Interfaces
- Scales
- Software

Let's explore each of these in a little more depth.

Data Neighborhood

The Data Neighborhood category shows you all of your currently configured *virtual channels* and provides utilities for testing and reconfiguring those virtual channels. Data Neighborhood also provides access to the DAQ Channel Wizard, which allows you to create new virtual channels.

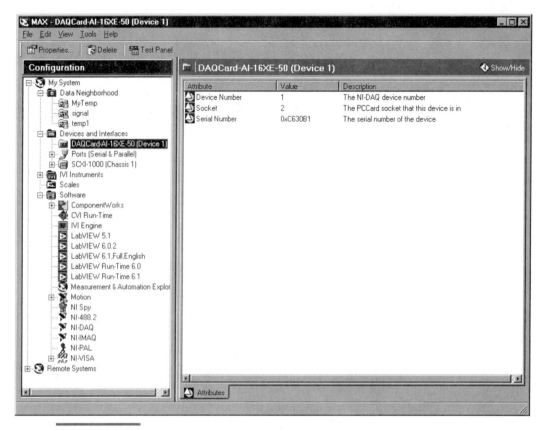

Figure 10.28
The MAX configuration utility.

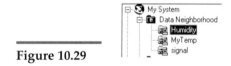

Figure 10.29

So what is a virtual channel? A virtual channel is a shortcut to a configured channel in your system. You can set up the configuration information for your channel and give the channel a descriptive name at the same time. Later, you can use the descriptive name to access that channel and its configuration information in LabVIEW. You can give your channel a description, decide what type of transducer your channel will use, set the range (to determine gain), choose the grounding mode, assign custom scaling for your virtual channel, and give the channel a descriptive name to replace the channel number all at the same time. For example, if channel 0 on your DAQ board was wired to a temperature sensor, you could create a virtual channel for channel 0 and call it Temperature Sensor. You can create virtual channels for analog input, analog output, and digital I/O. In this case, referring to a channel by a name (Temperature Sensor) instead of a number (0) helps you remember what the channel does.

You don't have to use virtual channels, of course. In your LabVIEW application you can refer to channels by their number (0, 1, 2 . . .). But configuring them first as a virtual channel in MAX is handy, because as we'll see in the next chapter, you can use a LabVIEW front panel or block diagram DAQ channel ring that gives you the names of all your virtual channels.

Figure 10.30

To create a virtual channel, you right-click on the **Data Neighborhood** icon in MAX, and choose **Create New....**This will pop up the **DAQ Channel Wizard**, which will guide you through the steps to set up your virtual channel.

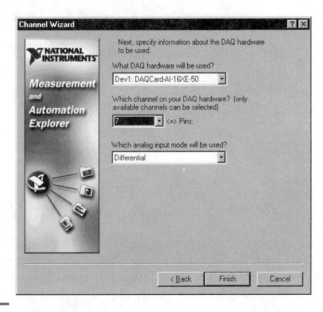

Figure 10.31
DAQ Channel Wizard.

Devices and Interfaces

The next category in MAX is called *Devices and Interfaces*. As the name implies, Devices and Interfaces shows you any currently installed and detected
National Instruments hardware, such as plug-in DAQ boards, SCXI and PXI
chasses, GPIB boards, etc. Devices and Interfaces also includes utilities for
configuring and testing your devices.

Figure 10.32

When you right-click on an installed board or device, the pop-up menu
gives you the options: **Properties**, **Test Panels**, and **Delete**.

The **Properties** panel is where you actually configure your DAQ board, as
shown in Figure 10.33.

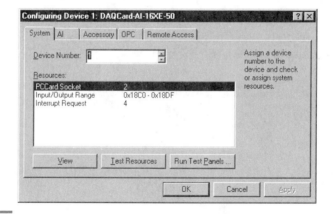

Figure 10.33

The panel has tabs across the top that you should use for configuring your
DAQ device. The tabs are as follows:

• **System**—This allows you to change your device number, and it provides two buttons for testing your DAQ device. The first button is the
"Test Resources" button. This button performs a basic test of the system
resources assigned to the device. The system resources tested are the
base I/O address, the interrupt request (IRQ), and the direct memory
access (DMA), which are explained in the following section:

- *Base I/O Address*: A DAQ device communicates with a computer primarily through its registers. NI-DAQ writes to configuration registers on the device to configure the device and reads data registers on the device to obtain the device's status or a signal measurement. The base I/O address setting determines where in the computer's I/O space the device's registers reside.

- *Interrupt Request (IRQ)*: Another way the DAQ device communicates with the computer is through processor interrupts, which give the processor the ability to respond quickly to its peripherals. Think of a processor interrupt as a doorbell. If your door did not have a doorbell, you would have to go to the door periodically to see if anyone happened to be there at that particular time. With a doorbell, you only need to go to the door when the doorbell rings and you are confident that someone is there waiting. In the case of a DAQ device, it is not efficient for the processor to continually check whether data are ready to be read from the device. A DAQ device can use an interrupt as a doorbell that signals the processor that it has data waiting to be read. Each interrupt request has a number assigned to it. For example, the device shown in the **Properties** window was given interrupt request 4.

- *Direct Memory Access (DMA)*: The third way the DAQ device can communicate with the computer is through direct memory access (DMA). DMA is a data transfer method in which data are transferred directly from the peripheral to computer memory, bypassing the processor. DMA is usually required to achieve maximum data transfer speed, making it useful for high-speed DAQ devices. DAQ devices that use the PCI bus have their own onboard DMA channels, and the PCI bus handles the sharing of that DMA.

The vast majority of the time, you should not have to worry about or set the base I/O address, IRQ, or DMA, as this is usually automatically done for you by Windows and NI-DAQ when you install the board. You should only change these settings if you have a hardware conflict or other particular reasons.

- **AO**—The **AO** tab configures the default polarity of your analog output signal and allows you to specify if you are using an external voltage reference for your DAC.

- **AI**—The **AI** tab configures your analog input signal, allowing you to specify the type of signal, its voltage limits, and what scale to map it to.

- **Accessory**—The **Accessory** tab specifies any accessories you are using with your DAQ device, such as a TBX-68 (terminal block with built-in cold-junction compensation). If NI-DAQ does not need to know about your accessory, it will not be on the list. In that case, choose **None**.

- **OPC**—The **OPC** tab allows you to set the AI recalibration period if you are using the NI-DAQ OPC server. OPC is *Open Process Control*, a protocol used in industrial automation. If you don't know what OPC is about, you can safely ignore this tab.

After your device has passed the basic resource test and you have configured the **System**, **AI**, **AO**, **Accessory**, and **OPC** tabs, you should return to the **System** tab and click the **Test Panels** button. The window that appears is shown in Figure 10.34. The **Test Panel** is a utility for testing the analog input, analog output, digital I/O, and counter functionality of your DAQ device.

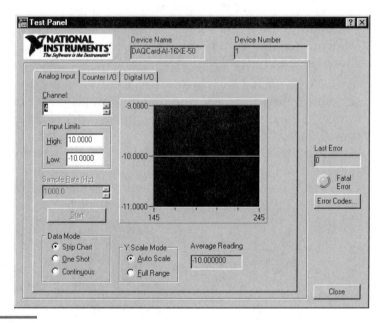

Figure 10.34

The **Test Panel** is a great utility for troubleshooting because it allows you to test the functionality of your device directly from NI-DAQ. If your device doesn't work in the **Test Panel**, it isn't going to work in LabVIEW. If you are ever having unexplainable trouble with data acquisition in a LabVIEW program, you can use the **Test Resources** button and the Test Panels to make sure the device is working properly.

Scales

Figure 10.35

Scales is the next feature in the MAX utility. Here you can create custom scales you can use to determine scaling information for existing virtual channels. This is sometimes necessary or useful particularly for sensors that are not linear or where you want to read the actual units directly instead of having to convert voltage or current to the desired unit, such as temperature. Each custom scale can have its own name and description to help you identify it. A custom scale can be one of three types: linear, polynomial, or table.

- *Linear*—A scale that uses the formula $y = mx + b$.
- *Polynomial*—A scale that uses the formula $y = a0 + a_1x + a_2x^2 + \ldots + a_nx^n$.
- *Table*—A scale where you enter the raw value and the corresponding scaled value in a table format.

Software

Figure 10.36

The final MAX category is **Software**. **Software** shows all of your currently installed versions of National Instruments software. The icon for each software package is also a shortcut that you can use to launch the software. For example, if you clicked on the LabVIEW icon in the previous, LabVIEW would launch. The **Software** category also includes a Software Update Agent. The purpose of the Software Update Agent is to check if your National Instruments software is the latest version. If your software isn't the latest version, the Software Update Agent will link you to a page on ni.com to download the latest version of your software.

10.6.3 DAQ Boards under MacOS and Linux

If you are using MacOS (Classic or OS X), Linux, Solaris, or HP-UX, the previous section on MAX doesn't apply, because, at least at press time, MAX was a Windows-only utility. However, National Instruments does provide driver support for many (but certainly not all) of its hardware products for other operating systems, including the MacOS and Linux.

The driver software package, NI-DAQ, is specific to each operating system and you should check the documentation for information on installing and configuring your hardware.

10.7 Using a GPIB Board

A GPIB board is used to control and communicate with one or more external instruments that have a GPIB interface. The General Purpose Interface Bus, invented by Hewlett-Packard (HP) in the 1960s, has become the most popular instrument communications standard. All HP instruments support it, as well as thousands of others. GPIB was also updated and standardized by the IEEE, and was duly named *IEEE 488.2*. Some nice features about GPIB are:

- It transfers data in parallel, one byte (eight bits) at a time.
- The hardware takes care of handshaking, timing, etc.
- Several instruments (up to 15) can be strung together on one bus.
- Data transfer is fast: 800 Kbytes/second or more.

Figure 10.37
GPIB boards are usually used to communicate with external instruments, such as an oscilloscope.

You can obtain a plug-in GPIB board for almost every platform and bus, as well as external interfaces that will convert your serial port, parallel port, Ethernet connection, or USB port to GPIB.

Another feature that makes GPIB popular is that the computer and instrument talk to each other using simple, intuitive ASCII commands. For example, a PC connected to an HP 3458A Digital Multimeter might say something like this:

PC: IDN? ; [Identity?—ie., who are you?]

HP: HP3458A

PC: RMEM 1; [Recall memory register 1]

HP: +4.23789

A multitude of GPIB instrument drivers, available free of charge, make communication with an external instrument as simple as using a few subVIs that do all the ASCII commands *for* you.

Installing a GPIB board is usually pretty straightforward. The GPIB boards from NI integrate with NI-MAX so that you can easily configure them under the **Devices and Interfaces** tree. The best advice is to just follow the installation procedures described in the manual that comes with the GPIB board.

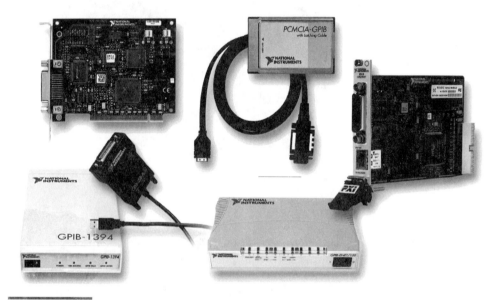

Figure 10.38
GPIB controller boards come in many different interface flavors: PCI, PC Card, Ethernet, USB, PXI, etc.

10.8 Getting Ready for Serial Communications

Serial communication has the advantage that it's cheap and simple. You rarely need any additional hardware, since many computers have at least one serial port available. It doesn't mean it will be easy to get the hardware hooked up right, however Although popular standards such as RS-232 and RS-485 specify the cabling, connectors, timing, etc., manufacturers of serial devices more often than not abuse and ignore these standards. If you just plug a cable from your serial port on your PC to the instrument, you have one chance in three that it will work perfectly the first time.

While RS-232 communication may eventually be replaced by newer standard serial protocols like USB, a lot of RS-232/485 instruments are still out there, and if you need to use them, then it will be important to become familiar with your serial instrument, its pinouts, and the parameters it uses, such as baud speed, parity, stop bits, etc. You should also know what the different lines of a serial port are for.

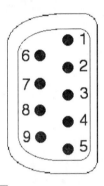

Pin	Function
1	RLSD
2	RxD
3	TxD
4	TDR
5	GND
6	DSR
7	RTS
8	CTS
9	RI

Figure 10.39
PC serial pinout (DB-9 connector). RxD: receive data; TxD: transmit data; DTR: data terminal ready; CTS: clear to send; GND: ground or common; DSR: data set ready; RTS: ready to send; RLSD: received line signal detect; RI: ring indicator.

In many cases, only a few of the lines are used. Chances are you'll end up using the following or fewer:

Transmit (TxD): Sends data *from* the PC *to* the instrument.

Receive (RxD): Sends data *to* the PC *from* the instrument.

Ground (GND): Ground reference. Never forget to hook this one up.

Clear-to-Send (CTS): The PC uses this line to tell the instrument when it's ready to receive data.

Ready-to-Send (RTS): The PC uses this line to advise the instrument it's ready to send data.

If you experience problems getting your serial device to communicate, try a few of these things:

- Swap the transmit and receive lines. You can use a special cable, called a "null modem" cable to do this.

- Check the baud speed, parity, stop bits, handshaking (if any), and any other serial parameters on your PC. Then check what these parameters are on the instrument. If *any* of these differ, the two parties won't talk to each other.

- The serial instrument will always require a power supply. Make sure it's on.

- Make sure the CTS and RTS lines are connected properly. Some instruments require their use; others don't.

- Be sure the serial port is not being used by *any* other application.

- Check to see if you are using the serial port you think you're using (remember, there are usually least two).

- Make sure you are sending the proper termination characters (EOL).

A handy way to know if your PC is set up correctly for serial communication is to get a second PC and connect their serial ports together. Then, using a dumb terminal program on each computer, you can verify that the data you type on one screen appear on the other and vice versa.

10.9 Wrap It Up!

Whew! This chapter has been a heavy one—if you read it end-to-end, you deserve to take a break.

We've covered basic signal and data acquisition theory. Different types of signal can be classified for measurement purposes into analog AC, analog DC, digital on–off, digital counter/pulse, and frequency. Signal sources can be *grounded* or *floating. Grounded signals* usually come from devices that are plugged in or somehow connected to the building ground. *Floating source* examples include many types of sensors, such as thermocouples or accelerometers. Depending on the signal source type, signal characteristics, and number of signals, three measurement systems can be used: *differential, referenced single-ended (RSE),* or *nonreferenced single-ended (NRSE).* The big no-no is a grounded source using a referenced single-ended system. The sampling rate of a data acquisition system (for AC signals) is very important. According to the *Nyquist theorem,* the sampling rate must be more than twice the maximum frequency component of the signal being measured.

Selecting a DAQ board or system that will do the right the job is the next step. Many kinds of DAQ boards exist for different platforms, applications, and budgets. National Instruments' SCXI and PXI systems provide ways to work with a very high number of channels, as well as perform elaborate *signal conditioning.* Installing a DAQ board is getting easier than before, but it still requires a little knowledge of how to set parameters in the configuration utility NI-MAX. The collection of software drivers for National Instruments' boards is called NI-DAQ, which NI-MAX relies upon to function properly.

Finally, we looked briefly at some hardware aspects of communicating to external instruments through *GPIB* and *serial* interfaces. The GPIB interface is a widely accepted standard for many instruments, and you can often obtain an instrument driver in LabVIEW for your particular instrument. Serial communication is cheap and conceptually simple but in practice requires much troubleshooting and tweaking.

Don't feel bad if much of the material in this chapter eluded you. DAQ is a complex subject and often requires the expertise of an instrumentation engineer. However, for relatively simple DAQ systems, if you have a willingness to experiment and a sense of adventure, you can assemble your own DAQ system at your PC. Remember to consult the manuals that come with your board and hardware for more details.

If you're interested in learning more about DAQ with LabVIEW, check out the book LabVIEW for Data Acquisition by Bruce Mihura (2001, Prentice Hall).

10.10 Solutions to Activities

10-1. 1, 3, 4, 2, 5, 4, 1, 2, 2, 2, 1

10-2.

   A. 1. Analog AC signals, small amplitude, floating.

   2. Differential (because of small amplitudes relative to ground).

   3. Nyquist frequency $= fn = 1/(2 \text{ ms}) = 1/(2 \times 10^{-3} \text{ ms}) = 500$ Hz. Sample at much more than 1,000 Hz, such as 5 kHz.

   4. Yes. Amplification (small signal amplitude), isolation (safety), and perhaps anti-aliasing filters (noise) are required.

5. Any E-series board should do.

B. 1. Analog input DC (resistance, thermocouple readings), analog output DC (voltage excitation), digital output on–off (relay to turn on machine), digital input counter (count machine cycles).

2. Referenced single-ended (RSE).

3. All DC measurements. 10 Hz should be a good sampling frequency.

4. Yes, thermocouple conversion.

5. SCXI system: because of high channel count and special conditioning needed.

OVERVIEW

In this chapter, we'll get to the heart of what LabVIEW is often commissioned to do: data acquisition and instrument control. You'll become familiar with some of the VIs on the **Data Acquisition** palette and the **Instrument I/O** palette. Analog input and output, digital input and output, GPIB instrument control, and serial communications are covered. We'll go through the basic steps necessary to get you started with some data acquisition in LabVIEW and point you in the right direction to do your own more advanced DAQ and instrument control programs.

GOALS

- Investigate the **I/O** palette on the front panel: waveform data type, DAQ channel names
- Become familiar with the basic VIs in the **Data Acquisition** palette
- Learn the sequence of VIs that you should use for simple DAQ measurements, both analog and digital
- Get started with some examples and DAQ wizards
- Become familiar with the basic VIs in the **Instrument I/O** palette
- Learn about VISA VIs for instrument control
- Examine some samples of GPIB instrument control
- Get an overview of serial communication

KEY TERMS

- Device
- Channels
- Port
- VISA
- Analog input
- Analog output
- Line
- Wizard
- Buffer
- Triggering
- GPIB
- IVI
- Sample waveform
- Serial
- Scan
- TaskID
- Waveform

DAQ and Instrument Control in LabVIEW

11

11.1 Definitions, Drivers, and Devices

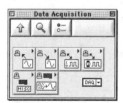

Figure 11.1

Figure 11.2

Take a good look at the palettes in Figures 11.1 and 11.2. The VIs in these palettes go a level beyond what any other function in LabVIEW does—they let you communicate with, read from, write to, measure, control, turn on and off, or blow up stuff in the external world via DAQ and GPIB boards. We'll be taking a whirlwind tour of the analog, digital, GPIB, serial, and

335

VISA functions available in LabVIEW. To do this effectively, you first need to understand a little bit about the interface between LabVIEW and the boards.

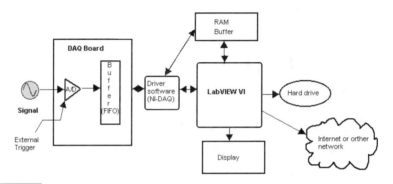

Figure 11.3

As you can see from Figure 11.3, initiating a DAQ operation involves Lab-VIEW calling *NI-DAQ* (which is the catch-all driver for the board in use), which in turn signals the hardware to initiate the I/O operation. DAQ boards use on-board *buffers* (called FIFOs, for "First-In, First-Out") and RAM buffers as an intermediate place to store the data they acquire. Also note that the software isn't the only place an initiation of an I/O operation takes place—an external piece of hardware can also trigger the operation.

Two important characteristics help classify the type of DAQ operation you are performing:

• Whether you use a buffer
• Whether you use an external trigger to start, stop, or synchronize an operation

11.1.1 Buffers

A *buffer*, as used in this context, is an area of memory in the PC (not the on-board FIFO) reserved for data to reside temporarily before it goes somewhere else. For example, you may want to acquire a few thousand data samples in one second. It would be difficult to display or graph all these data in the one same second. But by telling your board to acquire the data into a buffer, you can quickly store the data there first, and then later retrieve them for display or analysis. Remember, buffers are related to the speed and volume of the DAQ operation (generally analog I/O). If your board has DMA capability, analog input operations have fast hardware path to the RAM in your computer, meaning the data can be acquired directly into the computer's memory.

Not using a buffer means you must handle (graph, save to disk, analyze, whatever) each data point *one at a time* as it is acquired since there is no place to "keep" several points of data before you can handle them.

Use *buffered I/O* when:

- You need to acquire or generate many samples at a rate faster than is practical to display, store on a hard drive, or analyze in real time
- You need to acquire or generate AC data (>10 samples/second) continuously and be able to do analysis or display of some of the data on the fly
- The sampling period must be precise and uniform throughout the data samples

Use *nonbuffered* I/O when:

- The data set is small and short (for example, acquiring one data point from each of 2 channels every second)
- You need to reduce memory overhead (the buffer takes up memory)

We'll talk more about buffering soon.

11.1.2 Triggering

Triggering refers to any method by which you initiate, terminate, or synchronize a DAQ event. A trigger is usually a digital or analog signal whose condition is analyzed to determine the course of action. Software triggering is the easiest and most intuitive: You control the trigger directly from the software, such as by using a Boolean front panel control to start or stop data acquisition. Hardware triggering lets the circuitry in the board take care of the triggers, adding much more precision and control over the timing of your DAQ events. Hardware triggering can be further subdivided into *external* and *internal* triggering. An example of an internal trigger is the programming of the board to output a digital pulse when an analog-in channel reaches a certain voltage level. All National Instrument's DAQ boards have an external trigger pin, which is a digital input used for triggering. Many instruments provide a digital output (often called "trigger out") used specifically to trigger some other device or instrument, in this case, the DAQ board. Use *software triggering* when:

- The user needs to have explicit control over all DAQ operations *and*
- The timing of the event (such as when an analog input operation begins) needn't be very precise

  Use *hardware triggering* when:

- Timing a DAQ event needs to be very precise
- You want to reduce software overhead (for example, a While Loop that watches for a certain occurrence can be eliminated)
- The DAQ events must be synchronized with an external device

In the following sections, you will see how you can use the DAQ VIs to configure an I/O operation as buffered or nonbuffered and to configure the type of triggering.

11.2 Analog I/O

Figure 11.4

The examples in this section, and selected examples in subsequent sections, assume you've been able to install and configure a DAQ board. If you need some help doing this, consult Chapter 10 and the documentation that came with your board. The examples and exercises also assume you have some signal source to measure.

A few definitions are in order before we delve into the DAQ VIs. The following terms appear consistently in the inputs and outputs to the VIs, so it's critical to understand what they are if you are going to do any LabVIEW DAQ programming. One of the DAQ VIs with its inputs and outputs is shown for clarity in Figure 11.5.

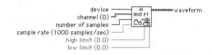

Figure 11.5
AI Acquire Waveform.vi.

The **device** is the "device number" that the board is assigned by NI-DAQ. You can find out what this is by looking up the properties of the board in the MAX configuration utility. This parameter tells LabVIEW what kind of board you're using, keeping the DAQ VI itself independent of the board type (e.g., if you later used a different compatible board and assigned it the same device number, all your VIs would work without modification).

The **sample** represents one A/D conversion: It's just one point—one numeric value corresponding to the real analog value at the time the measurement was taken.

The **channel** specifies the physical source of the data sample(s). A board with 16 analog input channels, for example, means you can acquire 16 sets of data points at the same time. In the LabVIEW DAQ VIs, a channel or set of channels are specified using the **DAQ Channel Name** data type. The **DAQ Channel Name** data type is very similar to a string data type. The **DAQ Channel Name** data type on the front panel can be created from the **I/O** palette.

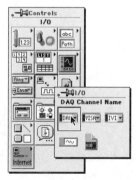

Figure 11.6

You use the **DAQ Channel Name** to specify what channels you are measuring. You can do this in several ways:

- If you configured virtual channels in MAX, you can click on the pop-up menu arrow of a **DAQ Channel Name** to see a list of your virtual channels and select the appropriate one.
- Or, you can type in the name of your virtual channel directly.
- Or, you can type in the channel number (or range of channel numbers) directly.

Figures 11.7 and 11.8 show the different ways you can represent a **DAQ Channel Name** and an array of **DAQ Channel Names**.

Figure 11.7
DAQ Channel Names.

Figure 11.8
Array of DAQ Channel Names.

The easiest way to work with DAQ is usually to go specify your virtual channel in MAX first, and then select it from the DAQ Channel Control. But if you want to directly type in the channel numbers, you can do so. You can specify one, several, or a range of channel numbers. For example:

Channels	*DAQ Channel Name string*
5	5
0–4	0:4
1, 8 and 10–13	1,8,10:13

A **scan** is a sample from each separate channel being used, assuming you are sampling multiple channels.

A **waveform** is a set of samples from *one* channel, collected over some period of time, and ordered by time. Usually, but not always, the time period between data points is constant for a given waveform.

Beginners often confuse scans and waveforms. A scan is a set of samples versus *channels* (one sample from each channel at just one instant in time); a waveform is a set of samples (from the same channel) versus *time*.

Notice also the waveform output is a waveform *data type*. We talked about the waveform data type in Chapter 8. You may want to review that section if you are confused about waveforms, because practically all the analog I/O functions in LabVIEW rely on it.

The **high limit** and **low limit** are the voltage limits you expect to have on your signal. By changing these inputs from their defaults of 10 and –10 V, you can set the *gain* of your DAQ system. For example, if you set them at 5 and –5 V, the gain is set to 2. If you set them at 1 and –1 V, the gain is 10. Thus, you wire these inputs if you know that the expected *range* of your input signal is different from the default. Use this formula to determine the gain applied:

```
Gain = 20/(|High Limit| - |Low Limit|)
```

Be aware that many DAQ boards support only certain predetermined gain values. If you set a theoretical gain that the board does not support, LabVIEW will automatically adjust it to the nearest preset gain. A typical board might have available gains of 0.5, 1, 2, 5, 10, 20, 50, and 100.

The **taskID** is a 32-bit integer (I32 type numeric) that some of the DAQ VIs use to identify a specific I/O operation that's going on. Many of the DAQ VIs require a **taskID in** and return a **taskID out** to pass on to the next VI. By wiring this taskID, you don't have to provide every VI with information about your board, sampling rate, limits, etc. An initial VI can take all this and pass a taskID out that will tell the other VIs what the configuration is. If you're confused by all this, just hold on and look at an example. It really isn't that hard.

11.2.1 Quick and Dirty Analog I/O: The Top Tier

If you can't wait to try out your DAQ board, try some of these easy VIs for starters. Even if you will move on to more complex data acquisition, it's a good idea to use the following examples as a means to test your board. All the VIs at the top tier of the **Analog Input** and **Analog Output** palette are "easy" VIs: They can be used as standalone VIs with minimal setup and configuration.

Here's what the "easy" DAQ VIs do:

Analog Input

Obtains one sample from the specified channel.

Figure 11.9
AI Sample Channel.vi.

Obtains one sample from each of the specified channels in the **channel** input. The samples are returned in the **samples** waveform array, ordered by channel.

Figure 11.10
AI Sample Channels.vi

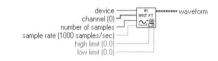

Obtains one waveform (a set of samples acquired over a period of time) from one channel, at the specified sampling rate. The samples are returned in the **waveform** output.

Figure 11.11
AI Acquire Waveform.vi.

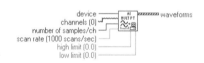

Obtains a waveform from each channel in the **channel** input. The samples are returned in the array of **waveforms**, ordered by channel and sample period. Each channel's data are stored in an individual waveform component.

Figure 11.12
AI Acquire Waveforms.vi.

Although by default the output of the previously mentioned analog input VIs is a waveform or array of waveforms, you can also wire a scalar (for single-point acquisition), a 1D array (for multiple points), or a 2D array (for multiple waveforms) to the output. You can do this because the outputs of these VIs are polymorphic; that is, they will adapt to other valid data types. In general though, you will find it easiest to just work with the waveform data types that are the default outputs.

Analog Output

Figure 11.13
AO Update Channel.vi.

Sets the specified voltage at the specified output channel. This voltage remains constant at the output channel until it is changed or the device is reset.

Figure 11.14
AO Update Channels.vi

Sets the specified voltages at the specified output channels. These voltages remain constant at the output channels until they are changed or the device is reset.

Figure 11.15
AO Generate Waveform.vi.

Generates a waveform at the specified output channel. The waveform data, in volts, should be provided as components of the **waveform** input. The **update rate** specifies the time between points (the **t0** and **dt** of the waveform are ignored).

Figure 11.16
AO Generate Waveforms.vi.

Same as **AO Generate Waveform**, except that multiple waveforms, one for each channel, can be generated simultaneously.

The **AO Generate Waveform** and **AO Generate Waveforms** functions can also accept a 1D array and 2D array, respectively, in the **waveform** input—these inputs are polymorphic. Often this is easier since **t0** and **dt** of the waveform are ignored.

Finally, you should know that the previous analog input and output VIs are *synchronous* with the I/O on the DAQ board; that is, the VI does not finish executing until all the data are read or written.

Try this simple example:

11.2.2 Activity 11-1: Analog Input

1. Connect a voltage source, such as a function generator (or even a 1.5 V battery, if that's all you have) to channel 0 on your DAQ board. Be sure you know whether you've configured your board for differential or single-ended measurements.

2. Build the front panel and block diagram shown. **AI Sample Channel** is found on the **Analog Input** palette. If your DAQ board is not device #1 in the MAX or NI-DAQ Utility, change the device constant to match your device number.

3. Save your VI as **Quick Analog In.vi**

4. Run the VI with <u>Acquire?</u> turned on, and turn it off after a few seconds.

5. Examine the data in the <u>Acquired samples</u> array.

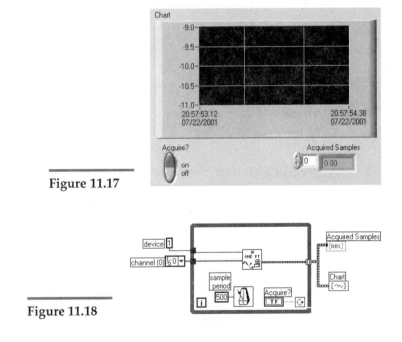

Figure 11.17

Figure 11.18

Notice that we wired the array of waveforms to a 1D array **"Acquired Samples"**; LabVIEW automatically takes care of extracting the **Y** components of the waveforms and placing them in that array.

The While Loop in Activity 11-1 is the trade-off for simplicity: It adds un-necessary software overhead. This is not a problem for a quick look at your data, but in a decent-sized block diagram, you really don't want the soft-ware trying to do all the sampling control; the DAQ boards and low-level drivers are designed to do that.

This method of DAQ works perfectly fine under many circumstances, provided

1. the sampling rate is slow (once per second or slower),

2. other time-consuming operating system (OS) events are not occur-ring while the VI is running, and

3. slight variations in the sampling times are acceptable.

Suppose you are running a VI using a LabVIEW timing function in a loop, sampling once per second, and you decide to drag the window around. Your VI will not acquire any data during the drag! Suppose you drag the window for five seconds. When you finally release the mouse button, the VI continues, but it gives *no indication* that the data were not sampled for a five-second period! When using these VIs that use the LabVIEW looping mecha-nism for timing, you should make sure that other programs are not running concurrently, as they may interfere with your timing, and you should avoid time-consuming OS events (mouse activity, disk activity, network activity, etc.) as the VI is running. To be extra cautious, you can use the LabVIEW timing functions such as **Get Tick Count** to gauge the accuracy of your loop timing.

Let's look at a second example for easy I/O. This example VI below lets you acquire several waveforms and display them on the graph.

11.2.3 Activity 11-2: More Analog Input

1. Connect four DC or low-frequency voltage sources to channels 0–3. If you don't have this many voltage sources, you can connect one source to all channels or make a resistor network to vary the ampli-tude of the voltage at each channel.

2. Build the front panel and block diagram shown in Figures 11.19 and 11.20. On the graph, select **Ignore Timestamp** from the pop-up menu

of the graph. The AI function returns an array of waveforms—one for each channel.

3. Set the <u>scan rate</u>, <u>channels</u>, and <u>number of scans to acquire</u> as shown.

4. Save your VI as **Acquire Multiple Channels.vi**.

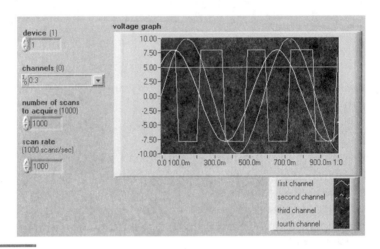

Figure 11.19

The block diagram is very simple. The **AI Acquire Waveforms** does everything for you. Note that this VI performs a nonbuffered, software-triggered ADC.

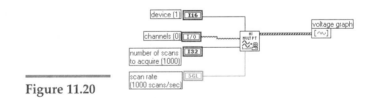

Figure 11.20

Note that although we used the string 0:3 to specify the channels; we could also have specified them as virtual channels if they were set up this way in the MAX utility. For example, if channel 0 was assigned to MySensor1 and channel 1 was MySensor2, etc., then we could have wired in an array of virtual channels instead.

An important limitation of multiple channel I/O needs to be mentioned. If you set a high scan rate for multiple channels, and observe the data from each channel over actual time (not over array index values), you'd notice a successive *phase delay* between each channel. Why is this? Most boards can only perform one A/D conversion at a time. That's why it's called scanning: The data at the input channels are digitized sequentially one channel at a time. A delay, called the *interchannel delay*, arises between each channel sample. By default, the interchannel delay is as small as possible but is highly dependent on the board.*

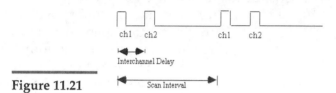

Figure 11.21

In DC and low-frequency signals, this phase delay in scans is generally not a problem. The interchannel delay may be so much smaller than the scan period that the board *appears* to virtually sample each channel simultaneously. For example, the interchannel delay may be in the microsecond range, and the sampling rate may be 1 scan/second (as shown in Figure 11.22). However, at higher frequencies, the delay can be very noticeable and may present measurement problems if you are depending on the signals being synchronized.

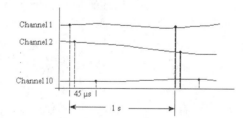

Figure 11.20

11.2.4 Better Analog I/O: The Middle Tier

If you are not going to be doing any DAQ with LabVIEW, you may want to skip the rest of the DAQ section.

* A few boards, such as the EISA-A2000, do support simultaneous sampling of channels. Other boards support either *interval scanning* or *round-robin scanning*. Boards that support interval scanning have a much smaller interchannel delay than those that use round-robin scanning.

One fundamental limitation of the easy VIs discussed previously is the redundancy of performing DAQ tasks. Every time you call **AI Sample Channel**, for example, you set up the hardware for a specific type of measurement, tell it the sampling rate, etc. Obviously, if you're going to take a lot of samples very often, you shouldn't need to "set up" the measurement on every iteration. The "top tier" VIs make for easy, quick programming, but they can add a lot of unnecessary software overhead and have little flexibility in applications where you are manipulating a lot of data.

The "middle tier" VIs offer more functionality, flexibility, and efficiency for developing your application. These VIs feature capabilities such as controlling intersampling rates, using external triggering, and performing continuous I/O operations. The following table describes briefly each of the middle-tier analog I/O VIs. You'll notice that some of these have a zillion inputs and outputs. Efficient use of the analog I/O VIs equates with only wiring the terminals you need. In most cases, you don't need to worry about the optional ("grayed out") terminals that show up in the "Detailed" Help window.

Analog Input

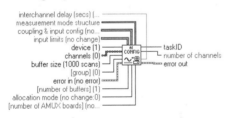

Figure 11.23
AI Config.vi.

AI Config configures the analog input operation for a specified set of channels, configures the hardware, and allocates a buffer in computer memory. **Device** is the device number of the DAQ board. **Channel** specifies the analog input channel numbers. **Input limits** specifies the range of the input signal and affects the gain applied by your hardware. **Buffer size** is specified in scans and controls how much computer memory **AI Config** reserves for the acquisition data. **Interchannel delay** sets the interchannel skew for interval scanning.

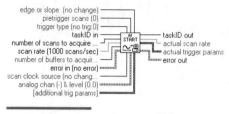

Figure 11.24
AI Start.vi.

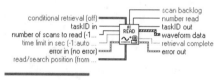

Figure 11.25
AI Read.vi.

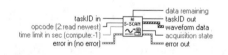

Figure 11.26
AI Single Scan.vi.

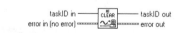

Figure 11.27
AI Clear.vi.

AI Start starts a buffered analog input operation. This VI controls the rate at which to acquire data, the number of points to acquire, and the use of any hardware trigger options. Two important inputs to **AI Start** are **scan rate (scans/sec)** (how many scans per second to acquire on each channel) and **number of scans to acquire** (how many times to scan through the channel list).

AI Read reads data from the buffer allocated by **AI Config**. This VI can control the number of points to read from the buffer, the location in the buffer to read from, and whether to return binary data or scaled voltage data. The output of this VI is an array of waveforms, one for each channel in the channel list.

AI Single Scan returns one scan of data. The **waveform data** output of this VI is the voltage values read from each channel in the channel list. You use this VI only in conjunction with **AI Config**; there is no need to use the **AI Start** and **AI Read** VIs when using **AI Single Scan.**

AI Clear clears the analog input operation, deallocates the buffer from computer memory, and frees any DAQ board resources such as counters.

The first VI you always use when setting up an analog input application is **AI Config**. **AI Config** produces a **taskID** and an error cluster. All other Analog Input VIs accept the **taskID** as an input to identify the device and channels on which to operate and then output the **taskID** when they complete. Because the **taskID** is an input and output to other Analog Input VIs, this parameter forms a data dependency among the DAQ VIs that control the execution flow of the diagram, making sure the DAQ VIs execute in the correct order.

Analog Output

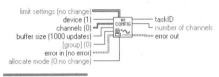

Figure 11.28
AO Config.vi.

AO Config configures the analog output operation for a specified set of channels, configures the hardware, and allocates a buffer in computer memory for the operation. **Device** is the device number of the DAQ board. **Channels** specifies the analog output channel numbers. **Limit settings** specifies the range of the output signals. The **task ID** output is used by all subsequent analog output VIs to identify the device and channels on which to operate.

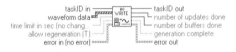

Figure 11.29
AO Write.vi.

AO Write writes the data in **waveform data** (an array of waveforms) into the buffer used for the analog output operation. The data should be one waveform per channel in the channel list.

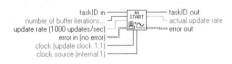

Figure 11.30
AO Start.vi.

AO Start starts a buffered analog output operation. **Update rate** is the number of updates to generate per second. If you wire a 0 to **number of buffer iterations**, the board will continuously output the data buffer until you run the AO Clear function.

Figure 11.31
AO Wait.vi.

AO Wait waits until the task's waveform generation is complete before returning. It checks the status of the task at regular intervals and waits asynchronously between intervals to free the processor for other operations. The wait interval is calculated by dividing

the **check every N updates** input by
the **update rate**.

AO Clear stops the analog output operation, deallocates the buffer from
computer memory, and releases any
DAQ board resources such as counters.

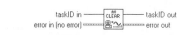

Figure 11.32
AO Clear.vi.

All the VIs shown in the previous two tables are meant to rely heavily on
dataflow to work together. As you probably already noticed, they follow a
logical sequence. The analog functions are tied together by the taskID and
error cluster nodes. The following table shows the intended sequence of VIs
for analog waveform acquisition and generation.

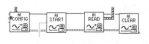

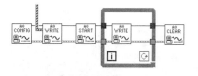

Figure 11.33
Acquisition.

Figure 11.34
Generation.

1. **AI Config**: Configure the channels and buffer	1. **AO Config**: Configure the channels and buffer
	2. **AO Write**: Write data to the buffer
2. **AI Start**: Start the acquisition	3. **AO Start**: Start the generation
3. **AI Read**: Read from the buffer	4. **AO Write**: Write new data to the buffer
	5. **AO Wait** [optional]: Wait for buffer to empty
4. **AI Clear**: Clear the buffer and deallocate resources	6. **AO Clear**: Clear the buffer and deallocate resources

Buffered Analog Input

Remember buffered I/O, from the beginning of the chapter? Here is where
you get to see it done. Previously, we used a While Loop to acquire multiple

points. Now we'll use the "intermediate" VIs to see how we can avoid using a While Loop by programming the hardware to acquire a set number of points at a certain sampling rate.

11.2.5 Activity 11-3: Buffered DAQ

This is an example of buffered DAQ. First, let's build the front panel shown in Figure 11.35, with a simple chart for viewing the data.

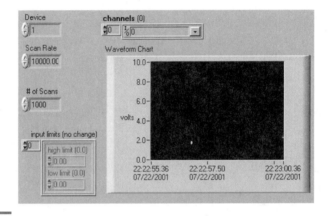

Figure 11.35

Next, connect a voltage source to channel 0. Then wire the block diagram (Figure 11.36) using the analog input VIs, in the order we mentioned before.

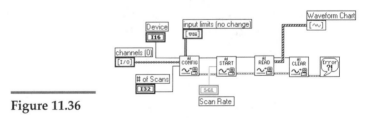

Figure 11.36

Save your VI as **Buffered Analog In.vi**.

Voilà—you've just built a (very simple) oscilloscope! There are some details we should note about the block diagram we just built:

• The <u>Input Limits</u> control is used to adjust the gain of the board. For example, if you have signals with 100 mV amplitude, the board will apply a gain of 100. However, if you have already set the limits for a virtual channel in the MAX utility, then you should not use the Input Limits control.

- The **AI Start** VI actually triggers the data acquisition. As soon as it's finished executing, the data have been loaded into the buffer. The buffer data are not "available" in LabVIEW until we decide to read them. Although we connected **AI Read** immediately after this VI, we could have read the data at any later time, provided the data in the buffer were not cleared or overwritten by some other process.

- **AI Clear** is important to use because it clears the buffer. Otherwise, a future **AI Read** might read some old data.

11.2.6 Activity 11-4: More Fun with DAQ

1. This is a modification of the previous exercise. Write a waveform display-and-store VI. This VI should have a front panel similar to the one in Figure 11.35, but with the added ability of storing all the data into a waveform file (see the waveform file functions from Chapter 8 and file I/O from Chapter 9 if you need some help). The user should be able to name the file. This same VI should be able to retrieve a file, and display on the same chart old data from this file.

Add a Boolean for "read/write" and a Case structure so only one operation can be done at a time.

2. Write a "waveform synthesizer" analog output VI. This VI should let you choose from a sine, triangle, square, and sawtooth waveform to output. You'll want to use the **Function Generator** function from the **Waveform>>Waveform Generation** palette. Use the analog output VIs in a similar fashion to the analog in VIs. As an optional feature, you could allow this VI to acquire a waveform from an external source and have this waveform be one of the possible analog output functions.

You'll find the solutions to these activities on the CD in CH11.LLB: **Acquire and Save.vi** and **Function Generator.vi**.

11.2.7 Nerd-Level Analog I/O

Yep, this is where we throw you some fancy concepts such as circular buffering (continuous data acquisition), hardware triggering, and disk streaming. Actually these concepts are pretty useful in practice. We'll cover each one briefly with an example.

Continuous Data Acquisition

Continuous data acquisition, or real-time data acquisition, returns data from an acquisition in progress without interrupting the acquisition. This approach usually involves a circular buffer scheme as shown in Figure 11.37. You specify the size of a large circular buffer. The DAQ board collects data and stores the data in this buffer. When the buffer is full, the board starts writing data at the beginning of the buffer (writing over the previously stored data, whether or not it has been read from LabVIEW). This process continues until the system acquires the specified number of samples, LabVIEW clears the operation, or an error occurs. Continuous data acquisition is useful for applications such as streaming data to disk and displaying data in real time.

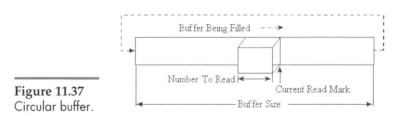

Figure 11.37
Circular buffer.

11.2.8 Activity 11-5: Continuous Acquisition

Build the front panel shown in Figure 11.38.

Save this VI as **Continuous Acquisition.vi**. Then wire the analog input VIs as shown in Figure 11.39.

Configure LabVIEW for continuous data acquisition by instructing **AI Start** to acquire data indefinitely. To do this, set <u>number of scans to acquire</u> in **AI Start** to 0. This acquisition is *asynchronous*, meaning that other Lab-

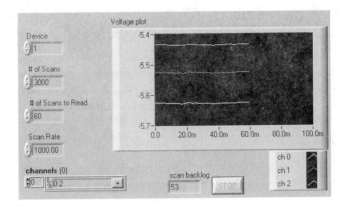

Figure 11.38

VIEW operations can execute during the acquisition. Figure 11.39 shows the block diagram for the front panel in Figure 11.38. **AI Read** is called in a looping structure to retrieve data from the buffer. From there, you can then send the data to the graph. **AI Clear** halts the acquisition, deallocates the buffers, and frees any board resources. Notice that the differences between this VI (which performs *continuous* acquisition) and the one for a waveform acquisition (the ordinary buffered DAQ VI in the previous section) are the While Loop, the 0 wired to the **number of scans to acquire** input of **AI Start**, and the **number of scans to read** parameter.

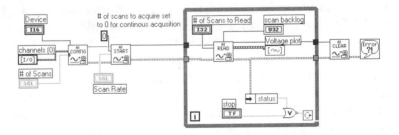

Figure 11.39

This VI is designed to run continuously, or "forever." To accomplish this without running out of memory, a fixed-length buffer is allocated and filled up with samples of data from beginning to end, then the data at the beginning are overwritten as the buffer is filled up again. As an example, suppose this VI were run with a buffer size of 10 and the sampling rate was once per second. Here are snapshots of how the buffer is being filled up with data:

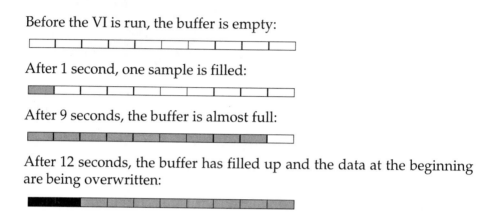

Before the VI is run, the buffer is empty:

After 1 second, one sample is filled:

After 9 seconds, the buffer is almost full:

After 12 seconds, the buffer has filled up and the data at the beginning are being overwritten:

This VI uses the "Intermediate" VIs to take advantage of LabVIEW's hardware timing and memory handling capabilities. If, while **Single Channel AI.vi** is running, an OS event ties up the microprocessor, the intermediate level VIs will use the DAQ board's buffer and DMA capability (if any) to continue collecting data without any microprocessor involvement. DMA allows the DAQ hardware to write directly to the computer's memory, even if the processor is tied up. If the OS event interrupts the microprocessor longer than the on-board FIFO buffer and DMA buffer can handle, only then will LabVIEW lose data samples. This can best be understood by referring to the previous four snapshots of the buffer. First, assume the DAQ hardware has no on-board FIFO, but it has DMA capability. Ideally, the DAQ hardware will write data into the buffer continuously, and LabVIEW will be continuously reading a few samples behind the last sample written to the buffer. Suppose the microprocessor gets tied up after one second, and LabVIEW has read the first sample. When the microprocessor is tied up, the DAQ hardware device can write to this buffer, but LabVIEW can't read from the buffer. If after 12 seconds, the microprocessor is still tied up, then LabVIEW has missed the data in the second slot from the left when it contained good data (light gray). The four intermediate AI VIs used in this previous example will indicate such an error.

Why do we need the <u>Scan Backlog</u> indicator? It's pretty useful to know if LabVIEW is keeping up with reading the data. If the buffer fills up faster than your VI can retrieve the data in it, you will start losing some of this data, since the buffer will be overwritten.

Hardware Triggering

There are two ways to begin a DAQ operation: 1. through a software trigger or 2. through a hardware trigger.

With software triggering, the DAQ operation begins when the software function that initiates the acquisition executes. For example, the "Easy" DAQ VIs use software triggering. As soon as LabVIEW executes the VI, the acquisition or generation starts. All DAQ boards support software triggering.

Another common method to begin a DAQ operation is to wait for a particular external event to occur. You usually begin the acquisition depending on the characteristics of a digital or analog signal such as the signal state, level, or amplitude. Hardware circuitry on the plug-in board uses this analog or digital event to start the board clocks that control the acquisition. Most DAQ boards support *digital* triggering to initiate an acquisition. Other boards also support *analog* triggering to initiate an acquisition. The input pin on your DAQ board that accepts the triggering signal is either EXTTRIG or START TRIG. All National Instruments E-series boards support digital triggers.

You can also use an external analog trigger to initiate the *read* of the data from the buffer in LabVIEW, rather than just *initiate* the DAQ operation. This sort of triggering is called a *conditional retrieval*. When performing a conditional retrieval, your DAQ board acquires data and stores it in the acquisition buffer using software triggering. However, the board does not retrieve data until it acquires a sample that meets certain level and slope conditions. (Careful here—It's easy to confuse conditional retrieval with hardware analog triggering. Systems using analog triggering do not store data in the acquisition buffer until the trigger condition occurs; systems using conditional retrieval do store the data in a buffer, but these data are not read into LabVIEW until the trigger occurs).

Figure 11.40 shows the conditional retrieval cluster that sets the criteria for retrieving data from the acquisition buffer. This cluster is an input to the **AI Read** VI. After the acquisition starts, the board continually samples the signal and compares it to the retrieval conditions. After the conditions are met, **AI Read** returns the amount of data specified in its **number of scans to read** input.

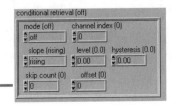

Figure 11.40

11.2.9 Activity 11-6: Triggering

Build an analog input VI that uses hardware triggering to start and stop the acquisition.

1. Connect a switch with a TTL signal to the EXT TRIG (or similar) pin on your DAQ board, and connect a couple of analog input signals.

2. Build the front panel shown in Figure 11.41. The <u>Trigger Type</u> control is a ring control that lets you specify the type of trigger:

   0: no triggering (default)

   1: analog trigger

   2: digital trigger A

   3: digital triggers A and B

   4: scan clock gating

You can use the one on the front panel of the example on the CD, or simply use a numeric control.

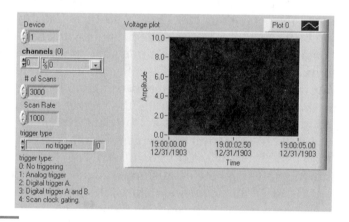

Figure 11.41

3. Make a block diagram like the one shown in Figure 11.42. Notice that the only difference between this one and the simple buffered analog

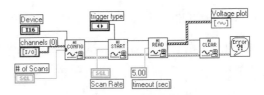

Figure 11.42

input is the connection of the **Trigger type** input on the **AI Start**.

4. Save this VI as **Triggered Analog In.vi**.

Streaming Data to File

You have already written exercises that send data to a spreadsheet file. In these exercises, LabVIEW converts the data to a spreadsheet format and stores them in an ASCII file after the acquisition completes. A different and sometimes more efficient approach is to write small pieces of the data to the hard disk while the acquisition is still in progress. This type of file I/O is called streaming. An advantage of streaming data to file is that it's fast;hence you can execute continuous acquisition applications and yet have a stored copy of all the sampled data.

With continuous applications, the speed at which LabVIEW can retrieve data from the acquisition buffer and then stream it to disk is crucial. You must be able to read and stream the data fast enough so that the board does not attempt to overwrite unread data in the circular buffer. To increase the efficiency of the data retrieval, you should avoid executing other functions, such as analysis functions, while the acquisition is in progress. Also, you can use the binary data output from **AI Read**, instead of the voltage data output, to increase the efficiency of the retrieval, as well as the streaming. When you configure **AI Read** to produce only binary data (specified by the **output units** input), it can return the data faster to the disk than if you used the **File I/O VIs** after acquiring the data into an array. One disadvantage to reading and streaming binary data is that users or other applications cannot easily read the file.

You can easily modify a continuous DAQ VI to incorporate the streaming feature. Although you may not have used binary files before, you can still build the VI in the following example. We will be discussing binary and other file types in Chapter 15.

11.2.10 Activity 11-7: Streaming to Disk

From LabVIEW, open the example **Cont Acq to File (binary)**, found under
LabVIEW\examples\daq\anlogin\strmdsk.llb. You will observe how it
streams to disk.

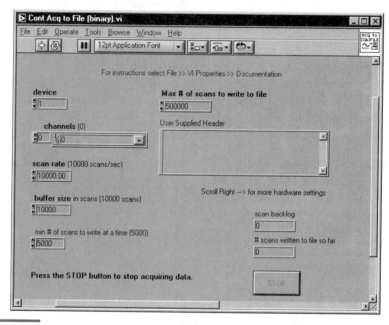

Figure 11.43

Run this VI for a few seconds—you may want to first enter header information in <u>User Supplied Header</u> that will go in your file.

To see the data written to disk, use the companion example VI **Display
Acq'd File (binary)**, found in the same path as the previous VI.

Run this VI, choosing the filename you used in the previous VI. Notice
that the first VI that streams data doesn't attempt to graph the data. The
point of streaming is to acquire data to the disk *fast*—and look at it later.

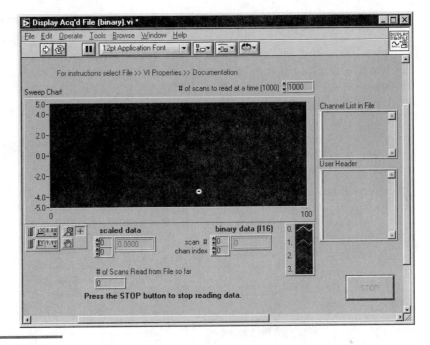

Figure 11.44

11.3 Digital I/O

Figure 11.45

Ah, the digital world, where everything—for the most part—is much simpler than its analog counterpart. 1 or 0, high or low, on or off; that's mostly about it. Actually, some knowledge of binary representation and arithmetic is required, along with knowing the following LabVIEW definitions.

A digital **line** is the equivalent of an analog channel: a path where a single digital signal is set or retrieved. Digital lines are usually either **input lines** or **output lines**, but sometimes they can be **bi-directional**. On most boards, digital lines must be configured as input or output; they can't act as both at the same time.

A **port** is a collection of digital lines that are configured in the same direction and can be used at the same time. The number of digital lines in each port depends on the board, but most ports consist of four or eight lines. For example, a multifunction DAQ board could have eight digital lines, configurable as one 8-line port, two 4-line ports, or even eight 1-line ports. Ports are specified as **digital channels**, just like analog channels.

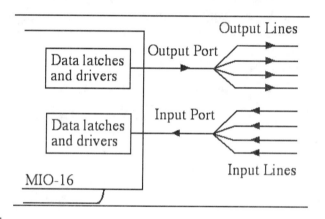

Figure 11.46

Port width is the number of lines in a port.

State refers to the one of two possible cases for a digital line: a Boolean TRUE (same as logical 1 or "on") or a Boolean FALSE (same as logical 0 or "off").

A **pattern** is a sequence of digital states, often expressed as a binary number, which describes the states of each of the lines on a port. For example, a four-line port might be set with the pattern "1101," meaning the first, third, and fourth lines are TRUE, and the second line is FALSE. The first bit, or least-significant bit (LSB), is the rightmost bit on the pattern. The last (fourth in this example), or most-significant bit (MSB), is the leftmost bit in the pattern. This pattern can also be converted from its binary equivalent to the decimal number 13.

Incidentally, National Instruments' DAQ boards use TTL positive logic, which means a logical low is somewhere in the 0–0.8 V range; a logical high is between 2.2 and 5.5 V.

11.3.1 Quick and Dirty Digital I/O: The Top Tier

For simple digital I/O, the top tier VIs on the Digital I/O palette are great: They are very easy and intuitive to use.

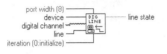

Figure 11.47
Read from Digital Line.vi.

Read from Digital Line reads the logical state of a digital line. **Device** is the DAQ board device number and **digital channel** specifies the port containing the line. **Line** specifies the digital line to read. **Line state** returns the state of the line: high (TRUE) or low (FALSE).

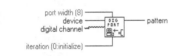

Figure 11.48
Read from Digital Port.vi.

Read from Digital Port reads the state of all lines in a port. **Digital channel** specifies the digital port to read. **Pattern** returns the digital line states as a decimal number, which, if you convert to a binary representation, makes it easy to see the state of each individual line in the port.

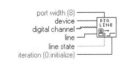

Figure 11.49
Write to Digital Line.vi.

Write to Digital Line sets a particular line on a port to a logical high or low state. **Device** is the DAQ board device number. **Digital channel** specifies the port containing the line. **Line** specifies the digital line to write to, and **line state** indicates the state to write to the line: high (TRUE) or low (FALSE).

Figure 11.50
Write to Digital Port.vi.

Write to Digital Port outputs a digital pattern to the specified port. **Digital channel** specifies the digital port to update. **Pattern** is the decimal equivalent of the digital or binary pattern to be written to the lines in the port.

All of the above VIs implement *immediate* or *nonlatched* digital I/O. This means that as soon as **Write to Digital Line** is executed, for example, that line is set to high or low immediately and remains in that state until another call to the VI changes it.

The **iteration** input on all these VIs needs some explanation. Since these are top-level digital I/O functions, by default they configure the digital ports for the appropriate direction and type when they are called. This configuration normally only needs to be done once for a series of reads or writes. If you use one of these functions repeatedly, such as in a loop, you can get rid of the extra overhead of configuring the ports over and over by wiring any nonzero number to the **iteration** input. Zero, the default value, tells the VI to initialize when it executes. The following example should help clarify the use of these VIs.

Suppose you had a high-voltage switch with a digital relay to let you know when it closed. The digital line from this relay is normally high (switch open). When the switch closes, the line goes to a logic low[†]. To create a simple VI that checks the status of digital line 3 on digital channel (port) 0 of your DAQ board (and thus determines the position of the switch), you might design something like the block diagram in Figure 11.51.

Figure 11.51

Notice how the loop's iteration terminal (☐) is wired to the **iteration** input on the digital VI. The first time around the loop, when i=0, the **Read from Digital Line** will perform the configuration and initialization of the digital port. Subsequently, after i>0, the VI will only be reading from the line, allowing the loop to execute in less time. Once the digital line is set to FALSE, when the switch closes, the loop terminates and our <u>switch closed</u> Boolean comes on.

[†] Logic low for "closed" and high for "open" may seem counter-intuitive since a closed switch usually means something has turned on, which we associate with a logic high. Nevertheless, most standard relays do use this "negative" logic.

11.3.2 Activity 11-8: Digital Output

Write a VI that has four Boolean LEDs on the front panel in an array. Make these controls (by popping up on them and selecting **Change to Control**). The objective of this panel is to turn on some real LEDs connected to the digital lines on your board. (Be sure and put a resistor in series with the LED.) The user should be able to turn the real LEDs on and off with the "virtual" LEDs. Also, make this program write to the whole digital port, and not to each individual line, to reduce overhead. Save your VI as **Digital Port.vi**.

Figure 11.52

The solution is fairly simple—but is made easier if you find the **Boolean Array to Number** function on the **Boolean** palette. You can take the Boolean array and, using this function, wire the **pattern** input painlessly.

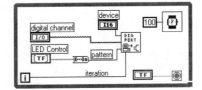

Figure 11.53

You'll notice that there are many other functions in the **Digital I/O** palette. For more advanced applications, consult the LabVIEW manuals and examples.

11.4 Instrument Control in LabVIEW: VISA, GPIB, and Serial

We could devote a whole book to the subject of instrument control with Lab-VIEW, and the intricacies of instrument control are beyond the scope of this chapter or book. However, we'll point you in the right direction with this last section.

In Chapter 10 we discussed some of the hardware protocols for instrument communication: in particular, GPIB and serial. How you communicate with the instrument may depend on the type of instrument. Common types of instruments include the following:

- GPIB
- Serial port
- VXI
- PXI
- Computer-based instruments

11.4.1 VISA

In an attempt at creating a standard software interface that is independent of the hardware protocol or bus, developers have come up with **VISA**. VISA, or Virtual Instrumentation Software Architecture, is a standard I/O Application Programming Interface (API) for instrumentation programming. VISA can control VXI, GPIB, PXI, or serial instruments, making the appropriate driver calls depending on the type of instrument being used.

LabVIEW provides VISA functions in the **Instrument Control>>VISA** palette (Figure 11.54).

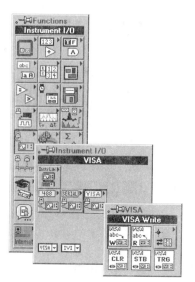

Figure 11.54

The **VISA** palette provides a number of functions including **VISA Write**, **VISA Read**, **CLR** (clear), **STB** (status byte), and **TRG** (trigger). These last three are standard functions on most instruments.

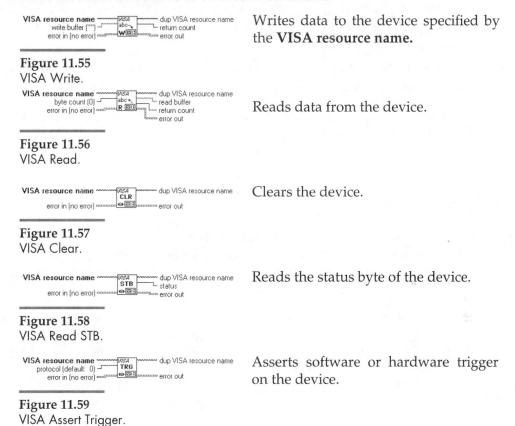

Writes data to the device specified by the **VISA resource name.**

Figure 11.55
VISA Write.

Reads data from the device.

Figure 11.56
VISA Read.

Clears the device.

Figure 11.57
VISA Clear.

Reads the status byte of the device.

Figure 11.58
VISA Read STB.

Asserts software or hardware trigger on the device.

Figure 11.59
VISA Assert Trigger.

The **VISA Resource name** is similar to a DAQ channel; it tells the VISA functions what instrument (or "resource") you are communicating with. You can set up VISA resource names in your **Devices and Interfaces** in the MAX utility. Then you can use a **VISA Resource Name Constant** from the **Instrument I/O** palette to specify the instrument.

For example, if you had a serial device connected on port 1 (COM1), and you wanted to write a command specific to that device, and read the 10 bytes that it returns, your block diagram might look like Figure 11.60.

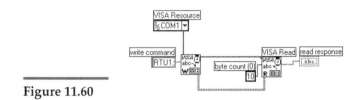

Figure 11.60

Of course, the command strings and lengths depend on the instrument's own particular protocol.

Ocassionally, instruments don't work very well with VISA communication, or you may have your own reason for not using VISA. In any case, LabVIEW gives you the lower-level tools to communicate directly using serial or GPIB protocols.

11.4.2 GPIB

The GPIB palette is in the **Instrument I/O** palette as shown in Figure 11.61. The majority of GPIB communication involves initializing, sending data commands, perhaps reading back a response, triggering the instrument, and closing the communication channel.

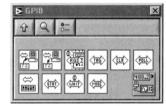

Figure 11.61

Remember, all GPIB instruments have an *address*, which is a number between 0 and 30 uniquely identifying the instrument. Sometimes instruments will also have a *secondary address*. In all the GPIB VIs, you must provide the address of the instrument (in string format) in order to talk to it. This makes it easy to use the same VIs to talk to different devices on the same GPIB bus.

A good way to test if you are communicating over GPIB with your instrument is to use the built-in example, **LabVIEW<->GPIB.vi** (found in `examples\instr\smplgpib.llb`).

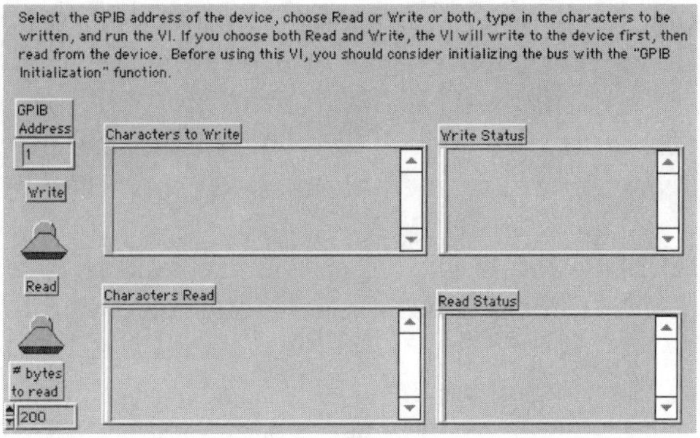

Select the GPIB address of the device, choose Read or Write or both, type in the characters to be written, and run the VI. If you choose both Read and Write, the VI will write to the device first, then read from the device. Before using this VI, you should consider initializing the bus with the "GPIB Initialization" function.

GPIB Address

1

Characters to Write

Write Status

Write

Read

Characters Read

Read Status

bytes to read

200

Figure 11.62

11.4.3 Serial Communications

The **Serial** palette, not surprisingly is also on the **Instrument I/O** palette (Figure 11.63).

Figure 11.63

Notice that these functions are actually just specialized VISA functions for serial ports. Like GPIB, there are functions for reading and writing. More specific to the serial port, you can do things like find out how many bytes are waiting to be read at the serial port or configure the serial port (e. g., baud rate).

In one sense, serial communication is the simplest to program. However, serial communication suffers from abused and ignored hardware standards, obscure and complex programming protocols, and relatively slow data transfer speeds. The difficulties people encounter with writing an application for a serial instrument are rarely LabVIEW-related! We'll take a quick look at how to use the serial port VIs to communicate with a serial device.

You should become familiar with some basic concepts of how serial communication works if you've never used it before. If you have used a modem, and know what things like baud rate, stop bits, and parity roughly mean, then you should know enough to get started. Otherwise, it might be a good idea to read some documentation about the serial port (any good book on RS-232).

A good place to get started if you are talking to a serial instrument is to use the **LabVIEW<->Serial** example included with the LabVIEW examples (Figure 11.64).

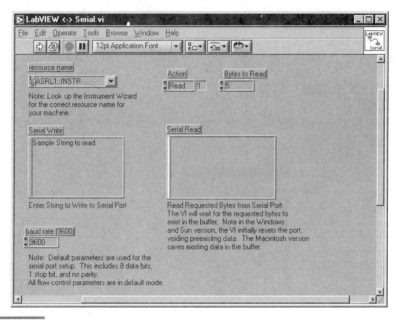

Figure 11.64

11.4.4 Instrument Drivers

Finally, no discussion of instrument control would be complete without talking about *instrument drivers*. An instrument driver is a collection of functions that implement the commands necessary to perform the instrument's operations. Because LabVIEW instrument drivers simplify instrument programming to high-level commands, you do not need to learn the mysterious, hard-to-remember low-level instrument-specific syntax needed to control your instruments. Instrument drivers are not necessary to use your instrument, of course; they are merely time savers to help you develop your project so that you do not need to study the instrument manual before writing a program.

Instrument drivers create the instrument commands and communicate with the instrument over the serial, GPIB, or VXI bus. In addition, instrument drivers receive, parse, and scale the response strings from instruments into scaled data that can be used in your test programs. With all of this work already done for you in the driver, instrument drivers can significantly reduce development time.

LabVIEW includes a couple of sample instrument drivers, for example, the instrument driver for the **HP34401A Multimeter**. If you look at the **Instrument Driver** palette (in the Instrument I/O palette) you can see what instrument driver functions look like (Figure 11.65).

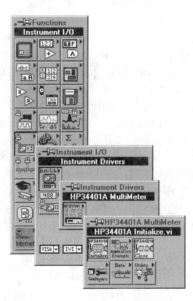

Figure 11.65

Although generally you'll need the instrument driver that is specifically for your instrument model,* your odds are fairly good: LabVIEW provides more than 700 LabVIEW instrument drivers from more than 50 vendors. A

---

* Actually, **IVI (interchangeable virtual instruments)** Instrument Drivers have begun to provide instrument drivers that will work with certain types of instruments, such as oscilloscopes, from multiple vendors and with multiple models. See http://ni.com for more details.

list is available on the National Instruments Developer Zone, `zone.ni.com/idnet`. You can use these instrument drivers to build complete systems quickly. Instrument drivers will drastically reduce your software development time because you can practically just start using your instrument from LabVIEW.

11.5 Wrap It Up!

This chapter has helped get you started with using LabVIEW to perform data acquisition and taken a peek at instrument control.

We took at look at the concepts of buffering and triggering. Buffering is a method for storing samples temporarily in a memory location (the buffer) and is used to achieve fast and accurate sampling rates. Triggering is a method to initiate and/or end data acquisition dependent on a trigger signal. The trigger can come from the software or from an external signal.

The **Analog Input** and **Analog Output** palettes contain VIs for analog I/O at different levels of complexity. The top tier VIs, or "easy" analog VIs, are useful for simple I/O. These easy VIs let you acquire one point (nonbuffered AI) or a whole waveform (buffered AI) from one or multiple channels. The middle tier VIs, or "intermediate" VIs provide more control over data acquisition. With these VIs, you use them in a specific sequence to configure, start, read, and clear the operation. The intermediate VIs allow you to perform buffered I/O, hardware triggering, and disk streaming. The bottom tier, advanced and utility VIs, are not generally suitable for use by beginners.

The **Digital I/O** palette contains some simple VIs for reading and writing to the digital lines of a DAQ board. You can read or write to either a single line or a whole port (four or eight lines, depending on the board).

For instrument control, the **VISA** palette should allow you to communicate with just about any type of instrument, regardless of the communication bus it uses. Sometimes you'll still want to use the **GPIB** or **Serial** palette to go one level lower. And if you can obtain an **instrument driver** that was designed for your instrument, you're already well on your way to building your final application.

OVERVIEW

This chapter will show you how to use some of the more advanced and powerful functions and structures of LabVIEW. LabVIEW has the capability to manipulate and store local and global variables—much like conventional programming languages. You will also see how you can make controls and indicators more flexible and useful by using their property nodes, which determine the behavior and appearance of front panel objects. In addition, this chapter covers some miscellaneous advanced functions such as system calls, calling and generating external code, and dialogs. Finally, you'll take a look at advanced data conversions and why you might need them.

GOALS

- Understand local and global variables, and know how and when to use them
- Be able to customize the appearance and behavior of your front panel objects using property nodes
- Learn about some miscellaneous Advanced functions
- Get an overview of what LabVIEW can do to interface with external code
- Know how to perform advanced conversions between different data types

KEY TERMS

- Local variable
- Global variable
- Read and write mode
- Race condition
- Property node
- DLL
- CIN
- Call Library
- ASCII string
- Binary string
- Typecasting

Advanced LabVIEW Functions and Structures

12

Every program has at least one bug and can be shortened by at least one instruction—from which, by induction, one can deduce that every program can be reduced to one instruction which doesn't work.

—from an unnamed computer science professor

12.1 Local and Global Variables

Local and global variables are, technically speaking, LabVIEW structures. If you've done any programming in conventional languages like C or Pascal, then you're already familiar with the concept of a local or global variable. Up until now, we have read data from or written to a front panel object via its terminal on the block diagram. However, a front panel object has only one terminal on the block diagram, and you may need to update or read a front panel object from several locations on the diagram or other VIs.

Local variables (locals for short) provide a way to access front panel objects from several places in the block diagram of a VI in instances where you can't or don't want to connect a wire to the object's terminal.

Global variables (globals for short) allow you to access values of any data type (or several types at the same time if you wish) among several VIs in cases where you can't wire the subVI nodes or when several VIs are running simultaneously. In many ways, global variables are similar to local variables, but instead of being limited to use in a single VI, global variables can pass values among several VIs.

This section will teach you some of the benefits of using locals and globals, as well as show you some common pitfalls to watch out for.

12.1.1 Local Variables

Local variables in LabVIEW are built-in objects that are accessible from the **Structures** subpalette in the **Functions** palette. When you select a local variable object, a node showing a "**?**" first appears to indicate that the local is undefined. By clicking on this node with the Operating tool, a list of all current controls and indicators will appear; selecting one of these will define the local. Or you can pop up on the local variable and choose **Select Item** to access this list. You can also create a local variable by popping up on an object's terminal and selecting **Create>Local Variable**.

Figure 12.1

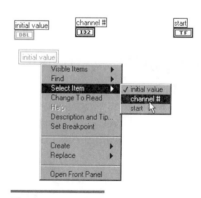

Figure 12.2

There are at least a couple of reasons why you might want to use locals in your VI:

- You can do things, such as control parallel loops with a single variable, that you otherwise couldn't do.
- Virtually any control can be used as an indicator, or any indicator as a control.

Controlling Parallel Loops

We've discussed previously how LabVIEW controls execution order through dataflow. The principle of dataflow is part of what makes LabVIEW so intuitive and easy to program. However, occasions may arise (and if you're going to develop any serious applications, the occasions *will* arise) when you will have to read from or write data to front panel controls and indicators without wiring directly to their corresponding terminals. A classical problem is shown in Figure 12.3. We want to end the execution of two independent While Loops with a single Boolean stop control.

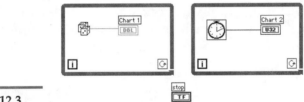

Figure 12.3

How can we do this? Some might say we could simply wire the stop button to the loop terminals. However, think about how often the loops will check the value of the stop button if it is wired from outside the loops.

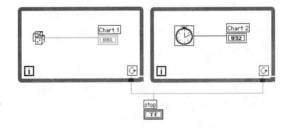

Figure 12.4

Wiring the stop button from outside the loops to both conditional terminals will not work, since controls outside the loops are not read again after execution of the loop begins. The loops in this case would execute only once if the <u>stop</u> button is FALSE when the loop starts, or execute forever if <u>stop</u> is TRUE.

So why not put the <u>stop</u> button inside one loop, as shown in Figure 12.5? Will this scheme work?

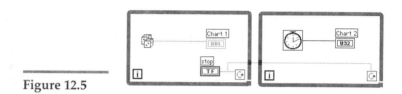

Figure 12.5

Putting the <u>stop</u> button inside one loop and stretching a wire to the other loop's conditional terminal won't do the trick either, for similar reasons. The second loop won't even begin until the first loop finishes executing (recall the notion of data dependency?).

The solution to this dilemma is—you guessed it—a local variable. Local variables create, in a sense, a "copy" of the data in another terminal on the diagram. The local variable always contains the up-to-date value of its associated front panel object. In this manner, you can access a control or indicator at more than one point in your diagram without having to connect its terminal with a wire.

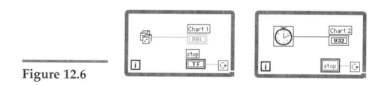

Figure 12.6

Referring to the previous example, we can now use one <u>stop</u> button for both loops by wiring the Boolean terminal to one conditional terminal and wiring its associated local variable to the other conditional terminal.

*There's one condition on creating a Boolean local variable: The front panel object can't be set to **Latch** mode (from the **Mechanical Action** option). Although it isn't obvious at first, a Boolean in **Latch** mode along with a local variable in read mode produces an ambiguous situation. Therefore, LabVIEW will give you the "broken arrow" if you create a local variable of a Boolean control set to **Latch** mode.*

Blurring the Control/Indicator Distinction

One of the really nice features about local variables is that they allow you to *write to* a control or *read from* an indicator, which is something you can't normally do with the regular terminals of an object. Locals have two modes: *read* and *write*. A local variable terminal can only be in one mode at a time, but you can create a second local terminal for the same variable in the other mode. Understanding the mode is pretty straightforward: In read mode, you can read the value from the local's terminal, just as you would from a normal control; in write mode, you can write data to the local's terminal, just as you would update a normal indicator. Just remember this formula for wiring locals:

READ mode = CONTROL

WRITE mode = INDICATOR

Another way to look at it is to consider a local in read mode the data "source," while a local in write mode is a data "sink."

You can set a local variable to either read or write mode by popping up on the local's terminal and selecting the **Change To...** option. A local variable in read mode has a heavier border around it than one in write mode (just like a control has a heavier border than an indicator), as shown in Figure 12.7. Pay close attention to these borders when you are wiring the locals, to make sure you are in the correct mode. If you attempt to write to a local variable in read mode, for example, you'll get a broken wire—and it may drive you crazy trying to figure out why.

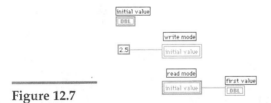

Figure 12.7

Last but not least, you must give a label to the control or indicator that the local refers to. That is, when creating a control or indicator, if you don't give it a name (which is possible in LabVIEW), you won't be able to create a local variable for it.

As a simple example, let's say you want to create a knob that represents a timer: The user can set the time and watch the knob turn as it counts down the time, just like those old-fashioned timers used in the kitchen. Obviously, the front panel object is going to have to be a control since the user will set the time, but it must also be able to accept block diagram values and "turn" accordingly as time goes by. Try building such a timer in the next activity.

Activity 12-1: Using Local Variables

1. Build a front panel with a knob and select the **Visible Items>>Digital Display** option, as shown in Figure 12.8.

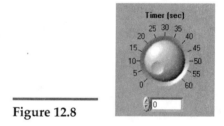

Figure 12.8

2. Create a local variable by selecting it from the **Structures** palette. You will get a local variable icon. Click on this icon with the operating tool and select <u>Timer (seconds)</u>. The local should be in write mode by default.

3. Build the simple block diagram shown in Figure 12.9.

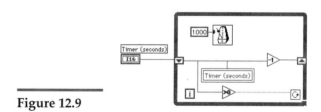

Figure 12.9

4. Save your VI as **Kitchen Timer.vi**.

In this example, the shift register is initialized with the value of <u>Timer</u> that was set at the front panel. Once loop execution begins, the shift register's value is decremented once per second and this value is passed to the <u>Timer</u> local variable, which is in write mode. The knob on the front panel rotates to reflect the changed value.

A nice feature to add to this example would be a sound to alert you that the timer reached zero, just like the old-fashioned ones (Ding!). Later in this chapter you will see how we can create simple sounds with LabVIEW.

Locals sound like a great structure to use, and they are. But you should watch out for a common pitfall when using locals: race conditions. A race condition occurs if two or more copies of a local variable in write mode can be written to at the same time.

There is a hazard to using locals and globals: accidentally creating a *race condition*. To demonstrate, build the simple example shown in Figure 12.10 and 12.11.

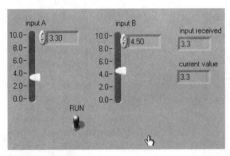

Figure 12.10

Figure 12.11

Notice the two While Loops controlled by <u>RUN</u> and a local variable <u>RUN</u>. However, the local variable <u>input received</u> is being written in the left loop as well as in the right loop. Now, execute this VI with <u>RUN</u> set to FALSE and different values on the two sliders. The loops will execute just once. What value appears at <u>current value</u>? We can't tell you, because it could be either value from <u>input A </u>or <u>input B!</u> There is nothing in LabVIEW that says execution order will be left-to-right, or top-to-bottom.

If you run the above VI with <u>RUN</u> turned on, you will likely see <u>current value</u> jump around between the two input values, as it should. To avoid race conditions such as this one, one must define the execution order by dataflow, sequence structures, or more elaborate schemes.

Another fact to keep in mind is that every read or write of a local creates a copy of the data in memory. So when using locals, remember to examine your diagram and the expected execution order to avoid race conditions, and use locals sparingly if you're trying to reduce your memory requirement.

Activity 12-2: More Fun with Locals

1. Another case where locals are very useful is in an application where you want a "Status" indicator to produce a more interactive VI. For example, you may want a string indicator that is updated with a comment or requests an input every time something happens in the application. Build a simple data acquisition VI similar to the ones you wrote in Chapter 11, but modify it to have a string indicator that tells the user:

   • when the program is waiting for input

   • when the program is acquiring data

   • when the program is graphing the data

   • when the program has stopped

   To help you get started, Figures 12.12–12.14 show the front panel and the first two frames of a Sequence structure.

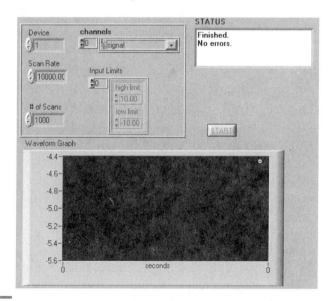

Figure 12.12

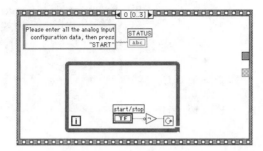

Figure 12.13

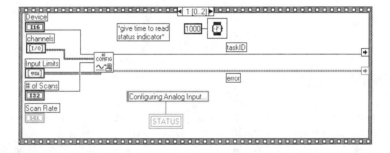

Figure 12.14

2. Save your VI as **Status Indicator.vi**.

Activity 12-3: More Fun with Locals

In many applications, you may want some type of "master" control that modifies the values on other controls. Suppose you wanted a simple panel to control your home stereo volumes. The computer presumably is connected to the stereo volume control in some way. In the VI shown in Figure 12.15, a simulated sound control panel has three slide controls: left channel, right channel, and master. The left and right channel can be set independently; moving the master slide needs to increment or decrement the left and right volumes proportionally.

Build the block diagram for the front panel shown in Figure 12.15. The fun part about this is that by moving the master slide, you should be able to watch the other two slides move in response.

Save your VI as **Master and Slave.vi**.

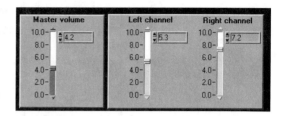

Figure 12.15

You will need to use shift registers for this exercise.

12.1.2 Global Variables

If you've never used globals in your programs before, congratulations! Global variables are perhaps the most misused and abused structure in programming. Globals are more often than not the cause of mysterious bugs, unexpected behavior, and awkward programming structures. Having said this, there are still occasions when you might want to and need to resort to globals. (It's not that globals are bad, they just need to be used with caution.)

Recall that you can use local variables to access front panel objects at various locations in your block diagram. Those local variables are accessible only in that single VI. Suppose you need to pass data among several VIs that run concurrently or whose subVI icons cannot be connected by wires in your diagram. You can use a global variable to accomplish this. In many ways, global variables are similar to local variables, but instead of being limited to use in a single VI, global variables can pass values between several VIs.

Consider the following example. Suppose you have two VIs running simultaneously. Each VI writes a data point from a signal to a waveform chart. The first VI also contains a Boolean <u>Power</u> button to terminate both VIs. Remember that when both loops were on a single diagram, we needed to use a local variable to terminate the loops. Now that each loop is in a separate VI, we must use a global variable to terminate the loops. Notice that the global terminal is similar to a local terminal except that a global terminal has a "world" icon inside it.

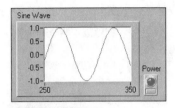

Figure 12.16
First VI front panel.

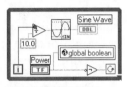

Figure 12.17
First VI block diagram.

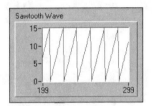

Figure 12.18
Second VI front panel.

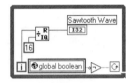

Figure 12.19
Second VI block diagram.

Creating Globals

Like locals, globals are a LabVIEW structure accessible from the **Structures** palette. And like locals, a single global terminal can be in write or read mode. *Unlike* locals, different VIs can independently call the same global variable. Globals are effectively a way to share data among several VIs without any having to wire the data from one VI to the next; globals store their data independently of individual VIs. If one VI writes a value to a global, any VI or subVI that reads the global will contain the updated value.

Once the global structure is selected from the palette, a 🔲 icon appears on the diagram. The icon symbolizes a global that has not been defined yet. By double-clicking on this icon, you will see a screen pop up that is virtually identical to a VI's front panel. You can think of globals as a special kind of VI—they can contain any type and any number of data structures on their front panels, but they have no corresponding diagram. Globals store variables without performing any kind of execution on these variables. Placing controls or indicators on a global's front panel is done in an identical fashion to a VI's front panel. An interesting tidbit about globals is that it makes no difference whether you choose a control or an indicator for a given data type, since you can both read and write to globals. Finally, be sure to give labels to each object in your global, or you won't be able to use them.

A global might contain, as in the following example, a numeric variable, a stop button, and a string control.

Figure 12.20

Save a global just like you save a VI (many people use a ".gbl" extension when naming globals just to keep track of them). To use a saved global in a diagram, you can choose **Select a VI...** in the **Functions** palette. A terminal showing one of the global's variables will appear on the diagram. To select the variable you want to use, pop up on the terminal and choose **Select Item>**, or simply click on the terminal using the Operating ✌ tool. You can select only one variable at a time on a global's terminal. To use another variable, or to use another element in the global, create another terminal (cloning by <ctrl>-dragging or <option>-dragging is easiest).

Figure 12.21

Just like locals, globals can be in a read or a write mode. To choose the mode, pop up on the terminal and select the **Change To...** option. Read globals have heavier borders than write globals. As with locals, globals in read mode behave like controls, and globals in write mode act like indicators. Again, a global in read mode is a data "source," while a global in write mode is a data "sink."

READ mode = CONTROL

WRITE mode = INDICATOR

Here are some important tips on using global variables:

1. Always initialize your globals in your diagram. The initial value of a global should always be clear from your code. Globals don't preserve any of their default values unless you quit and restart LabVIEW.

2. Never read from and write to global variables in the same place, that is where either one could occur before the other (this is the famous "race condition").

3. Since globals can store several different data types, group global data together in one global instead of several global variables.

It's important that you pay attention to the names you give the variables in your globals. All the VIs that call the global will reference the variable by the same name; therefore, be especially careful to avoid giving identical names to controls or indicators.

An Example

Let's look at a problem similar to the two independent While Loops. Suppose that now, instead of having two independent While Loops on our diagram, we have two independent subVIs that need to run concurrently.

Figures 12.22–12.25 shows two subVIs and their respective front panels. These two VIs, **Generate Time** and **Plot** are designed to be running at the same time. **Generate Time** just continuously obtains the tick count from the internal clock in milliseconds starting at the time the VI is run. **Plot** generates random numbers once a second until a stop button is pressed, after which it takes all the tick count values from **Generate Time** and plots the random numbers versus the time at which the numbers were generated.

Figure 12.22   **Figure 12.23**

Figure 12.24

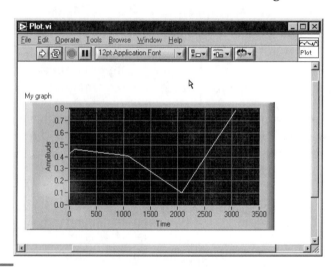

Figure 12.25

The way these two subVIs exchange data is through the use of a global. We want **Plot** to obtain an array of time values provided by **Generate Time**, and more importantly, we want both subVIs to be stopped by a single Boolean control.

So, first we create a global with the two necessary variables. Remember, to create a new global, select the **Global Variable** structure from the **Structures** palette, and double-click on the "world" icon to define the global's components. In this case, we define a numeric component <u>Time (ms)</u> and a Boolean <u>Stop</u>. The name of the global variable is **The Global**.

Figure 12.26

Then we use the global's variables at the appropriate places in both subVIs.

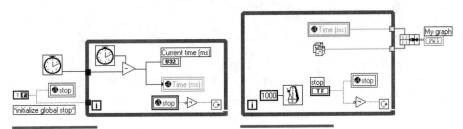

Figure 12.27   **Figure 12.28**

Notice how the <u>stop</u> Boolean variable is used: A stop button from **Plot** writes to the global variable <u>stop</u>, which is in turn used to stop the While Loop in **Generate Time**. When the <u>stop</u> button is pressed in **Plot**, it will break the loop in **Generate Time** as well. Similarly, the time values from **Generate Time** are passed to the global variable <u>Time</u>, which is called by the **Plot** VI to build an array.

Hopefully, these two VIs have given you an example of how globals work. We possibly could have avoided using globals for what we wanted to accomplish in this program, but the simple example is good for illustrative purposes.

If you looked at a block diagram that called the two subVIs, you'd see another problem with using globals: There are no wires anywhere! Globals obscure dataflow since we can't see how the two subVIs are related. Even when you see a global variable on the block diagram, you don't know where else it is being written to. Fortunately, version 4.0 of LabVIEW addressed this inconvenience by including a feature that searches for the instances of a global.

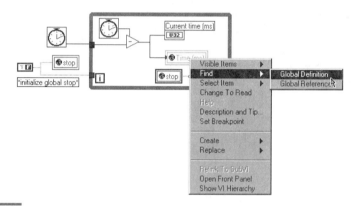

Figure 12.29

If you pop up on a global's terminal, you can select **Find>Global Definition**, which will take you to the front panel where the global is defined. The other option is **Find>Global References**, which will provide you with a list of all the VIs that contain the global. For more information about LabVIEW's search capabilities, see Chapter 13.

12.2 Property Nodes

With property nodes (what used to be called "attribute nodes" in LabVIEW 5.1 and earlier), you can start making your program more powerful and a lot more fun. Property nodes allow you to programmatically control the properties of a front panel object: things such as color, visibility, position, numerical format, etc. The key word here is *programmatically*, that is, changing the properties of a front panel object according to an algorithm in your diagram. For example, you could change the color of a dial to go through blue, green, and red as its numerical value increases. Or you could selectively present the user with different controls, each set of them appearing or disappearing according to what buttons were pressed. You could even animate your screen by having a custom control move around to symbolize some physical process.

To create a property node, pop up on either the front panel object or its terminal, and select **Create>Property node**. A terminal with the same name as the variable will appear on the diagram. To see what options you can set in a variable's property node, click on the node with the Operating tool or pop up on the node and choose **Select Item>**. Now you have the choice of which property or properties you wish to select. Each object has a set of *base properties* and, usually, an additional set of properties specific to that object.

Figure 12.30

Just like with local variables, you can either read or write the property of an object (although a few properties are read-only). To change the mode of a property, pop up on it and select the **Change to Write/Read** option. The small arrow inside the property node's terminal tells you which mode it's in. A property node in write mode has the arrow on the left, indicating that data are flowing into the node, *writing* a new property. A property node in read mode has the arrow on the right, *reading* the current property and providing these data. The same analogy we used for locals, a control (read mode) and an indicator (write mode), holds for property nodes.

An interesting feature of property nodes is that you can use one terminal on the block diagrams for several properties (but always affecting the same control or indicator). To add an additional property, you can use the Positioning tool to *resize* the terminal and get the number of properties you need, much in the same way multiple inputs are added to functions like **Bundle**, **Build Array**, etc. Figure 12.31 shows two properties on the same terminal for the numeric control gain.

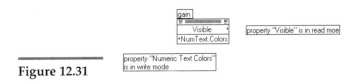

Figure 12.31

Let's look at a simple example. Suppose you wanted to have a front panel that would hide certain specialized controls except for those occasions when they were needed. In the front panel, shown in Figure 12.32, we see a tension gauge and a Boolean alarm switch. We include a button that says show/hide advanced controls..., hinting at the possibility that if you pressed it, some really obscure and intricate options will pop up.

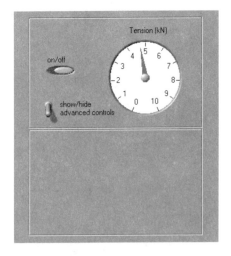

Figure 12.32

In this example, we've included two more controls, gain and offset, which are made invisible by setting their property nodes' option "Visible" to

FALSE unless the button is pressed. If the <u>show/hide advanced controls...</u> button is pressed, then ta-dah! . . . the two knobs become visible.

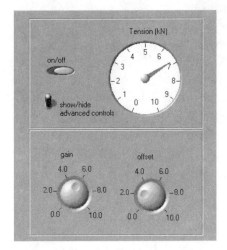

Figure 12.33

The entire block diagram would be encompassed in a While Loop like the one shown in Figure 12.34 to make the button control the visibility of the two knobs and thus give the "pop up" effect.

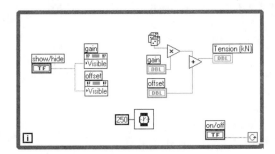

Figure 12.34

Often you will want to use more than one option in an object's property node. Remember, instead of creating another property node, you can select several options at a time by enlarging the terminal with the Positioning tool (much like you enlarge cluster and array terminals). You will see each new option appear in sequence; you can later change these if you like by clicking on any item with the Operating tool.

What do some of the base options in a property node refer to?

- **Visible**: Sets or reads the visibility status of the object. It is visible when TRUE and hidden when FALSE. This is often a better choice than coloring an object transparent, since transparent objects can accidentally be selected.

- **Disabled**: Sets or reads the user access status of a control. A value of 0 enables the control so that the user can access it; a value of 1 disables the control without any visible indication; and a value of 2 disables the control and "grays it out."

- **Key Focus**: When TRUE, the control is the currently selected key focus, which means that the cursor is active in this field. Key focus is generally changed by tabbing through fields. This is useful for building a mouseless application. See Chapter 13 for more information on Key Focus.

- **Position**: A cluster of two numbers that respectively define the top and left pixel position of the front panel object.

- **Size**: A cluster of two numbers that respectively define the height and width in pixels of the entire front panel object.

- **Blinking**: When TRUE, the front panel object blinks.

- **Format and Precision**: Sets or reads the format and precision properties for numeric controls and indicators. The input cluster contains two integers: one for format and one for precision. These are the same properties you can set from the pop-up menu of the numeric object.

- **Color**: Depending on the type of object, you may have several color options. The input is one of those neat color boxes that sets the color of the text, background, etc., depending on the object.

You'll also notice that every object has properties called **ClassID**, **Owner**, **Owning VI**. Don't worry about them at this point. They are for very advanced and relatively obscure LabVIEW programming scenarios.

The Help window really is helpful when using property nodes. If you move the cursor onto the terminal of a property node, the Help window will show you what the property means and what kind of data it expects. You can also pop up on the property node terminal and choose **Create Constant** to get the correct data type—this comes in very handy when the input is a cluster.

Almost all controls or indicators have the base properties. Most of them have many more, especially tables and graphs (which can have over 100 properties!). We won't even begin to go into most of these properties, partly because you may never care about many of them and you can always look up the details in the manuals. The best way to learn about property nodes is to create some to go with your application and begin to play around with them. You'll find that property nodes are very handy for making your program more dynamic, flexible, and user-friendly (always good for impressing your nontechnical manager).

Another Example

Graphs and charts have zillions of options in their property nodes, as you can see by popping up on the terminal of a chart's property node (Figure 12.35).

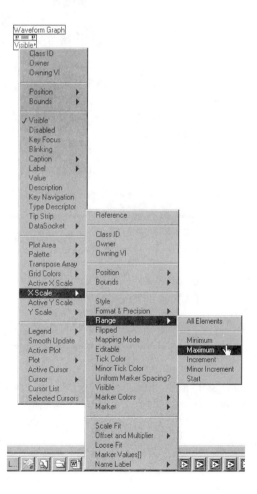

Figure 12.35

This next example, which can be found in the property node examples in the full version of LabVIEW, shows just one of the many aspects of a graph you can control programmatically. **Chart Property node** lets you select one of three display types for the chart: Strip, Scope, and Sweep (if you need to, review Chapter 8 to see what they do).

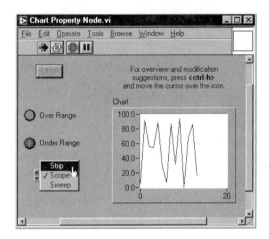

Figure 12.36

You can select the chart mode and watch it change even while the VI is running. The way this is done is through the **Update Mode** option on a chart's property node.

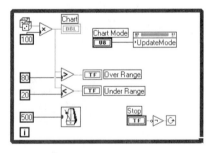

Figure 12.37
Inside the loop is a property node that alters the update mode of the chart according to the setting specified by the Chart Mode ring control of the front panel.

Activity 12-4: Using Property Nodes with Charts

Write a VI that graphs three channels of data (either real-time data through a DAQ board or random data). Let the user turn on or off any of the plots with three buttons. Since this is a chart and not a graph, the data will accumulate along with the old data, even if the VI is closed and reopened. Add a "CLEAR" button that clears the chart.

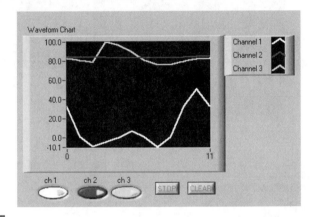

Figure 12.38

 LabVIEW includes a "transparent" color. To set a color property, you can create a cluster of colorbox constants. The colorbox is actually a numeric type, found buried deep in the Numeric>Additional Numeric Constants palette. Pop up on a colorbox constant to select its color (remember that "T" is the transparent color).

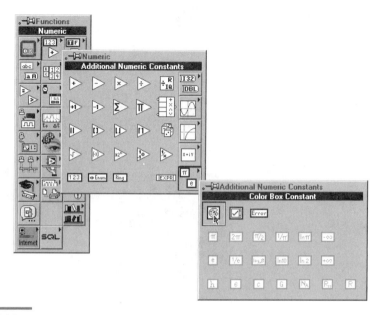

Figure 12.39

When using a multiplot graph or chart, you can only affect the properties of one plot at a time. The plots are numbered 0,1, . . ., n for property node purposes. A specific property, called **Active Plot**, is used to select the plot for the properties you are modifying or reading. In this exercise, you will need to build some case statements like those shown in Figures 12.40 and 12.41.

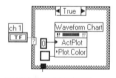

Figure 12.40

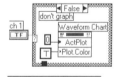

Figure 12.41

*To clear a chart, use the **History Data** property of the chart. Then wire an empty array to this property node.*

Save your VI as **Property Nodes-Graph.vi**.

Activity 12-5: Using Property Nodes to Create Dynamic Menus

Ring controls, along with property nodes, can create some powerful interactive VIs. Examine the following example, called **Ring Control Menus**.

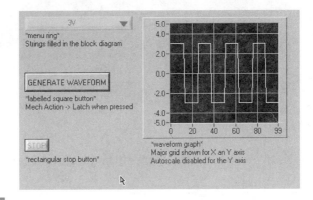

Figure 12.42

A pull-down menu ring, initially empty, will be used to query the user for the type of output. Once <u>GENERATE WAVEFORM</u> is pressed, the ring control will say "Choose Waveform" and present several options: sine, square, or sawtooth wave.

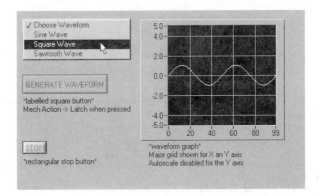

Figure 12.43

Next, the same ring control *will change* to let the user choose the output amplitude. Finally, the waveform is graphed and generated at the DAQ board.

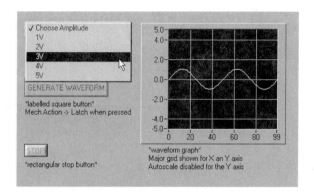

Figure 12.44

Figure 12.45 gives a peek at part of the block diagram.

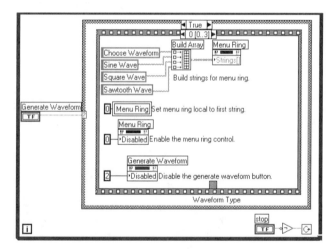

Figure 12.45

12.3 Other LabVIEW Goodies

A "miscellaneous" category is always hard to avoid—like the ever-growing one in my budget. Many miscellaneous functions are located in the **Advanced** palette. We'll cover some of the functions in that menu as well as some others that don't quite fit anywhere else but nevertheless can be quite useful.

12.3.1 Dialogs

Can LabVIEW carry on a conversation with you? For now, you can make LabVIEW pop-up dialog windows that have a message and often some response buttons, such as "OK" and "Cancel," just like in other applications. We've mentioned dialog functions before, but you may find the review useful.

The dialog functions are accessible from the **Time & Dialog** palette. Lab-VIEW provides you with two types of pop-up dialog boxes: one-button and two-button. In each case, you can write the message that will appear in the window and specify labels on the buttons. The two-button dialog function returns a Boolean value indicating which button was pressed. In any case, LabVIEW halts execution of the VI until the user responds to the dialog box. The dialog box is said to be *modal*, which means that even though windows (such as other VIs' front panels) will continue to run and be updated, the user can't select any of them or do anything else via the mouse or keyboard until he or she has dealt with the dialog box.

As an example, suppose you wanted to add a dialog box to confirm a user's choice on critical selections. In the example shown in Figures 12.46 and 12.47, a dialog box is presented when the user presses the computer's self-destruct button.

Figure 12.46

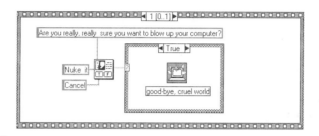

Figure 12.47

Dialog boxes are a nifty feature to make your application have the look and feel of a true Windows or Mac executable, but don't overdo them. Remember, LabVIEW programs already provide the interactive graphical interface, so having unnecessary dialogs popping up can make your application more cumbersome and messy. It's best to reserve the dialogs for very important notifications and/or confirmations of a user's choice. Also, note that LabVIEW does not let you have more than two response buttons in a dialog box. For more elaborate "dialogs," you're best off writing a subVI whose front panel contains the dialog options you'd like. You can make this subVI pop up and even customize its appearance by popping up on the VI's icon, selecting **VI Properties...>>Window Appearance** and selecting **Dialog**.

12.3.2 Saying "NO" Harshly

LabVIEW gives you a couple of functions to abort the execution of your code immediately: **Stop** and **Quit LabVIEW**. Like the occurrence functions, these functions are in the **Application Control** palette.

 Stop has a Boolean input. When TRUE (which is the unwired default input), it halts execution of the VI and all its subVIs, just as if you had pressed the stop button on the Toolbar.

 Quit LabVIEW does just that, when its input is TRUE (also the default input). Be careful with this one!

Some of us reminisce about a third function that was included back when LabVIEW was in version 2.2: **Shutdown**. On certain Macintosh models, this function would effectively turn off the whole computer, monitor and all. We never understood how it could be useful, but it was fun to leave a colorful VI running with one big button that said "DO NOT PRESS THIS BUTTON" on someone else's Mac, and then just wait to see who would be too curious . . . The button was, of course, connected to the operating system's Shutdown function. It was a good way to amuse ourselves with the less experienced LabVIEW users.

On a serious note, however, you should use these functions with caution. It is considered bad programming etiquette (in any language) to include a "hard stop." Strive to always have a graceful way to exit your program.

 Make sure your VI completes all final tasks (closing files, finishing the execution of subVIs, etc.) before you call the Stop function. Otherwise, you may encounter unpredictable file behavior or I/O errors, and your results can be corrupted.

12.3.3 Sound

A very nice or very obnoxious feature to have in many applications is sound, depending on how you use it. An audible alarm can be very useful in situations where an operator cannot look at the screen continuously during some test. A beep for every warning or notification, however, can be downright annoying. So use sounds with some thought to how often they will go off.

You can access the sound function from the **Graphics & Sound >> Sound** subpalette.

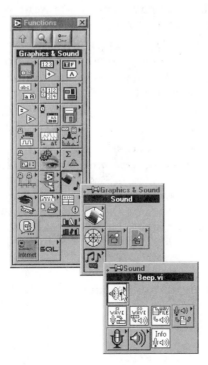

Figure 12.48

At a very simple level, on all operating systems, LabVIEW gives you access to the operating system's beep, through the **Beep** function.

Figure 12.49
Beep.vi.

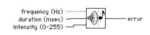

 In the Mac operating system,
you can set the frequency, intensity, and duration of the sound.
 In Windows, these inputs are ignored and only the standard system beep is used.

On Windows and the MacOS only, LabVIEW has the ability to record and playback sound files (wave and aiff format). You can use the functions **Sound Play Wave File**, **Sound Read Waveform**, and **Sound Write Waveform** to add sound files in LabVIEW. Browse the sound examples with LabVIEW to get a feel for how they work.

12.4 Calling Code from Other Languages

What happens if you already have some code written in a conventional language (such as C, Pascal, FORTRAN, Basic) that you'd like to use? Or if for some reason you just miss typing in all those semicolons in your familiar text-based code? LabVIEW does give you some options for interfacing with code from other languages. If you are thinking about writing *all* your code in C or C++, you should check out LabWindows/CVI (available from National Instruments), a programming environment very similar to LabVIEW; the main difference is that C code replaces the graphical block diagram. But if you'd like to (dare we say it?) actually have *fun* programming, then stick to LabVIEW and use conventional code only when you have to.

 You may want to skip this section if you aren't familiar with writing code in other programming languages such as C, or if you don't expect to need to interface LabVIEW and external code.

You have three options for calling external code:

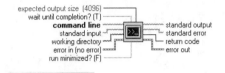

Figure 12.50
System Exec.vi.

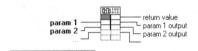

Figure 12.51
Call Library Function.

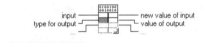

Figure 12.52
Code Interface Node.

1. The simplest option is to launch a separate executable or program that does what you need. On **Windows, Unix,** and **MacOS X** systems, you do this with the **System Exec** function (**Communication** palette). On **MacOS 9.x** and earlier, you can use **AppleEvents** functions (**Communication** palette).

2. If you have compiled functions— **Dynamic Link Libraries (DLLs)** on Windows, **Shared Libraries** on Unix, and **Code Fragments** on the MacOS—you can use the **Call Library Function** in LabVIEW (**Advanced** palette).

3. Finally, if you want to write your own C code and embed this into LabVIEW, you should use the **Code Interface Node (CIN)** (Advanced palette).

In Windows, LabVIEW can also export a VI into an executable (.exe) or a DLL (.dll) that other applications in turn can use. We'll discuss this capability briefly in the next chapter.

12.4.1 Using CINs

Using a CIN involves basically the following steps:

1. Place the CIN on the block diagram (from the **Advanced** palette).

2. The CIN has terminals for passing the inputs and outputs from the code. By default, the CIN node has only one pair of terminals You can resize the node to include the number of parameters you need.

Figure 12.53

By default, each pair of terminals is an input–output: The left terminal is an input, and the right terminal is an output. However, if a function returns more outputs than inputs (or has no input parameters at all), you can change the terminal type to **Output-Only** by popping up on that terminal and choosing this option.

3. Wire the inputs and outputs to the CIN. You can use any LabVIEW data type when wiring the CIN terminals (of course, it *must* correspond to the C parameter data type for the function you're calling). The order of the terminal pairs on the CIN corresponds to the order of the parameters in the code. In the example depicted in Figures 12.54 and 12.55, we call CIN that filters the <u>Raw Data</u> input array and passes the output into the <u>Filtered Data</u> array. Notice also how the terminals, once wired, indicate the type of data you are passing.

Figure 12.54

4. Create a .c file by selecting this option from the pop-up menu. The .c file created by LabVIEW, in the style of C programming language, is a template in which you write the C code. With a little effort, you can paste in your existing code if you have it.

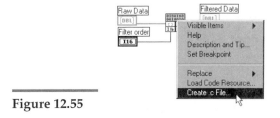

Figure 12.55

5. Compile the CIN source code. This step could be tricky, depending on what platform, compiler, and tools you are using. You have to first compile the code with a compiler that LabVIEW supports, and then, using a LabVIEW utility, you must modify the object code so LabVIEW can use it.

6. Load the object code into memory by selecting **Load Code Re-source...** from the pop-up menu. Select the `.lsb` file you created when you compiled the source code.

Once all these steps are completed successfully, you're calling C code in the CIN as if it were a subVI, with one major exception: Calls to CINs execute synchronously, meaning that, unlike most other tasks in LabVIEW, they do not share processor time with other LabVIEW tasks. For example, if you have a CIN and a For Loop at the same level in the block diagram, the For Loop halts execution until the CIN finishes executing. This is an important consideration if your timing requirements are tight.

There is one important final fact about CINs: They are most definitely *not* portable across platforms. If you compile your LabVIEW code with a CIN on a Power Mac and then try to run it under Windows, it won't work. The way around this is to rebuild the CIN using a C compiler for the platform that LabVIEW is running. The C compilers that have been tested with LabVIEW include:

Windows	Microsoft Visual C++, Symantec C
MacOS	Metrowerks CodeWarrior, Apple MPW
Linux, Solaris, HP-UX	GNU C++ (gcc) compiler

There are many complex issues involved in LabVIEW communicating with external code. Since many of these are highly dependent on the processor, operating system, and compiler you're using, we won't attempt to go any further into discussing CINs or Call Library functions. If you'd like more details, contact National Instruments and request their application notes on this subject or look at the *Using External Code in LabVIEW* manual included in the LabVIEW manual set.

12.5 Fitting Square Pegs into Round Holes: Advanced Conversions and Typecasting

Remember *polymorphism*, discussed early in this book? It's one of LabVIEW's best features, allowing you to mix data types in most functions without even thinking about it. (Any compiler for a traditional programming language would scream at you if you tried something like adding a constant directly to an array—but not LabVIEW.) LabVIEW normally takes care of doing conversions internally when it encounters an input of a different but compatible data type than it expected at a function.

If you're going to develop an application that incorporates instrument control, interapplication communication, or networking, chances are you'll be using mostly string data. Often you'll need to convert your numeric data, such as an array of floating-point numbers, to a string. We talked about this a little in Chapter 9. It's important to distinguish now between two kinds of data strings: *ASCII strings* and *binary strings*.

ASCII strings use a separate character to represent each digit in a number. Thus, the number 145, converted to an ASCII string, consists of the characters 1, 4, and 5.* For multiple numbers (as in arrays), a delimiter such as the <space> character is also used. This kind of string representation for numbers is very common in GPIB instrument control, for example.

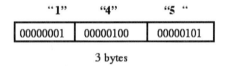

3 bytes

Binary strings are a bit more subtle—for one thing, you can't tell what data they represent just by reading them as ASCII characters. The actual bit pattern (binary representation) of a number is used to represent it. Thus, the number 145 with I8 representation is just a single byte in a binary string (which, incidentally, corresponds on my computer's ASCII set to the "ë" character—not that most humans could tell that means 145). Binary strings are common in applications where a minimal overhead is desired, because generally it is much faster to convert data to binary strings and because they take up less memory.

1 byte

Ultimately, all data in computers are stored as binary numbers. So how does LabVIEW know if the data are a string, Boolean, double-precision

---

* Unicode, the increasingly popular alternative to ASCII, uses 4 bytes (32 bits) to represent characters, allowing a much broader range of international characters and symbols. At the time of this writing, LabVIEW did not have support for Unicode.

floating-point number, or an array of integers? All data in LabVIEW have two components: the *data* themselves, and the data's *type descriptor*. The data type descriptor is an array of I16 integers that constitute code that identifies the representation of the data (integer, string, double-precision, Boolean, etc.). This type descriptor contains information about the length (in bytes) of the data plus additional information about the type and structure of the data. For a list of data type codes, see *Application Note 154, LabVIEW Data Types*, included with the LabVIEW PDF manuals.

In a typical conversion function, such as changing a number to a decimal string, the type descriptor is changed and the data are modified in some way. The common conversion functions you use most of the time convert numbers to ASCII strings, such as in the example shown in Figure 12.56.

Figure 12.56

 The topics in the rest of this section can be very confusing to beginners. If you do not have any need to use binary string conversions in your application, you can safely skip the rest of this section.

In some cases, you may want to convert data to a *binary* string. Binary strings take up less memory, are sometimes faster to work with, and may be required by your system (such as a TCP/IP command, or an instrument command). LabVIEW provides a way to convert data to binary strings using the **Flatten To String** function. You can access this function from the **Data Manipulation** subpalette under the **Advanced** palette.

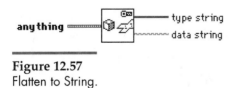

Figure 12.57
Flatten to String.

The input to **Flatten To String** can be of any LabVIEW data type, including complex types such as clusters. The function returns two outputs: the binary **data string**, representing the data, and **type string** (which is not a string but actually an array of I16 yet is called a string because of nerdy programmer's terminology). The **type string** gives you the information in the data type descriptor.

A flattened binary string contains not only the data in compact form but *header* information as well. The header information included at the beginning of the binary string contains information about the length, type, structure, etc. of the data. For example, the diagram in Figure 12.58 shows an array of DBL numeric types flattened to a string. The <u>type string</u>, which is always an array of I16, also has some of this header information.

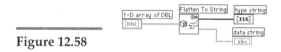

Figure 12.58

The key to using these confusing binary flattened strings is that most of the time you don't need to manipulate or view the strings directly—you just pass these strings to a file, network, instrument, etc., for little overhead. Later, to read the data, you will need to unflatten the string.

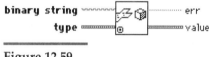

Figure 12.59
Unflatten From String.

To read back binary strings, use the inverse function, **Unflatten From String**. The **type** input is a dummy LabVIEW data type of the same kind that the **binary string** represents. The **value** output will contain the data in the **binary string**, but will be of the same data type as **type.err** is TRUE if the conversion was unsuccessful.

Figure 12.60 shows how we convert our flattened binary string back to a DBL array. We wire the binary string input, as well as a dummy (empty) 1D array of DBL to specify the data type, and we get back our original array of DBL. Notice how we flattened and unflattened without ever needing to "peek" or "manipulate" the binary strings directly.

Figure 12.60

For very fast, efficient conversions let's take a final look at another, very powerful LabVIEW function: **Type Cast** (Figure 12.61).

Figure 12.61
Type Cast.

This function allows you to change the data type descriptor, without changing the data at all. There is no conversion of the data in any way, just of the type descriptor. You can take almost any type of data (strings, Booleans, numbers, clusters, and arrays) and call it anything else. One advantage of using this function is that it saves memory since, unlike other conversion functions, it doesn't create another copy of the data in memory.

A common use of type casting is shown in Figure 12.62, where some instrument returns a series of readings as a binary string, and they need to be manipulated as an array of numbers. We are assuming that we know ahead of time that the binary string is representing an array of I16 integers.

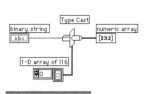

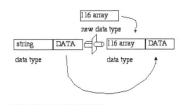

Figure 12.62   **Figure 12.63**

Type casting from scalars or arrays of scalars to a string works like the **Flatten To String** function, but it has the distinct feature that it doesn't add or remove any header information, unlike the binary strings created by the Flatten and Unflatten functions. The **type** input on the **Type Cast** function is strictly a "dummy" variable used to define the type—any actual data in this variable are ignored.

You need to be very careful with this function, however, because you have to understand exactly how the data is represented so that your type casting yields a meaningful result. As mentioned previously, binary strings often contain header information (such as the output from the **Flatten To String** function). **Type Cast** does not add or remove any header information in binary strings. If you don't remove headers yourself, they will be interpreted as data and you'll get garbage as a result.

Make sure you understand how data to be type cast are represented. Type casting does not do any kind of conversion or error checking on the data. Byte ordering (MSB first or last) is platform-dependent, so understand what order your data is in.

You might try experimenting with the **Type Cast** function to see what kind of results you get. To show you how bizarre you can get with typecasting, we took a cluster containing numeric controls (type I16) and cast the data to a string, a numerical array, and a Boolean (Figure 12.64).

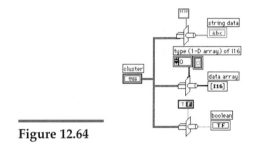

Figure 12.64

The front panel (Figure 12.65) shows you what results we obtained with two numbers.

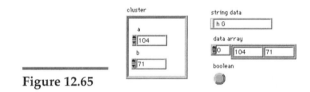

Figure 12.65

The string returned two characters corresponding to the ASCII values of 104 and 71. The array simply reorganized the numbers into array elements instead of cluster elements. And the Boolean? How would a Boolean variable interpret cluster data? Since we didn't know, we tried this. We found out that the Boolean is TRUE if the numerical value is negative; otherwise it's positive. In the case of a cluster, it apparently ORs the interpretation of the cluster elements, so if any element had a negative number, the Boolean interprets it as TRUE. Pretty strange, eh?

12.6 Wrap It Up!

In this chapter, we mined some of LabVIEW's gemstones: local variables, global variables, and property nodes. We also looked at some of the advanced functions: dialogs, sound, calling external code, binary string conversion, and typecasting. The power and versatility offered by these structures and functions allow you to go a level deeper into programming applications in LabVIEW.

Local variables allow you to create a block diagram "copy" of a front panel object, which can be read from or written to in different places. Global variables are similar to locals, but they store their data independently of any particular VI, allowing you to share variables among separate VIs without wiring them together. Locals are very useful for controlling parallel loops or updating a front panel object from multiple locations in the diagram. Globals are a powerful structure, but care is required in their use because they are prone to causing problems.

Property nodes give you immense control over the appearance and behavior of indicators and controls, allowing you to change their properties programmatically. Every control and indicator has a set of base properties (such as color, visibility, etc.). Many objects, such as graphs, have numerous properties that you can set or read accordingly.

With LabVIEW, you can use external C routines by using a CIN (code interface node). A CIN lets you call C routines compiled with external compilers. Additionally, you can directly call DLLs (Dynamic Link Libraries) in Windows.

The functions **Flatten To String**, **Unflatten From String**, and **Type Cast** are powerful conversion utilities for working with different data types. These functions allow you to convert LabVIEW data types to and from binary strings. Binary strings are often needed in many applications.

OVERVIEW

By tweaking the numerous environment options, you can customize the appearance and functionality of LabVIEW so that it works best for you. You will also learn the VI Property options, for example, how to set up your VI so that you can select front panel controls through the keyboard instead of the mouse. We'll look at the powerful VI Server and how you can use it to dynamically control VIs. LabVIEW's numeric types give you the option of representing them as binary, octal, hexadecimal, or decimal representation. In addition, you can include built-in units as part of a numeric variable and allow LabVIEW to perform automatic conversions. You will see how to instantly transform a section of your diagram into a subVI. LabVIEW has some useful development tools in the **Tools** menu, such the **Find** function, which lets you quickly search for any object, subVI, or text.

GOALS

- Use the **Options** to customize your LabVIEW environment
- Become familiar with the **VI Properties** options that allow you to make choices on the appearance and execution of your VI
- Be introduced to the **VI Server** and what it does
- Be able to set up keyboard access of controls
- Learn how to represent numbers with other radices and assign them units
- Create a subVI from a selection of the block diagram
- Discover the utilities under the **Tools** menu

KEY TERMS

- Options
- VI Properties
- VI Server
- Application Class
- VI Class
- Methods and Properties
- Priority
- Reentrant Execution
- Key Focus
- Radix
- Unit
- Find
- Profile Window
- History
- VI Hierarchy

Advanced LabVIEW Features

13

"The time has come," the Walrus said,
"To talk of many things:
Of shoes—and ships—and sealing wax—
Of cabbages and kings—
And why the sea is boiling hot—
And whether pigs have wings."

—Lewis Carroll (1832–98), English writer, mathematician. *Through the Looking Glass* (1872)

13.1 Options, Options...

Maybe you've already used the **Options...** command under the **Tools** menu. In any case, it's a good idea to skim through all the options available, since often here is where you will find the solution to that persistent LabVIEW problem you've had for a month. You can specify things such as search paths, default front panel colors and fonts, time and date formats, etc.

Figure 13.1 shows some of the option categories. Each of the categories shown brings up several options when selected. Although a comprehensive preference list would be too long to show here, some preference options are more commonly used and turn out to be quite useful:

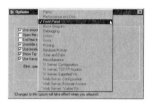

Figure 13.1

- In the **Paths** options, you can specify LabVIEW's default directory. Normally, this is just the LabVIEW directory. However, if you keep your project VIs somewhere else, you can change the default path so that when you start LabVIEW, it's easier to open your project VIs.

- The **Front Panel** options have some important options. For example, "Use smooth updates during drawing" turns off the annoying flickering you see on a graph or chart that is updated regularly. "Use transparent name labels" is another popular option because many people don't like the 3D box on the control and indicator labels. Finally, you can set the blink speed here (for the blink attribute of an object)—you'd think this feature would be a programmatic option, but it's buried here instead.

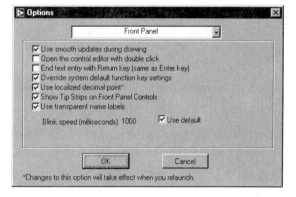

Figure 13.2

- The **Colors** preferences lets you set the default colors for the front panel, block diagram, blink colors, etc. If you just must have that magenta background all the time, this is the place to set it.

- The **Printing** preferences allow you to choose among bitmap, standard, and Postscript printing. Depending on the printer you have, you may need to choose a different option to get the best print results.

Take a look around at the rest of the categories. You can specify font styles, time and date styles, levels of resource sharing between LabVIEW and other applications, etc. You'll notice also that there are options for the **VI Server**, which we'll talk about shortly, and the **Web Server**, which is discussed in Chapter 14.

13.2 Configuring Your VI

Often you may want your program to bring up a new window when a certain button is pressed or a certain event occurs. For example, you may have a "main" front panel with several buttons that present options. Pressing one of the buttons would lead the user to a new screen, which may in turn contain an "Exit" button that would close the window and return to the main front panel. Since this pop-up window will usually be the front panel of a subVI, you also may want to customize what features of this window should appear, such as the **Toolbar** or the clickable box that closes the window. You can set several options that govern the window appearance and the execution of your VIs in two different places: **VI Properties...** (from the pop-up menu of the VI's icon pane, which you will find in the upper right-hand corner of the front panel window) and **SubVI Node Setup...** (from the pop-up menu on a subVI's icon in the block diagram). A very important distinction should be made at this point: The setup options under **SubVI Node Setup** for a subVI affect *only that particular instance of the called subVI*, while the **VI Setup** options *always* take effect, whether it's run as a top-level VI or called as a subVI. We'll start first with the few and simple **SubVI Node Setup** options.

13.2.1 SubVI Node Setup Options

When you select the setup option from the pop-up menu on a subVI, you get the dialog box shown in Figure 13.3.

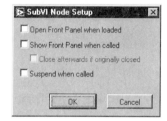

Figure 13.3

You can select any of the following options:

- **Open Front Panel when loaded**—The subVI's front panel pops open when the VI is loaded into memory (such as when you open a VI that calls this subVI).

- **Show Front Panel when called**—The VI front panel pops open when the subVI is executed. You'll find this option and the next to be pretty useful for creating interactive VIs.

- **Close afterward if originally closed**—This causes the subVI front panel to close when its execution is complete, giving that "pop-up" window effect. This option is only available if the previous option is selected.

- **Suspend when called**—This has the same effect as setting a breakpoint; that is, it causes the VI to suspend execution when the subVI is called. You can use this as a debugging tool to examine the inputs to your subVI when it is called but before it executes.

Remember, all the subVI setup options apply only to the particular subVI node you set them on. They do not affect any other nodes of the same subVI anywhere else.

13.2.2 Activity 13-1: Using SubVIs

This activity will let you use the **SubVI Setup Options** to create a "login" shell that can be used with any application.

1. Make a simple front panel, as shown in Figure 13.4, that will call a "pop-up" subVI when the button <u>Change User</u> is pressed. Call this top-level VI **Shell.vi**.

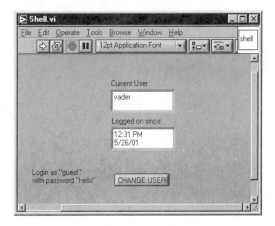

Figure 13.4

2. Use the subVI **Login.vi** (which you will find on the CD) to build the block diagram shown in Figures 13.5 and 13.6.

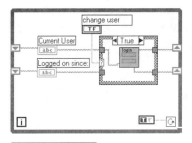

Figure 13.5

Figure 13.6

3. Pop up on the **Login** subVI, and choose **SubVI Node Setup** as shown in Figure 13.7. When this VI is run, it will display the last value of the <u>Current user</u> (from the uninitialized shift register), until the **Login** VI is called and changes it.

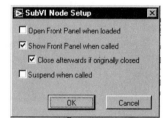

Figure 13.7

Login *must be closed before you run this example. Otherwise it will not close after it finishes executing—hence the phrase "Close afterwards if originally closed."*

13.2.3 VI Properties Options

VI Properties are a bit more numerous. The dialog box that appears when you choose **VI Properties...** by popping up on the icon pane gives you a number of property sets: **General, Memory Usage, Documentation, Revision History, Security, Window Size, Window Appearance, Execution,** and **Print Options.** Some of these are read-only properties (such as memory usage); others are configurable. Let's examine these in more detail briefly.

General

This VI property set shows you the file path to the VI, the icon, etc.

Memory Usage

This property pane shows you how much RAM your VI is using.

Documentation

Here you can type in documentation for your VI. In the **VI Description** field—which as good programming practice, you should always fill in—you can type a description of your VI. This same description will show up in the Help window if you hover the mouse button over your VI's icon. You can also optionally enter a **Help Tag** and a **Help Path** to an external Help file

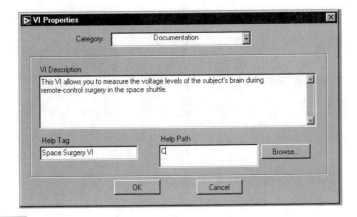

Figure 13.8

Revision History

This property set allows you to set up a very simple source code history. It gives you the option of seeing what revision your VI is in and any comments from previous revisions. This is only useful if you have been typing in comments in this window every time you save the VI.

Security

By default, anyone who has your VI can view and modify the source code (the block diagram). In some cases you don't want others to either modify it by accident or even see the block diagram. Here you can specify a password that will be required for anyone to view or modify the block diagram.

If you password-protect a VI, be sure you remember or write down the password somewhere. There isn't any way to access the VI's code if you forget the password.

13.2.4 Window Appearance

The **Window Appearance** options, shown in Figure 13.9, let you control many details of your VI's window appearance. You can choose among the predefined setups of "top-level application," "dialog," or "default," or you can customize the look.

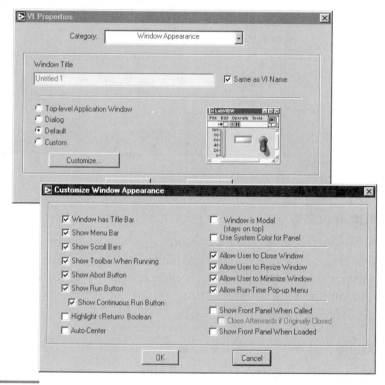

Figure 13.9

Most of the options are self-explanatory; however, a few comments are in order:

- Be careful with options such as disabling the **Show Abort Button**. A VI with this option can't be hard-stopped, even with the keyboard short-cut, once it's running! Only use this for VIs you have tested and which you know will exit properly from the code.

- The **Dialog Box** option gives the VI an appearance of the standard OS system dialog box and prevents accessing other LabVIEW windows.

- The **Hilite <Return> Boolean** is not so self-explanatory. This option will highlight (put a black border around) the Boolean control that has been assigned the <Return> or <Enter> key, as discussed in the next section.

Window Size

This property allows you to force the VI to be a minimum pixel size. You can also specify settings for letting front panel objects automatically scale if the front panel is resized.

13.2.5 Execution

The Execution options allow you to set some very advanced execution tweaks:

- **Priority**: Don't mess with this; you shouldn't need it. If you really want to know what this is about, read the LabVIEW manuals.
- **Preferred Execution System**: This is LabVIEW's attempt at multithreading. It allows you to choose one of several threads for your VI to run it. You probably won't need this either.
- **Allow Debugging**, checked by default, gives you access to LabVIEW's debugging tools when the VI is running. You can uncheck it to obtain a slight 1–2% increase in speed and decrease in memory usage of the VI.
- **Run When Opened** does just that.
- **Suspend When Called** is a rarely used debugging option.
- **Reentrant Execution** deserves a little more explanation.

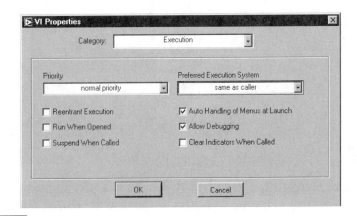

Figure 13.10

Reentrant Execution

Reentrant execution is an important concept. Normally, when you have two or more instances of a call to the same subVI that occur in parallel, LabVIEW assigns the same data storage space to both VIs, which they share. In some cases, though, you need separate data storage blocks for each subVI node. In the block diagram shown in Figure 13.11, we are using the subVI **Running Average** on each data channel.

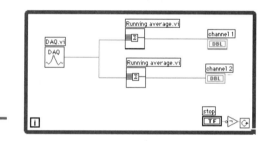

Figure 13.11

Running Average uses *uninitialized* shift registers as memory elements. Remember that uninitialized shift registers keep their last value even after the VI is stopped and run again (see Chapter 6 for a discussion of uninitialized shift registers). If this VI were left in the default mode, it would give unexpected results because the call to either subVI node would contain the shift register data from the last call to the other node (since the subVIs will normally take turns executing). By choosing the "Reentrant execution" option, each subVI node is allocated an independent data storage space, just as if they were two completely different subVIs.

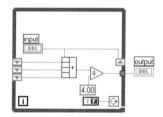

Figure 13.12
Running Average.vi block diagram.

Activity 13-2: Reviewing Recycled Reentrantcy

As an interesting activity, run this VI we just mentioned, **Running Average.vi**, which is on the CD, first to see how it works. Then run the **Reentrant**

VI (also on the CD), first as it is, and then selecting the "Reentrant Execution" option for the **Running Average** VI to compare the differences.

Print Options

These settings allow you to specify options such as headers and margins for printing.

13.3 The VI Server

The VI Server is a powerful feature in LabVIEW that gives you the capability to programmatically access features in LabVIEW such as opening and running VIs, changing the color or value of a front panel object, printing programmatically, etc.

This is quite an advanced topic, so don't feel bad if much of this VI Server introduction confuses you. Almost everyone struggles to understand how to use VI Server functions at first. The good news is that for simple LabVIEW applications, you won't ever need it. Once you get warmed up and comfortable with LabVIEW, though, you'll want to play with all the fun and powerful things you can do via the VI Server—eventually you'll wonder how you ever programmed in LabVIEW without it.

You don't need to understand or use the VI Server for any of the other material in this book; so feel free to gloss over it if you want to move ahead.

Don't let the name confuse you: "VI Server" is much more than just some type of networking server built into LabVIEW (although it is that as well). The VI Server functionality is really a way of introducing *object-oriented programming* into LabVIEW.

For example, with VI Server, you can programmatically:

- Load a VI into memory, run it, and then unload the VI without the need to have it statically linked as a subVI in your block diagram
- Dynamically run a subVI that gets called at runtime, by only knowing its name and connector pane structure (known as calling a VI by reference)

- Change properties of a particular VI, such as the size and position of the front panel window, whether it is editable, etc.
- Make LabVIEW windows move to the front of the screen
- From the block diagram, call a subVI without waiting for it to finish executing (one of the few places where you can get away with not obeying the normal dataflow paradigm!)
- Dynamically change the attributes (properties), such as color and size, of a front panel object

In addition to all this cool stuff you can do, there's more: Everything we just mentioned works with *network transparency*. Network transparency means you can do all of the mentioned manipulation of a VI or of Lab-VIEW itself *on another machine across the network (including the Internet)* in just the same way as if it were on your own machine. This means that, for example, you could have a data acquisition VI running at remote sites, while your local analysis VI gets information from the remote machines without having to write any special networking and without using complex TCP/IP functions.

Figures 13.13 and 13.14 show you an example of how trivial it is to run a VI remotely. (Don't worry about what the functions are if you haven't seen them before; they're from the **Application Control** palette and we'll examine them in detail shortly.)

Figure 13.13
This tells the local machine to run `My Cool VI.vi`.

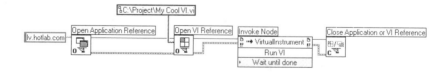

Figure 13.14
This tells the LabVIEW across the Internet at the address `lv.hotlab.com` to run `My Cool VI.vi`. The path to the VI refers to the remote drive.

The VI Server exposes its functionality in LabVIEW through block diagram functions like those in Figures 13.13 and 13.14. But it also allows its functionality to be accessed in Windows from external programs through an ActiveX automation client (e.g., a Visual Basic program or macro) and from a remote LabVIEW VI over TCP/IP. The diagram in Figure 13.15 illustrates this architecture. As you can see in the figure, the VI Server functionality is accessible by block diagram functions, by external ActiveX programs, or by a remote LabVIEW VI over TCP/IP.

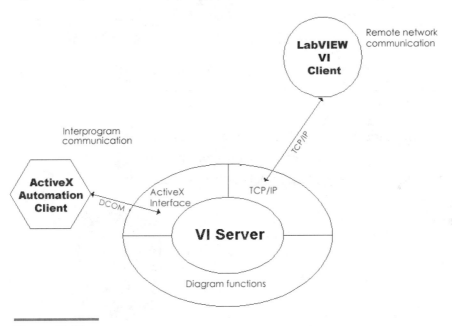

Figure 13.15

13.3.1 Mechanisms for Accessing the VI Server

To enable the VI Server in LabVIEW, go to the **Tools** menu and choose **Options...>VI Server: configuration**.

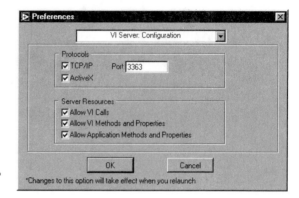

Figure 13.16

The functions for using the VI Server are in the **Application Control** palette (Figure 13.17). There are three classes of "objects" in LabVIEW that the VI Server functions let you manipulate:

- Application Class—the LabVIEW environment itself
- VI Class—a specific VI in memory or on a disk
- Control Class—a front panel control or indicator

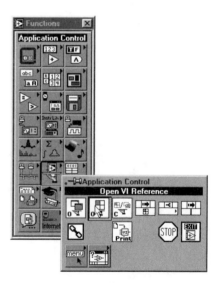

Figure 13.17

The VI Server block diagram functions let you work with all three classes; these are briefly explained in the following table:

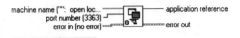

Figure 13.18
Open Application Reference.

Returns a reference to a VI Server application running on the specified **machine name**. If you specify an empty string for **machine name**, then it returns a reference to the local LabVIEW application in which this function is running. If you do specify a **machine name**, then it attempts to establish a TCP connection with a remote VI Server on that machine on the specified port. It returns an **application reference** refnum that can be used by the other VI Server functions.

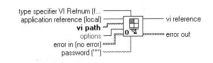

Figure 13.19
Open VI Reference.

Returns a **vi reference** to a VI specified in **vi path** by a name string or a path to the VI's location on disk. If the **application reference** is unwired, it always refers to a VI on the local machine; to refer to a VI on a remote machine, it should have an application reference first to that remote machine.

Figure 13.20
Close Application or VI Reference.

Closes the reference to an Application or VI object.

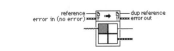

Figure 13.21
Call By Reference Node.

The **Call By Reference Node** is very similar to a subVI node in that you can use either to call a VI. The function connectors adapt to reflect the connector pane that matches that of the VI whose reference is being passed. However, a subVI node is statically linked to a particular VI that you determine when you drop the node on the diagram. With the **Call By Reference Node**, the VI that is called is determined dynamically at runtime by the value of the VI reference wired to the reference input

Figure 13.22
Property Node.

Figure 13.23
Invoke Node.

at the top of the node. The VI the **Call By Reference Node** ultimately calls might be on a different computer.

Sets (writes) or gets (reads) VI and application property information. To select the VI or application class, pop up on the node and select the **Select LabVIEW Class** submenu. To set an application class, select **Application**. To set a VI class, select **Generic VI**, or wire the VI or application refnum to **reference** and the node choices change accordingly.

Invokes a method or action on a VI or Application object. Most methods have parameters associated with them. To select the method, pop up anywhere on the node and select **Methods**. Once you select the method, the associated parameters will appear below the method name. You can set and get the parameter values. Parameters with a white background are required inputs and the parameters with a gray background are recommended inputs.

Now let's work with a simple example of how to use the VI Server.

Activity 13.3: Using the Application Class Properties

In this activity, we will use the VI Server to find out what operating system we are running, where LabVIEW is installed, and the username of who is logged on.

1. With a new blank VI, create the block diagram shown in Figure 13.24 by making an **Application Property Node** (from the **Application Control** palette). You will be programmatically reading the directory LabVIEW is in, the OS name, and the LabVIEW username.

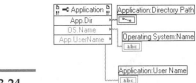

Figure 13.24

*When you are referencing the local Application class, you can omit the **Open Application Reference** function that normally would precede the **Invoke Node** function.*

2. On the front panel, run the VI and observe the results.

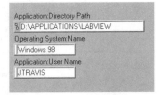

Figure 13.25

3. Save your VI as **AppClassDemo.vi**.

Now let's look at how you can use the VI Server to perform some "magic": allow one VI to change the value of another VI's front panel control, without wiring the two together.

Activity 13-4: Using the VI Class Methods and Properties

In this activity, you will manipulate one VI and toggle one of its controls through the VI Server interface from a "master" VI.

1. Create the simple **Chart.vi** as shown in Figure 13.26.

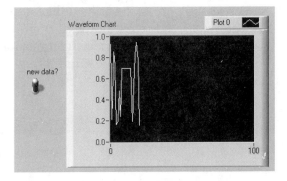

Figure 13.26

2. Now open the VI on the CD in `Ch13.11b`, **Master.vi** and run it. Make sure that **Chart.vi** is still open and you are not running it.

3. Run **Master.vi**. It will do the following:

   i.   Open the front panel of **Chart.vi**

   ii. Center the front panel on the screen

   iii. Run the **Chart.vi**

   iv. Allows the "new data?" control to be changed while running

Study the block diagram to gain insight into how it accomplishes this.

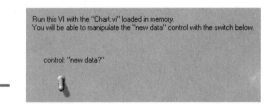

Figure 13.27

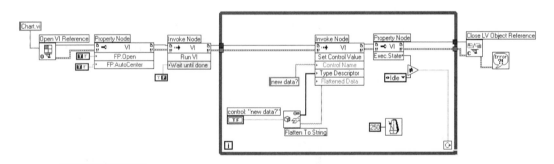

Figure 13.28

As we mentioned at the beginning of this section, don't worry too much if the VI Server seems perplexing—it's unlike anything you've encountered in LabVIEW before, and it breaks all the normal rules of how LabVIEW VIs interact with each other. Once you get comfortable with LabVIEW programming, you can always return to explore the numerous VI Server examples built into LabVIEW.

13.3.2 Keyboard Navigation

If you are one of those people who think mice are unfriendly (or your end users think so), there's good news—you can set up your VI to allow users to "navigate" the controls with the <tab> key. Without any special setup, you can always use <tab> to pick the control that will receive input (you can't pick indicators this way, since indicators don't accept inputs). A "selected" control—called the *key focus*—has a rectangular border enclosing it. Once a control is the key focus, you can use the appropriate keys to enter its value. The following tips may be useful:

- If you directly type the value into the selected control, you must hit the <enter> key when you are done to make your entry valid.

- For numerical and ring controls, you can also use the arrow keys to advance to the desired number. Pressing <shift> with the <up> or <down> arrow key advances the value faster. You can set the minimum increment from the **Data Range...** option in the pop-up menu of the control.

- For Boolean controls, the <return> key toggles the Boolean value.

- Tabbing from control to control normally follows the order in which you created the controls

For VIs with several controls, you may wish to set your own tabbing navigation order, that is, determine once a control is selected, which control will be selected next when the <tab> key is pressed. This sequence is known as *panel order* in LabVIEW. To change the panel order, choose **Set Tabbing Order...** from the **Edit** menu. This works in the same way as the **Cluster Order...** option, discussed in Chapter 7. The front panel in Figure 13.29 shows what it looks like when **Tabbing Order** is selected.

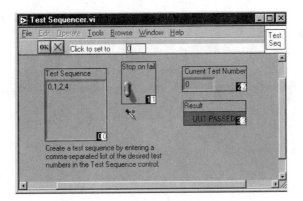

Figure 13.29

On your front panel, all controls will be boxed in with two numbers in the lowercase corner of each box. The number in the white background represents the previous panel order; the number in the black background represents the new panel order you are assigning to your diagram. To create the new panel order, click successively on each control in the desired order, or type the sequence number in the menu bar, and then click the OK button. To cancel all changes, click the X button.

You can also assign certain "special" keys to a control. Choosing the **Advanced>>Key Navigation...** option from the pop-up menu of a control brings up a dialog box like that shown in Figure 13.30.

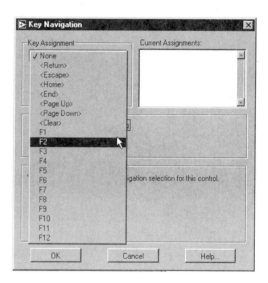

Figure 13.30

You can assign function keys (<F1>, <F2>, etc.) to a control, as well as function keys with modifiers (such as the <ctrl> key). Pressing the selected key will set the key focus on that control without having to "tab" to it. You will find Key Navigation useful if you have many controls but have a few you use more often.

Finally, you can also programmatically set or disable a control's key focus. The **Key Focus** property of a control is a Boolean that, when true, means the control is "selected" and ready for keyboard input.

Figure 13.31

Activity 13-5: A Login VI

Build a **Login** VI with a nice front panel as shown in Figure 13.32. This VI should make the Key Focus first appear in the <u>LOGIN</u> box. The VI should detect when the user has pressed <return> or <enter> and move the Key Focus to the <u>PASSWORD</u> box. The password characters should not be seen. Finally, the VI should appear in the middle of the screen with the same appearance as a dialog box from the OS.

Figure 13.32

1. *You will need to use the Key Focus attribute.*
2. *Pop up on the UNDERLINE_PASSWORD control and select the "Password" option for the string.*
3. *Pop up on the UNDERLINE_LOGIN and UNDERLINE_PASSWORD control and select the "Limit to single line" option.*
4. *How will this VI know when a user has finished the input string to the UNDERLINE_LOGIN? You will likely need a local variable to periodically check whether the input contains a carriage return character yet.*

The solution is found on the CD.

13.4 Radices and Units

A useful feature of LabVIEW numeric types is that you can manipulate them as more than just pure numbers. LabVIEW numeric controls and indicators can optionally have a representation (called a *radix*) and/or an associated *unit*. Normally, you represent numbers in decimal format. But for some applications, you may need to view numbers in a binary or hexadecimal representation. Similarly, you may want to treat certain numerical variables as having an associated unit (such as feet, calories, or degrees Fahrenheit)— especially if you plan to do unit conversions.

13.4.1 Radices

LabVIEW's numeric displays are very flexible. You can choose to display numbers in decimal, hexadecimal, octal, or binary format by selecting **Visible Items>>Radix** from the numeric's pop-up menu. Click with the Operating tool (do *not* pop up) on the tiny letter that appears, and you'll pull down a radix selection menu.

Figure 13.33

If your numeric does not have integer representation, all options except Decimal are disabled.

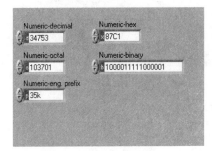

Figure 13.34

A user who is working with digital I/O, for example, might find it very useful to represent a numeric control in binary, since it would be easy to see the one-to-one correspondence of each digit to each digital line.

13.4.2 Units

Any floating-point numeric control can have physical units, such as meters or kilometers/second, associated with it. You can access the unit label by selecting **Visible Items > Unit Label** from an object's pop-up menu.

Once the unit label is displayed, you can enter a unit using standard abbreviations such as m for meters, ft for feet, etc. If you enter an invalid unit, LabVIEW flags it with a ? in the label. If you want to know which units LabVIEW recognizes, enter a simple unit such as m, and then pop up on the unit label and select **Unit....** A dialog box will pop up showing you information about LabVIEW's units.

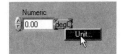

Figure 13.35

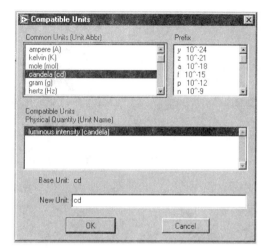

Figure 13.36

Any numeric value with an associated unit is restricted to a floating-point data type.

LabVIEW preserves the associated unit throughout any operations performed on the data. A wire connected to a source with a unit can only connect to a destination with a compatible unit, as shown in Figure 13.37.

Figure 13.37

When we say "compatible," we mean that the destination unit has to make sense from a physical or mathematical point of view. The destination unit does not have to belong to the same classification system (e.g., SI versus English units), as shown in Figure 13.38. Thus, you can transparently convert compatible units from different systems in LabVIEW.

You cannot connect signals with incompatible units, as shown in the following illustration. If you try, the error window (discussed soon) will contain a `Wire: unit conflict` error.

(A)   (B)

Figure 13.38

Some functions are ambiguous with respect to units and cannot be used if a unit is assigned to the data you are trying to operate on. For example, if you are using distance units, the Add One function cannot tell whether to add one meter, one kilometer, or one foot.

Finally, you need to be wary of doing arithmetic that combines numbers with and without units. For example, you can multiply a unitless constant times a number with a unit, but you cannot add a number with a unit to a number without a unit. LabVIEW allows you to "add" a unit to a unitless number using the **Convert Unit** function (from the **Conversion** palette in the **Numeric** palette), shown in Figure 13.39.

Figure 13.39
Convert Unit: Converts a physical number (number with units) to a pure number (number without units) or vice versa. If input is a pure number, output takes on specified units. If input is a physical number, output is input measured in specified units.

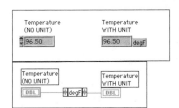

Figure 13.40

To specify the unit, just type the unit inside the middle box of the terminal, or as with numeric types, pop up on the terminal and select **Unit...** for a list of valid units. The **Convert Unit** function also allows you to "delete" a unit from a number.

13.5 Automatically Creating a SubVI from a Section of the Block Diagram

One key to writing your program right the first time with the least amount of bugs is *modularity*. The modules in LabVIEW are, of course, subVIs. When you're writing a large application, it's essential to break up the tasks and assign them to subVIs. The **Edit>Create SubVI** menu option makes converting sections of your block diagram into subVIs much simpler. What if you're working on some code and decide you should make part of it a subVI? No problem! Select the section of a VI's block diagram and choose **Edit>Create SubVI**. LabVIEW converts your selection into a subVI and automatically creates controls and indicators for the new subVI. The subVI replaces the selected portion of the block diagram in your existing VI, and LabVIEW automatically wires the subVI to the existing wires. Choosing this feature is also an excellent option to use, for example, if you need to repeat part of your block diagram in another VI.

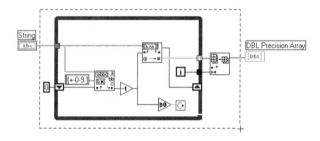

Figure 13.41

Figure 13.42

Despite how easy it is to use, let's look at some rules for using the **Create SubVI** option.

Because converting some block diagrams into subVIs would change the behavior of your program, there are some instances where you cannot use this feature. These potential problems are listed next.

- You cannot convert selections that would create a subVI with more than 28 inputs and outputs, because that is the maximum number of inputs and outputs on a connector pane. We'd feel sorry for the user who would have to wire even 15 inputs! In a case like this, select a smaller section or group data into arrays and clusters before selecting a region of the block diagram to convert.

- You cannot convert selections in which items inside and outside a structure are selected, yet the structure itself has not been selected.

- When your selection includes local variables but does not include the corresponding control, one local variable for the control remains in the calling VI and passes the data to or from the subVI. Within the subVI, the first local variable for a control becomes a read or write for the control and subsequent local variables refer to the control in the subVI. Yikes! What does that mean? The bottom line is: Be *very* careful when you create a subVI from the selection if it includes locals.

- When your selection includes local variables or front panel terminals inside a loop, the value that they are measuring may be changed elsewhere on the block diagram as the loop runs. Thus, when you convert the loop into a subVI, there is a possibility that the functionality of the selected code changes. If you have selected some but not all of the local variables or front panel terminals, LabVIEW displays a warning that allows you to choose between continuing and canceling the operation.

The **Create SubVI** can be a really handy tool in cases where you start with what you thought was a small project (hah!) and you soon realize your program is going to be more complex than you originally thought. However, don't just squeeze a bunch of diagram mess into a subVI to increase your workspace—subVIs should always have a clear, well-defined task. When considering whether to make a part of your diagram a subVI, you might ask yourself, "Could I ever use this piece of code or function somewhere else?" We'll come back to subVIs and modular programming again in Chapter 16, where we look at good programming techniques.

13.6 A Few More Utilities in LabVIEW

13.6.1 The Hierarchy Window

Generally, you will find the Hierarchy Window useful when you are working on a VI of fair to high complexity—such as one that contains 10 or more subVIs. The Hierarchy Window can help keep track of where subVIs are called and who calls them.

With a VI's front panel open, select **Browser>>Show VI Hierarchy** to bring up the Hierarchy Window for that VI. The Hierarchy Window displays a graphical representation of the calling hierarchy for all VIs in memory, including type definitions and globals.

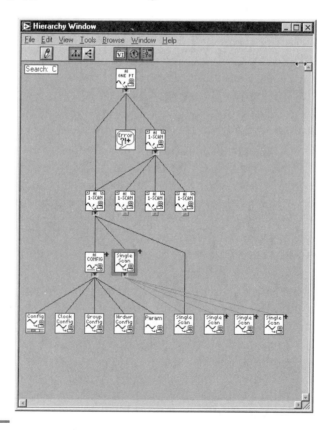

Figure 13.43

The buttons in the Hierarchy Window Toolbar let you configure a number of aspects of the display. For example, the layout can be displayed horizontally or vertically; you can hide or show as many levels of hierarchy as you wish; and you can include or exclude globals or type definitions. A nice feature in this window is the one that allows you to double-click on a subVI icon to open its front panel.

13.6.2 Searching for Objects in the Virtual Haystack

LabVIEW has a good search engine so that you can quickly find text, objects, or VIs in your project. LabVIEW's **Find...** function, under the **Edit** menu, can quickly locate any LabVIEW function, subVI, global variable, attribute node, front panel terminal, or text in your block diagram. You can limit your search scope to a single VI or a set of VIs, or you can include all VIs in memory.

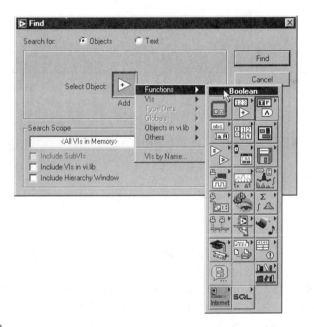

Figure 13.44

You can choose to search for objects or text by clicking in one of the **Search for** checkboxes. If you select objects, click on the **Select Object** button to access a pop-up menu that allows you to choose the type of object you want to search for. If you select text, you can type in the text and click on the **More Options...** button to further limit the scope of the search to sections of VIs and object labels.

If LabVIEW only finds a single match, it opens the corresponding VI and highlights the match. If LabVIEW finds multiple matches, it displays a **Search Results** dialog box, as shown in Figure 13.45.

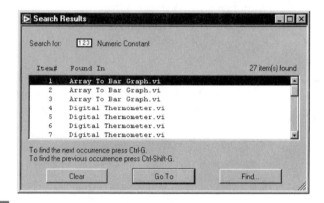

Figure 13.45

When multiple items are found, you can double-click on any the items displayed in the **Search Results** window to have LabVIEW show you where that item resides.

More Tools

Under the **Tools** menu, you will find an assortment of other LabVIEW utilities, such as a Web page builder, a VI Profiler (which measures the time and memory a VI consumes), and more. Depending on the version of LabVIEW you have installed (evaluation, basic, professional, full development) and any add-on toolkits you may have (Internet Toolkit, SQL Toolkit, etc.), you will see the corresponding utilities in this menu.

13.7 Wrap It Up!

This chapter covered a potpourri of interesting and useful LabVIEW features. We covered aspects such as **Options**, **VI Properties** and **SubVI Node Setup** options, the VI Server, key focus, radices, units, creating subVIs from selections, the VI Hierarchy, and the Find function.

The **Options...** command allows you to control myriad options under several categories. These options give you control over the LabVIEW environment and let you personalize aspects such as fonts, colors, printing types, paths, etc.

VI Properties and **SubVI Node Setup** contain useful selections to customize the aspects of a particular VI at the top level or as a subVI. With **VI Properties**, you can hide or show the Toolbar buttons, center the window, control execution options, etc. The **SubVI Node Setup** allows you to "pop up" the window of a subVI as its called and close it afterwards.

The VI Server was introduced as a powerful object-oriented approach to controlling VIs and LabVIEW.

Key focus is a Boolean attribute of a control that decides whether the control is "selected" and ready for input from the keyboard. You can set the key focus programmatically, or simply tab through all the controls. You can also assign special keys, such as function keys to toggle a particular Boolean control.

Radices and units are special built-in features of LabVIEW numerical types. A numerical type whose radix is shown can be changed to a decimal, binary, octal, or hexadecimal representation. The optional built-in units on numeric types allow you to perform mathematical operations on numbers with units and provide automatic conversion for different unit systems.

The modularity of LabVIEW is highlighted by the ability to select any part of the block diagram, and under certain conditions, turn it into a subVI with the appropriate inputs and outputs.

LabVIEW provides some advanced programming tools, such as a Hierarchy Window, a **Find** utility, and more.

OVERVIEW

In this chapter, you'll learn about the different ways you can connect LabVIEW VIs to the Internet, your network, and other programs. We'll see how you can easily use LabVIEW's built-in Web server to publish front panels to the Web. You'll become familiar with the DataSocket protocol and how it can be used to share data among LabVIEW VIs across the network. With the VI Server, you can begin to programmatically control many features of VIs and of LabVIEW in an object-oriented manner. ActiveX connectivity (Windows) and AppleEvents (MacOS) provide you with a way to have LabVIEW interact with external programs. Finally, we'll see where LabVIEW fits in the big picture of the "enterprise."

GOALS

- Become familiar with basic network and Internet protocols (URLs, TCP/IP, HTTP, etc.)
- Know how to use the LabVIEW Web Server to publish VIs to the Web
- Use DataSocket to publish and subscribe to data sources
- Discover the ActiveX functionality in LabVIEW
- Become familiar with LabVIEW's lower-level networking capabilities: TCP and UDP VIs
- Understand the role of databases, reports, Web publishing, and VIs

KEY TERMS

- Internet
- Client–server
- URL
- HTTP
- Web server
- DataSocket
- dstp
- ActiveX
- AppleEvents
- TCP/IP
- UDP
- Enterprise
- Databases

Connectivity in LabVIEW

14

14.1 LabVIEW, Networking, and the Internet

One of the main purposes of *virtual instrumentation*, as we have been showing in this book, is to create more powerful, flexible, and cost-effective instrumentation systems built around a PC using software as the engine and interface. A virtual instrument can easily export and share its data and information with other software applications since they often reside on the same computer.

When National Instruments introduced LabVIEW over a decade ago, it created a unique software tool for creating virtual instruments. The appeal of LabVIEW is largely tied to its graphical programming nature, since it is very easy to prototype and develop an application in a fraction of the time it would take to produce the same in a language like C++. There is an interesting parallel between the success and popularity of LabVIEW and the popularity of the Web: In both cases, it was not so much the underlying technology that was so innovative, but rather the well-designed graphical interface that made it accessible. After all, almost everything you could do in LabVIEW could be done in C or assembly code years before LabVIEW was popular. The Internet's origin goes back to the 1960s. What it took to create these revolutionary tools, though, was the development of an intuitive, human-friendly interface. For LabVIEW, that meant programming by wiring graphical objects together, like building a breadboard circuit. For the Web, it meant a "Web browser" application that involves little more than just pointing and clicking on images or words that were of interest and were hyperlinked to other places on the Web.

So you might say that using LabVIEW to create Internet-enabled applications brings some of the best user interface designs together. The possibilities are exciting for creating easy-to-use and intuitive networked applications that take virtual instrumentation to another level.

In this chapter, you'll learn about some of the tools LabVIEW has that allow you to publish VIs to the Web, share data across a network, communicate with other applications, and more. If you want to expand deeper into the topic of Internet-enabled virtual instrumentation, check out another book I have written, *Internet Applications in LabVIEW* (Prentice Hall, 2000). You can find more info at my website, http://jeffreytravis.com/books.

14.1.1 The Internet and the Client–Server Model

Before we get into some practical examples, let's look briefly at some of the core technologies that make it possible for Internet-enabling virtual bio-instrumentation systems.

The very term *Internet* is a truncated form of *interconnected network*. One basic definition of network is "two or more electronic devices linked in some way to permit the exchange of information." In information technology a network is a series of points or nodes interconnected by communication paths. The Internet consists of thousands of interconnected networks.

The Web is a client–server system, with your Web browser being the client and any given website's server being the server.

In the client–server model, common in networked applications, one set of processes (called clients) requests services from another set of processes (servers). If you are wanting to monitor a LabVIEW application from your laptop, you are requesting that the information stored on the server be viewable from a client—in this case, your laptop. As is the case in this scenario, a server usually waits for a client to initiate a connection. In many cases a server can service multiple clients at a time.

You are already familiar with the way the Web functions as a client–server model. You likely are also familiar with e-mail and ftp (file transfer protocol) services that use the client–server model to transport information.

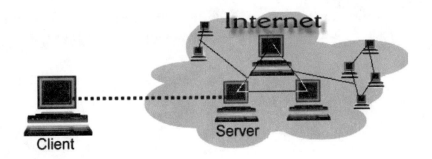

Figure 14.1

So how does that information get transferred from the server to the client?

The simple answer is that information is accessed, broken down into smaller pieces, sent electronically along various network pathways, re-ordered, and reassembled. That, however, is accomplished through a slightly more complicated series of *communication protocols*: a set of rules or a "language" that allows two computers or devices to communicate with each other. Protocols exist at several *layers*, with high-level protocols (such as the Web's *HTTP* protocol) being built on lower-level protocols (*HTTP* relies on *TCP/IP*), which in turn may be relying on lower protocols (*TCP/IP* needs a physical layer protocol, such as *Ethernet*), in chained fashion. The most common communication protocols that you will encounter in building networked VIs are TCP/IP, UDP, HTTP, DSTP (DataSocket Transfer Protocol), and WAP.

14.1.2 Choosing a Technical Solution with LabVIEW

When designing an Internet-enabled instrumentation system, you will have a wide array of choices and tools at your disposal when you use LabVIEW as the primary data acquisition and analysis software. LabVIEW has a number of built-in capabilities for Internet connectivity, including:

- TCP/IP and UDP functions
- A built-in Web server that can create front panel images on the fly
- LabVIEW's VI Server: a powerful framework for VIs or ActiveX applications to communicate over a network
- The DataSocket protocol for sharing live data over a LAN or the Internet

In addition, with the use of add-on tools and other software development packages, LabVIEW can also be used with additional technologies such as:

- Java applets to remotely control or monitor a VI
- ActiveX controls
- CGI support in the G Web server
- E-mail, ftp, and telnet

14.2 An Overview of How the Web Works

Probably the most common application of networked VIs is the ability to either monitor or control VIs through a Web browser. Before we see how to do this, though, it's helpful to first examine how the Web works.

14.2.1 Anatomy of a URL

One of the most important things to understand when working on the Web is the format for URLs, or *Uniform Resource Locators*. A URL basically lets you specify where a document is on the Web, by providing the Internet address and path of the document. You see URLs all the time whenever you type an address into your Web browser; for example, http://www.ni.com/labview/index.html is an example of a URL.

The basic syntax for URLs of most Web documents is

```
http://host/path
```

where `http` specifies that we are using the HyperText Transfer Protocol (HTTP), `host` is the Internet address of the Website (e.g., *www.ni.com*), and `path` (e.g., *labview/index.html*) points to the document requested on the server.

Most Web browsers let you omit parts of a URL and use default values; for example, if you leave out the "`http://`" at the beginning of the URL, the browser still uses HTTP to search for the document. In addition, if the `path` points to a directory instead of a file, the Website usually has a default document inside the specified directory that will be displayed. The Web server, and not the browser, controls what the name of the default document should be—usually it's something like `index.html` or `default.htm`.

Although HTTP is usually the scheme used for retrieving documents over the Web, there are other possible schemes for accessing data. The more generalized syntax for URLs is

```
scheme://host[:port]/path[/extra-path-info][?query-info]
```

Thevarious elements are defined as follows:

• *scheme* is the protocol used to connect to the host. For websites, the scheme is `http`; for ftp sites, it is `ftp`.

Figure 14.2
Protocols used in a URL.

The schemes that are often available include:

Scheme	Meaning
http://	HyperText Transfer Protocol; for websites
ftp://	File Transfer Protocol; for ftp sites (uploading and downloading files)
gopher://	Gopher sites; an older, less used type of server that indexes documents
news://	News servers
file:///	A scheme that does not use TCP/IP but points to a document through your local file system
telnet://	Not really a URL (since it doesn't point to a document) but can be used to initiate a telnet session with the specified host.
dstp://	DataSocket Transfer Protocol; a proprietary protocol from National Instruments for sending instrumentation data over a network

A nonhypertext URL might look like

```
news://comp.lang.labview
```

Looking up this URL would launch your news reader client and go to the comp.lang.labview newsgroup.

- *host* is the Internet address of the host.
- *port* is an optional port number for the requested service. Normally the port can be omitted if the service is running on its default port. For example, Web servers usually run on port 80, so the URL

```
http://www.travis.to:80
```

is the same as

```
http://www.travis.to
```

However, if you have another Web server running, say, on port 8172, the URL would need to be specified as

```
http://www.travis.to:8172
```

- *path* is the path to the document or data source.
- *extra-path-info* is optional information sometimes used by CGI programs.
- *query-info* is optional parameters used by CGI programs; it always has the "?" character before it.

Most Web browsers such as Netscape Communicator or Internet Explorer allow you to retrieve documents within the browser using the http, ftp, gopher, and file schemes. Often you can configure the browser to launch an external application to handle a telnet or news URL.

14.2.2 URL Encoding

When you build applications that dynamically create a URL, it's important to understand the syntax rules for URLs, collectively called *URL encoding*.

Besides the syntax to a URL we just saw, certain special characters in a URL are not allowed to be typed as normal ASCII characters. These include spaces, question marks, the @ symbol, etc., because all of these have a special meaning within a URL. These characters must instead be represented as "%" followed by the two-digit hexadecimal ASCII value of the character.

For example, if you are accessing a file with a space in it, such as "Come&Get It.html", the URL encoding would look like

`http://myserver/Come%26Get%20It.html`

The %26 replaces the "&" character and the "%20" replaces the space (0 × 26 is the hex value for "&" and 0 × 20 is the hex value for a space).

> As a special case, the space " " character can also be encoded as a plus sign ("+") in addition to its hexadecimal notation. So the document "Hello World.html" could be encoded as `http://myserver/Hello%20World.html` or `http://myserver/ Hello+World.html`.

14.2.3 Web Browsers, Web Servers, and HTTP

The World Wide Web (WWW) essentially works with a client–server architecture, as shown in Figure 14.3.

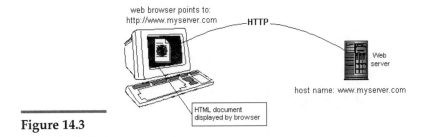

Figure 14.3

Web browsers are the clients, Web servers are the servers, and HTTP is the communication protocol among them. Let's take a brief look at each of these pieces that make up the Web.

Web Browsers

A Web browser is a client application that implements the HTTP protocol, can connect to Web servers, and can retrieve HTML documents and images. The modern Web browsers do much more than this: they usually have e-mail clients as part of the package, and they can handle content such as audio, video, Java applets, and ActiveX controls (more about these later). The two most popular Web browsers are *Netscape Navigator* (or *Netscape Communicator*, depending on the version of the application) and *Microsoft Internet Explorer*. Virtually all PCs have a Web browser pre-installed.

14.3 Publishing and Controlling VIs on the Web

Often you'll want other people to be able to access your data or your VIs through a Web browser. The main point to know is whether you want people to just monitor (read-only) or control (send data back to the VI) from the browser.

- *To monitor (not control) a VI, you can use LabVIEW's built-in Web server, explained shortly.*
- *To control a VI over the Web, you can use a variety of methods. One of the easiest and most painless ways to do it is to use an open-source toolkit called LabVNC. Lab-VNC is free and included on the CD with this book; you can also download the latest version at my website, http://jeffreytravis.com.* *

Let's look at the LabVIEW Web server in a little more detail. With the Lab-VIEW Web server, you can *dynamically* create Web pages with images of a VI's front panel on the fly, without the need for any special coding in your block diagram.

* The capability to control LabVIEW over the Web is forthcoming in a future version of LabVIEW, which may already be available by the time you read this book (at press time, Lab-VIEW 6.0 only supported remote monitoring).

14.3.1 Configuring LabVIEW's Built-in Web Server

Don't confuse the VI Server with the Web Server in LabVIEW. They're unrelated and independent of each other. The Web server allows a remote Web browser to see a VI image in the browser. The VI Server lets local and remote VIs call methods and properties of your VIs.

To enable the LabVIEW Web server, do the following:

1. To avoid system conflicts, make sure no other Web server (e.g., Microsoft Personal Web Server or the Internet Toolkit's server) is running on the same port before you enable the LabVIEW Web Server.

2. Go to **Tools>>Options>>Web Server:Configuration** (as shown in Figure 14.4).

3. Check the box "Enable Web Server."

4. If you desire, change the root directory. This is the directory on your hard drive where your Web documents will reside.

5. Click OK.

6. Quit LabVIEW and then restart LabVIEW (you need to do this to enable the Web Server).

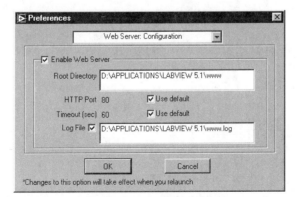

Figure 14.4

You can also control LabVIEW Web Server properties through the VI Server interface, using the Property Node for the Application class. This means that you can dynamically and programmatically turn the server on or off, set permissions for the access list, etc. With the VI Server interface, you don't have to relaunch LabVIEW.

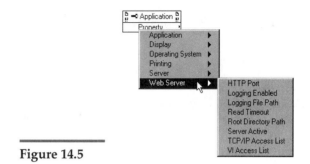

Figure 14.5

14.3.2 Publishing to HTML with LabVIEW's Web Server

With the LabVIEW Web Server, there is nothing to program on the block diagram. It dynamically creates images of a front panel in response to a Web browser request. The VI to be displayed in a Web browser must already be loaded into memory in LabVIEW to be served by the LabVIEW Web server. The LabVIEW Web server can provide either a static ("snapshot") image of a VI's front panel or a dynamic animated image ("monitor") of the VI's front panel.

The following activity shows how easy it is to view the **Temperature System Demo.vi** through a Web browser.

Activity 14-1: Using LabVIEW's Built-in Web Server

1. Make sure you've turned on the built-in Web Server, as described earlier.

2. Open the VI Temperature System Demo.vi (from the LabVIEW examples, in `/examples/apps/tempsys.llb`).

3. Open your Web browser.

4. To see a static image of the VI, type in your Web browser address window (on the same machine)

   `http://127.0.0.1/.snap?Temperature+System+Demo.vi`

5. To see an autorefreshing image, you will need to use a browser that supports "push" features on images. With Netscape's browser, type this URL:

   `http://127.0.0.1/.monitor?Temperature+System+Demo.vi`

You will see the image animate features like the graph. The push feature only works well on Netscape, and not on Internet Explorer. On Internet Explorer, the whole page will autorefresh continuously, instead of displaying pixel animation.

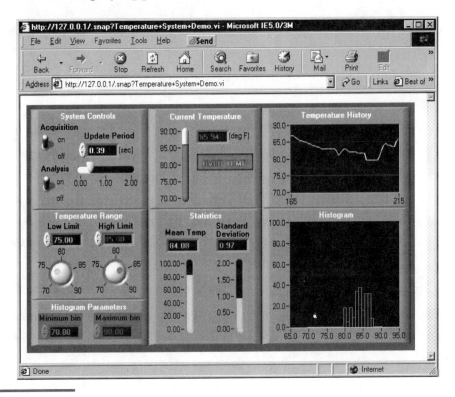

Figure 14.6

Let's examine the URL from the previous activity a little more closely:

127.0.0.1

This is just the special IP address for the local machine (if you were viewing the VI over the Web from a remote host, you would have to type in the real IP address, of course).

.snap?

The .snap part of the URL is a special command that tells the LabVIEW

	Web Server to take a snapshot of the VI that will follow the ? character.
`.monitor?`	The other option, `.monitor`, is a special command that tells the Lab-VIEW Web Server to monitor and provide an animated image of the VI that will follow the ? character.
`Temperature+System+Demo.vi`	This is just the *URL encoding* for "Temperature System Demo.vi." In any URL, spaces must be replaced with the "+" sign or "%20", and any special characters must also be re-placed (e.g., and / must use the ap-propriate URL encoding).

Note that these dynamic documents returned by the LabVIEW Web server are simply front panel images; they are not HTML documents. How-ever, it is very straightforward to create your own HTML documents that embed dynamic images of a VI's front panel; you do this by simply specify-ing the URL as the image source in an < IMG> tag.

The LabVIEW Web server only lets you monitor a LabVIEW VI; you cannot control the VI. To control a VI, use a third-party tool like LabVNC (on the CD).

In conjunction with the LabVIEW Web Server, you can use the simple **Web Publishing Tool** (from the **Tools** menu). This little utility builds a sim-ple HTML file for you where you can add text around your VI image.

In addition to the built-in Web server in LabVIEW, you can purchase the Internet Toolkit for LabVIEW from National Instruments. This tool set gives you an enhanced Web server, with all the same features as the built-in Web server, adding support for secure directories (user/password/groups), CGI capability, e-mail, ftp, and telnet VIs.

LabVIEW 6i	**+ Internet Toolkit**
http (Web) server **dstp** datasocket	**ftp** **smtp** (e-mail) **telnet** **http** with added capabilities, such as CGI and password-protected directories

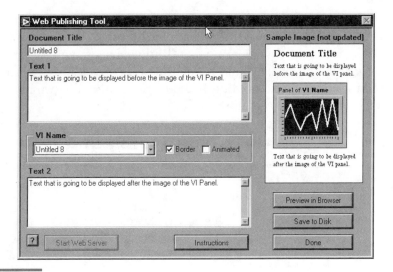

Figure 14.7

14.4 Sharing Data over the Network: DataSocket

Sometimes you may want to share data among several computers on the network, perhaps all of them running LabVIEW. While the Web publishing we just saw is useful for a graphical glimpse of the VI, you can't share data very effectively that way.

DataSocket is an interface provided in LabVIEW that lets you publish (write) and subscribe (read) LabVIEW-formatted data over the network. In addition to LabVIEW, the DataSocket protocol is supported for other software environments, like C++, Visual Basic, and Java. For more information about these, see http://ni.com/datasocket.

Essentially, DataSocket is a technology that allows you to send and receive data over a network from a variety of software platforms (including LabVIEW) without worrying about the low-level implementation details. With DataSocket, for example, you can send formatted data back and forth between two LabVIEW machines on a network, or between a LabVIEW and a LabWindows program, or even among several Web browser clients and LabVIEW.

Because one of the key components of DataSocket is the *DataSocket Server*, a small application that is external to the programming environment, your programs don't have to worry about how to manage TCP/IP connections, nor does performance in your specific application vary with the number of clients connected. In addition, DataSocket can handle several data types, including integers, floats, strings, and Booleans, as well as arrays of these. By letting DataSocket handle typecasting and conversions internally, you don't have to worry about sending header information or formatting your data in a special way to be transmitted over the network.

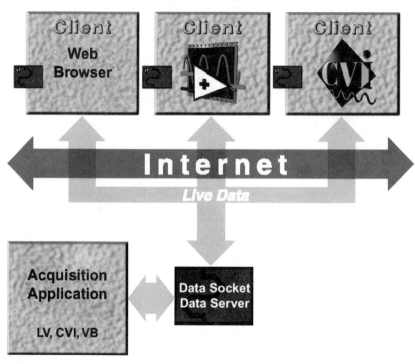

Figure 14.8

DataSocket has two main "pieces" that work together:

1. The *DataSocket Server*
2. The *DataSocket API* (Application Programming Interface) for clients

The DataSocket Server is a standalone application (at time of press, available for Windows only) that runs on a computer and will handle client connections. The client connections may write data to the server (known as DataSocket *publishers*) or read data (DataSocket *subscribers*) from any one of the publishers. The DataSocket server automatically handles the underlying network connections and data packet transmission, making it transparent to the clients. We will examine the DataSocket server in more detail shortly.

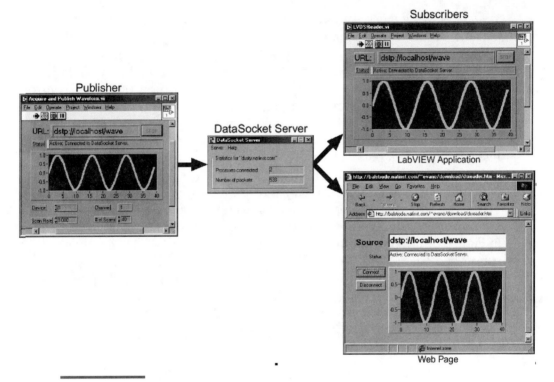

Figure 14.9

The DataSocket server is automatically installed with LabVIEW for Windows. Because the DataSocket server is implemented with an ActiveX interface, the DataSocket server is only available on the Windows platform (so, while unfortunately MacOS and Unix users can't currently run a DataSocket *server* on their machines, it is possible to create DataSocket *clients* on other platforms).

To start the DataSocket server, simply go to **Start>>Program Files>>National Instruments DataSocket>>DataSocket Server**.

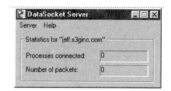

Figure 14.10

DataSocket is very easy to use: You can use DataSocket connections directly from the front panel without any block diagram programming!

Each front panel control or indicator can publish or subscribe to data through its own DataSocket connection. Because front panel DataSocket connections publish only the data, not a graphic of the front panel control, the VIs that subscribe through a DataSocket connection can perform their own operations on the data.

You can use front panel DataSocket connections to publish or subscribe to live data in a front panel object. When you share the data of a front panel object with other users, you publish data. When users retrieve the published data and view it on their front panel, users subscribe to the data.

To set a front panel's DataSocket connection, pop up on it and select **Data Operation>>DataSocket Connection....** You will get the dialog box shown in Figure 14.11. You can then choose to have this item's data be published, or subscribed to, or both. You will always need to specify a valid DataSocket server (which could be running on your own machine or some other one on the network).

Figure 14.11

Let's see how easy it is to share data among LabVIEW VIs in the following activity.

Activity 14-2: DataSocket from the Front Panel

In this exercise, you will see how easy it is to share data across the network using DataSocket. You won't even need to touch the block diagram!

This exercise works best if you have access to two networked computers, both running LabVIEW. If you don't, that's OK, you can just treat computer "A" and computer "B" as the same computer. You should also know the IP address or machine names of your computers (on Windows, you can get this information by typing "ipconfig" at the command prompt; it will display your IP address).

If you are only using one computer, you can use the local IP address of "localhost." This "special" IP address always points to the local computer.

1. Computer A should be running the DataSocket Server. Turn on the DataSocket Server by going to the **Program Files>>National Instruments DataSocket>>DataSocket Server**.

2. On computer A, create a VI with a knob on the front panel and nothing else. Pop up on the knob and select **Data Operations>> DataSocket Connection...**. Type

   ```
   dstp://<computer-ip-address>/knob
   ```

   in the URL, and select "Publish." Then click "Attach."

3. On computer B, create a VI with a chart on the front panel and nothing else. Popup on the chart and select **Data Operations>> DataSocket Connection...**. Type

   ```
   dstp://<computerA-ip-address>/knob
   ```

   in the URL, and select "Subscribe." Then click "Attach."

Continuous Run Button

4. Now, on both computers A and B, hit the "continuous run" arrow on your VIs. If everything works correctly, the small rectangular LED on each control should turn green (if it's red, check your URL or make sure the DataSocket server is running).

On computer A, tweak and twiddle the knob. Notice how the chart value changes correspondingly. Notice how there is *no* block diagram code!

Figure 4.12
Computer A front panel.

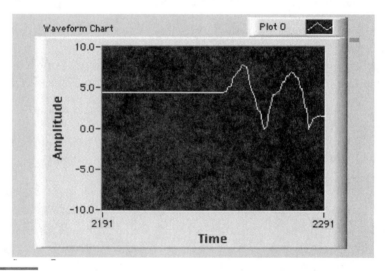

Figure 14.13
Computer B front panel.

14.4.1 DataSocket VIs

Although we just saw how you can do "DataSocket voodoo" on the front panel, it generally is a good idea to explicitly code any network functions on your block diagram. To do this, you can use the simple **DataSocket Read** and **DataSocket Write** VIs. The **DataSocket** function palette is part of Lab-VIEW; you can find it in **Functions>>Communication>>DataSocket** (Figure 14.14).

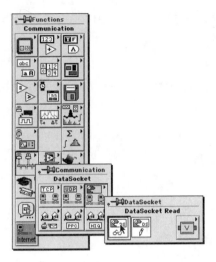

Figure 14.14

The DataSocket palette contains the **DataSocket Read**, **DataSocket Write**, and the **Variant** subpalette of functions.

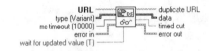

Figure 14.15
DataSocket Read.

The **URL** identifies the data source to read.

The **type (Variant)** specifies the type of data to be read and defines the type of the data ouput terminal. The default type is a Variant, which can be any type.

The **ms timeout** specifies how long to wait for a value update. This time is ignored if **wait for updated value** is false and an initial value has arrived.

Figure 14.16
DataSocket Write.

The **URL** identifies the data source to write to (normally a dstp server).

The **data** input is a polymorphic input that takes the data to be written.

The Variant subpalette of the DataSocket palette contains functions you're not likely to use frequently, unless you are trying to communicate with another software system that uses the **Variant** data type (e.g., Visual Basic with Componentworks).

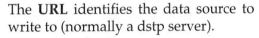

The next activity uses some of these functions to build a simple data publisher and subscriber.

Activity 14-3: Create a Simple Data Publisher and Subscriber with DataSocket

Like the previous activity, you may benefit more from trying this on two different computers; however, it should also work fine if both VIs are on the same computer.

Using the DataSocket VIs, you will create: 1. a data publisher VI that will write a waveform to a DataSocket server and 2. a subscriber VI that will pick what type of waveform it wants to read and then display the waveform read from the DataSocket server.

1. To create the data publisher, create a VI like the one shown in Figures 14.17, 14.18, and 14.19.

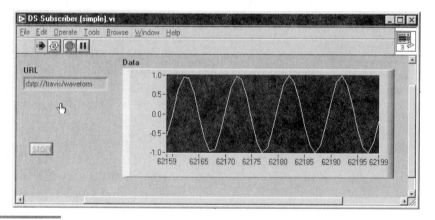

Figure 14.17

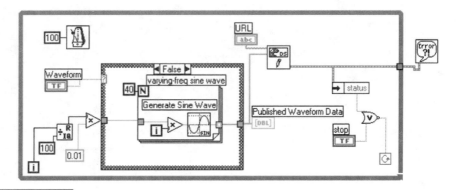

Figure 14.18

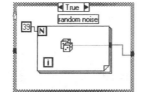

Figure 14.19

2. Save this VI as **DS Publisher (simple).vi**.

3. Create a reader VI like the one shown in Figures 14.20 and 14.21. Notice how you need to create a block diagram constant to let the **DataSocket Read** function know what kind of data you are expecting (in this case, an array of DBL).

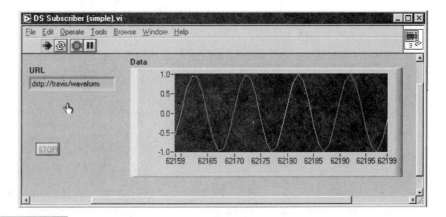

Figure 14.20

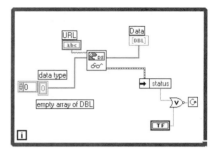

Figure 14.21

4. Save this VI as **DS Subscriber (simple).vi**.

5. On the machine where you have **DS Publisher (simple).vi**, start the **DataSocket Server** by going to **Program Files>>National Instruments DataSocket>>DataSocket Server.**

6. In the **URL** textbox of **DS Publisher (simple).vi**, type

   ```
   dstp://localhost/waveform
   ```

7. This creates a data item "waveform" on the local DataSocket server.

8. Run **DS Publisher (simple).vi**. You should see the DataSocket server show one process connected as the publisher sends it data.

Figure 14.22

9. If you are going to be running the **DS Subscriber (simple).vi** on the same machine as the publisher VI, enter in the **URL** string control

   ```
   dstp://localhost/waveform
   ```

   If you are running **DS Subscriber (simple).vi** on another networked machine, make the URL point to the machine where the publisher VI is running, for example

   ```
   dstp://remote.server.machine/waveform
   ```

10. Run the **DS Subscriber (simple).vi** and note how the waveform on the publisher appears on the subscriber. (If you are using two ma-

chines, you should see the DataSocket Server indicate two processes connected).

11. Experiment with changing the waveform on the publisher VI. Experiment with starting and stopping either the publisher or subscriber VI. Notice how robust the system is: Neither the publisher nor the subscriber is affected, and neither needs to "reset" if the other disconnects.

14.5 Connectivity to Other Programs and Devices

If you want to use LabVIEW to communicate with low-level protocols, you can generally do so, but be aware that low-level networking takes a good deal of design and often more coding. You can use TCP/IP functions, for example, to communicate with any other computer, device, or program that also supports TCP/IP. On Windows, you can use ActiveX automation to have LabVIEW interact with or use other ActiveX-enabled applications (e.g., Microsoft Excel, Internet Explorer).

LabVIEW supports the following additional communication protocols (all accessible from the **Communication** palette):

- TCP/IP
- UDP
- ActiveX automation (including ActiveX events)—Windows only
- AppleEvents and PPC—MacOS only

14.5.1 TCP/IP

TCP/IP is the underlying protocol for the Internet and most internal networks.

Incidentally, TCP stands for *Transmission Control Protocol* and IP stands for *Internet Protocol*. IP divides your data into manageable packets called *datagrams* and figures out how to get them from A to B. Problem is, IP isn't polite and won't do any handshaking with the remote computer, which can cause problems. And like bulk mail with the U.S. Postal Service, IP won't guarantee delivery of the datagrams. So they then came up with TCP, which added on to IP, provides handshaking, and guarantees delivery of the datagrams in the right order (more like a private express carrier, to follow the analogy).

TCP is a connection-based protocol, which means you must establish a connection using a strict protocol before transferring data. When you connect to a site, you have to specify its IP address and a port at that address. The IP address is a 32-bit number that is often represented as a string of four numbers separated by dots, like 128.39.0.119. The port is a number between 0 and 65535. You can open more than one connection simultaneously. If you're familiar with Unix or have used Internet applications, then this should all be old hat to you.

LabVIEW has a set of VIs, found under the **TCP** subpalette of the **Communications** palette, that let you perform TCP-related commands, such as opening a connection at a specified IP address, listening for a TCP connection, reading and writing data, etc. They are all fairly easy to use if your network is configured properly.

Figure 14.23
The TCP Palette.

A good example for getting started with building your networked VIs are the examples found in the full version of LabVIEW: **Simple Data Client.vi** and **Simple Data Server.vi.** Their diagrams are shown in Figures 14.24 and 14.25.

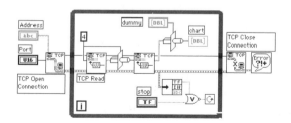

Figure 14.24
Simple Data Client.

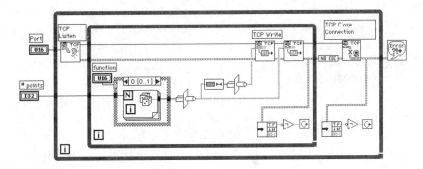

Figure 14.25
Simple Data Server.

You can learn quite a bit about writing client–server VIs by examining these diagrams. The basic process for the client is:

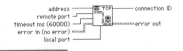

Figure 14.26
TCP Open Connection.vi.

1. Request a TCP connection. You can set a timeout to avoid hanging your VI if the server doesn't respond.

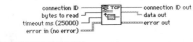

Figure 14.27
TCP Read.vi

2. Read (or write, in other cases) data. Data are always passed as a string type.

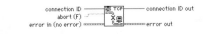

Figure 14.28
TCP Close Connection.vi.

3. Close the TCP connection.

The basic process for a server is:

Figure 14.29
TCP Listen.vi.

1. Wait for a connection.

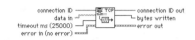

Figure 14.30
TCP Write.vi.

2. Write (or read, in other cases) data. Data are always passed as a string type.

3. Close the connection.

Because all data over a TCP/IP network have to be passed as a string, you will need to convert your data to the LabVIEW string type. The easiest way to do this, as in the previous examples, is to use the **Type Cast** function. Just ensure that both the server and client know exactly what kind of data they're passing. If the server, for example, typecasts extended-precision floats to a string, and the client tries to typecast the string into a double-precision number, the result will be garbage!

Networked applications are ideal when you need to write a program to control a large distributed system, such as in process control applications. You can find a lot more information about TCP/IP in LabVIEW from my other book, *Internet Applications in LabVIEW*.

14.5.2 UDP

UDP, which stands for *Universal Datagram Protocol*, is a lot like TCP, relying on IP addresses, except it is a "connection-less" protocol. What this means is that a server can broadcast data to a large number of clients without actually having to manage or be aware of connections among the clients. That's why for UDP there is no concept of a "listener" like there is in TCP.

LabVIEW provides support for UDP through the **UDP** VIs palette (Figure 14.31).

Figure 14.31
The UDP palette.

14.5.3 ActiveX

ActiveX is a framework from Microsoft that allows Windows applications to communicate with one another, as well as embed functionality from one external application into another. For example, with ActiveX you can embed a Microsoft Excel spreadsheet into a Microsoft Word document (this used to be called OLE).

ActiveX is a very broad topic, and we'll only point you to a few things LabVIEW can do it with it. Needless to say, only LabVIEW for Windows supports ActiveX since it's a Windows protocol. The ActiveX technologies supported by LabVIEW are grouped into two categories, *Automation* and *Container* functionalities.

- *ActiveX Automation*: For example, LabVIEW acting as a client can launch Excel, open a Workbook, etc. You can use a Visual Basic script to control LabVIEW acting as a automation server.

- *ActiveX Container*: The technology allows an application to contain (embed) components from some other software packages. For example, a LabVIEW front panel can contain a Microsoft Excel worksheet.

The ActiveX Automation capability allows LabVIEW to be either a server or client; the ActiveX container capability always handles LabVIEW as an ActiveX client.

LabVIEW as an ActiveX Automation Server

LabVIEW can expose properties and methods of the LabVIEW application itself and of specific VIs to other ActiveX-enabled applications (e.g., Microsoft Excel, Visual Basic, etc.). It does this through the VI Server interface. If you recall the VI Server discussion from Chapter 13, LabVIEW's VI Server capabilities are accessible by TCP/IP (for other LabVIEW applications only), block diagram functions, and ActiveX.

To enable the ActiveX access to the VI Server in LabVIEW, you must check the appropriate box in **Tools>>Options>>VI Server:Configuration,** as shown in the Figure 14.32.

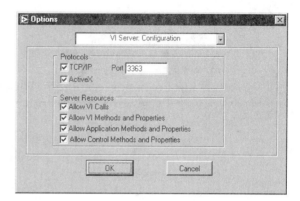

Figure 14.32

By enabling the ActiveX server, you can write external applications in other languages, such as Visual Basic or C++, that can interface with Lab-VIEW. For example, you could write a Visual Basic application that called a VI, showed its front panel, centered it on the screen, ran the VI, and closed it—doing all this programmatically through the ActiveX/VI Server interface. For more information on this capability, consult the LabVIEW manuals.

LabVIEW as an ActiveX Automation Client (Container)

LabVIEW can also be an ActiveX automation client, meaning that it can access the properties and methods of other ActiveX-enabled applications and ActiveX controls.

On the front panel, LabVIEW provides an **ActiveX** palette that allows you to create an ActiveX **Container,** an **OLE Variant**, or an **Automation Refnum**.

Figure 14.32A
The ActiveX control palette.

Figure 14.33
ActiveX functions palette.

On the block diagram, you have functions for working with ActiveX.

With the **ActiveX Container**, you can embed ActiveX controls directly into the front panel of your VI.

The **Automation Refnum** is a LabVIEW type that provides a reference to an external application. With this reference, you can use the **Automation** functions to read and write properties, call methods, and handle events particular to your external application. For example, from LabVIEW, you could open an Excel spreadsheet (on your local or a remote computer) and have it run some calculations.

The **OLE Variant** represents a special data type that is used by ActiveX applications (if you've used Visual Basic, you know what variants are). OLE Variants are used to pass data to an ActiveX control or application. To use the data in LabVIEW, you must convert the Variant to LabVIEW-type data with the **To G Data** function.

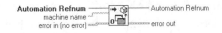

Figure 14.34
Automation Open.

Let's look briefly at the main block diagram functions for ActiveX:

Opens an **Automation Refnum** that refers to a specific ActiveX object. You select the class of the object by popping up on the function and choosing **Select ActiveX Class**. Once you open a refnum it can be passed to other ActiveX functions. If you have DCOM configured correctly, you can open a reference to a remote object specified by the IP address in **machine name**.

Closes the **Automation Refnum**.

Figure 14.35
Automation Close.

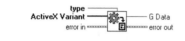

Figure 14.36
To G Data.

Converts an **ActiveX Variant** (same as an **OLE Variant**) to G-type data. The type of data is specified by **type**. This function works similarly to the typecast function.

Figure 14.37
Invoke Node.

Invokes a node on the object specified by the reference (the **Automation Refnum**). The methods that show up are particular to the ActiveX object and are not part of LabVIEW.

Figure 14.38
Property Node.

Gets or sets properties, specific to the ActiveX object specified by the **reference** (the **Automation Refnum**).

There are also ActiveX functions for handling ActiveX events (in the **Communication>>ActiveX>>ActiveX Event** palette); for more information on these, consult the LabVIEW documentation.

Probably one of the most useful ways to use ActiveX and LabVIEW is to take advantage of **ActiveX Containers**. **ActiveX Containers** allow you to embed objects or documents from one application into another. For example, you can insert a calendar control, a movie player, or in this case, a Web browser into an ActiveX container on a LabVIEW front panel. Do the following activity to see how you can embed a Web browser on your front panel.

Activity 14-4: Embedding an ActiveX Web Browser in a VI (Windows Only)

1. Open a new VI in LabVIEW. From the **Controls>>ActiveX** palette, select the **Container** object. Resize it to be very large on your front panel.

Figure 14.39

2. Pop up in the container and select **Insert ActiveX object**. From the **Select ActiveX Object** dialog box, select **Create Control**. A list of all available objects is displayed. Next, scroll through the list and select the **Microsoft Web Browser** control.

3. Switch to the block diagram. You can now use automation functions with this refnum. Use the **Invoke Node** function from the **Communications>>ActiveX** palette. Once you wire the automation reference of the control container to the **Invoke Node** function, it will display the automation object (IwebBrowser2) it is accessing. You now can select from a list of methods it exposes by popping up on the Method terminal. Select the "Navigate" method.

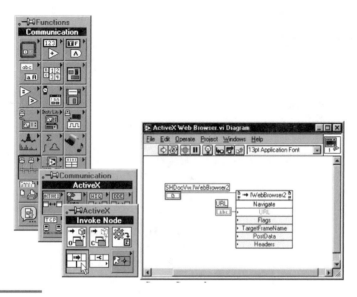

Figure 14.40

4. Pop up on the URL input and select **Create Control**. This will give you a string input for typing the URL you want to connect to.

5. Return to the front panel, type in a URL, and run the VI. You will see the Web page appear in the container. In the example shown in Figure 14.41, http://rayodyne.com was typed.

Unless you already understand ActiveX, this example may appear somewhat magical. How is it possible to have a Web browser as a LabVIEW front panel control? The answer is that the Web browser is not really part of LabVIEW; in fact, LabVIEW is doing very little. Instead, the fact that both LabVIEW and Microsoft Internet Explorer (IE) can communicate via ActiveX is what allows you to embed the IE browser into LabVIEW.

There are several things to note about using this ActiveX Web browser control:

• It is very easy to use and set up with minimal block diagram programming.

• You have access to many other Web browser properties and methods, such as resizing the browser, moving forward and backward, refreshing the Web screen, etc.

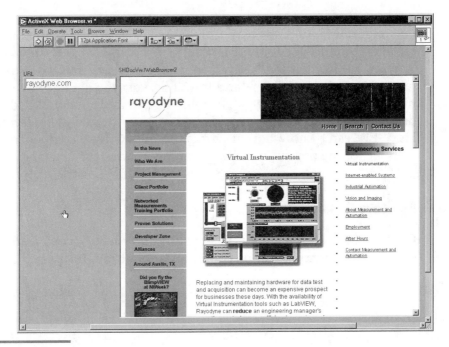

Figure 14.41

- You can easily and quickly access both static HTML documents (via the file:// URL) and Internet websites (via the http:// URL) from Lab-VIEW.
- This will only work on LabVIEW for Windows; it is not available for other platforms.

This type of example is great for an application where you wish to link to an online web-based help system, for example.

Finally, note that there are two ways to specify the ActiveX class to link to an **Automation Refnum** in LabVIEW:

1. Placing an **ActiveX container** or an **Automation Refnum** on the front panel, right-clicking, and selecting **Insert ActiveX Object...**
2. Using **Automation Open** function on the block diagram, or placing an **Automation Refnum** constant on the front panel, right-clicking, and selecting **Select ActiveX Class...**

The first method is used when you want to embed an ActiveX control (or document) into the LabVIEW front panel. The second method is used more for interapplication communication.

You'll notice that the ActiveX functions are very similar to the VI Server functions; so if you understand the VI Server, you should have no problem using the ActiveX functions. The main difference is that you will need to understand the methods and properties of the external ActiveX object (since LabVIEW can't know how it works). Many ActiveX components include a help file for the objects, which you can reference if it's available.

14.5.4 AppleEvents and PPC

For LabVIEW on the MacOS Classic (9.1 and earlier), you can use some VIs that provide interapplication communication through **AppleEvents**.

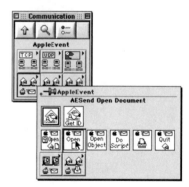

Figure 14.42

AppleEvents are a MacOS-specific protocol that applications use to communicate with each other. AppleEvents send messages to other applications or to the operating system itself to open a document, request data, print, etc. An application can send a message to itself, another application on the same computer, or another application on a remote computer.

You cannot use AppleEvent VIs to communicate with computers that aren't running MacOS (such as a PC running Windows). If you need to communicate with other platforms, use a protocol common to all of them, such as TCP/IP.

The actual low-level AppleEvent messages are quite intricate: You can use them with LabVIEW, but you'd better know the MacOS very well and have a good reference handy. For simpler stuff, though, LabVIEW comes with higher-level VIs for sending some of the popular commands to applications: telling the Finder to open a document, for example. The AppleEvent VIs are located in the **AppleEvent** subpalette of the **Communications** palette.

You can also send system commands with AppleEvents. For example, you might want a VI that could open and close a user-selectable application for viewing the text data file you just saved.

Figure 14.43

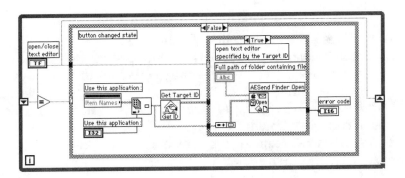

Figure 14.44

You can find a neat example of how LabVIEW can communicate with Microsoft Excel in the full version of LabVIEW under `>examples:comm:AE examples.llb`.

Another Mac protocol, Program-to-Program Communication (PPC), is a low-level form of Apple interapplication communication (IAC), which allows applications to send and receive blocks of data. LabVIEW provides you with VIs to perform PPC, found on the **PPC** subpalette of the **Communication palette**. PPC is a more complicated and advanced protocol than AppleEvents, so we won't say more about it here. If you know how to use PPC, then you can easily learn how to use LabVIEW's PPC VIs by looking at some examples or perusing the manuals.

14.6 Enterprise Connectivity — The Big Picture

For applications that require storing and processing lots of data, or where you need to be able to quickly sort and retrieve the data, you may need to design systems from the ground up with experts who understand the different tiers of an enterprise-level network. This often includes working with SQL databases such as those of Oracle or Sybase, tying together the LabVIEW labs to a common network, and implementing remote client access. Figure 14.45 illustrates this architecture.

For example, suppose you had several test labs using LabVIEW to do functional testing on widget products. Adding Internet capability would allow you to monitor the current test status of any of the labs using a Web browser. However, by tying a database into the network, you could not only obtain historical test information from the same Web browser but you could examine all kinds of interesting relationships. For example, you could compare average pass/fail times among different labs or run a query comparing historical results with current real-time tests.

So, of particular importance in the enterprise system is the back-end database. In many cases, you will want to store your data acquisition test results in a database, rather than on your local drive. Databases allow far more efficient storage and retrieval of large amounts of data, particularly for situations of concurrent publishers and subscribers. In addition, databases allow programmers to create all types of queries that can examine relationships among elements in the database.

Although there are a variety of database vendors and brands, from the simple Microsoft Access to high-end Oracle systems, most of them can work with a common language known as *SQL*. SQL stands for Structured Query Language, and it is principally used to write queries that are sent to the database in order to retrieve the desired data. It's possible to use SQL statements with a variety of different databases, partly thanks to *ODBC* (Open DataBase Connectivity), a "glue" layer that abstracts SQL statements from an application and lets them work with a variety of database drivers.

Another way databases are used is to allow Web access to the data remotely. By writing Web server scripts (using CGI or ASP, for example), a remote user with a Web browser can perform queries and pull up data or charts that are displayed right on the browser. If your LabVIEW-based application handles a large amount of data and you'd like to have Web access to not only the real-time processes but also historical data, consider tying a database to the LabVIEW system (as a publisher of data) and a Web server

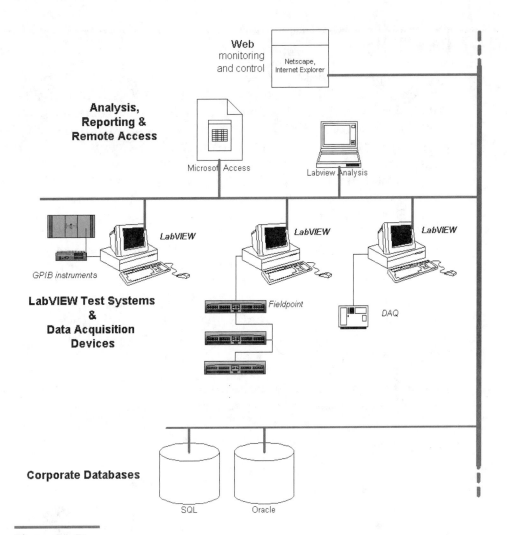

Figure 14.45
Enterprise connectivity.

script (as the subscriber of data). Your Web scripts could even be written in LabVIEW as CGI VIs!

National Instruments sells a *Database Connectivity Toolkit* for LabVIEW that has support for any ODBC-compliant database, allowing LabVIEW to read and write to almost any SQL database. You can also use the free, open-source *LabSQL* for connecting to almost any database. A copy of LabSQL is on the enclosed CD, or you can download it from the LabVIEW Open Source Tools (LOST) site, http://jeffreytravis.com/lost.

14.7 Wrap It Up!

In this chapter we looked at the role of the Internet and networks and the different connectivity features LabVIEW offers: Web-enabled VIs, DataSocket, TCP/IP, UDP, ActiveX, AppleEvents, and databases.

We saw how LabVIEW's built-in Web server allows you to easily monitor your VIs over the Web. With the LabVIEW Web Server you can simply type a URL in the Web browser and instantly see an image of the VI. To remotely control a VI through the Web browser, you can use a free tool like LabVNC, or you can create your own interface with Java, ActiveX, or CGI.

DataSocket is a protocol for sharing data among NI-based applications. With DataSocket, you can easily hook any front panel control or indicator into the DataSocket Server, so that other VIs or applications can publish or subscribe to that control. You can also use the DataSocket block diagram functions to share data.

Lower-level protocols, such as TCP/IP, UDP, and ActiveX on Windows and AppleEvents on the Mac, are supported in LabVIEW. If you know about these protocols, you can use them for writing network or interapplication communication VIs.

Finally, we saw the importance of databases in the enterprise. Often you'll want to use a database to write and retrieve your LabVIEW data. You can use tools such as LabSQL or the Database Connectivity Toolset to accomplish this.

OVERVIEW

In this chapter, you'll learn about some advanced file functions in LabVIEW. Knowing how to use the different kinds of files in LabVIEW (text, datalog, binary, and waveform) will help you pick the best file type for your application. When it's time to produce a printed report, we'll look at LabVIEW's options for printing and the report functions it has. This chapter also teaches you how to programmatically print from a VI.

GOALS

- Become familiar with the different types of files LabVIEW works with
- Use LabVIEW's low-level file I/O functions for text, datalog, and binary
- Understand how you can save and read waveform data from the different file types
- Learn about the different ways you can print from LabVIEW
- Discover the report function for creating printed or HTML reports from LabVIEW
- Know how to print programmatically

KEY TERMS

- Binary file
- Datalog file
- ASCII file
- Waveforms and files
- Refnum
- Printing
- HTML
- Reports

Advanced File I/O, Printing, and Reports

15

15.1 Advanced File I/O

In Chapter 9, you saw how to save your data in a text file, whether it was plain old text or a spreadsheet format, using the VIs from the File I/O palette. These files, stored in ASCII, or text, format, have the advantage of being the most easily portable type of file. Virtually any computer running on any operating system can read or write a text file. However, text files do have some drawbacks: They are the least space efficient (most bytes per piece of information), and they may require a lot of conversion and processor time if the data you want to store are not text (e.g., a graph). LabVIEW provides you with the ability to store and retrieve two other kinds of files: *datalog files* and *binary files*.

Datalog files are a special kind of binary file used by LabVIEW to store front panel information or any LabVIEW data you may want to save. For storing all the front panel information, you can think of datalog files as a sort of screen dump of your VI. When you create a datalog file, it records all the values in all the controls and indicators at the time you saved it. You can later load this datalog file into your VI to see the stored values displayed in your front panel. You can even save several "sets" of values from the same

487

front panel in one datalog file. Datalog files can only be made and read by LabVIEW. They are fairly easy to use because you can manipulate datalog files from the menus in LabVIEW without writing code. You can also create datalog files that record some specific LabVIEW data type, such as a cluster or a string.

Binary files (also called *byte stream* files) normally contain a byte-for-byte image of the data as they were stored in memory. You can't just "read" a binary file with a text editor, or with any other program unless you know exactly how the file is formatted—just as you need to know the data type when using binary strings (see Chapter 12). The advantages of a binary file are that you have the least overhead, since no conversion is required, and you save a lot of disk space compared to ASCII files. For example, storing an array of 100 numbers in a 8-bit binary file takes up about 100 bytes, while a text file might require over 400 bytes. That's because each 8-bit integer takes up only one byte (in binary format), but the same number in text format can take up to 3 or 4 bytes (one byte for each ASCII digit).

Table 15.1 *Summary of File Types.*

ASCII	Datalog	Binary
• Easiest to use • Compatible with other applications; easy to view and manipulate • Require the most disk space and conversions • Suitable for small to medium-size data sets that will be used in other applications (e.g., spreadsheets)	• Easy to use interactively; requires more elaborate programming to use it in applications • Can only be used within LabVIEW • Best for storing data in LabVIEW object or the whole front panel	• Requires strict programming protocols • Most efficient use of disk space and processor time • Fast (disk streaming) • With care, can be read by other programs • Useful for applications that will record large files in real time

15.1.1 Giving Directions to Find Your File

To locate a file within a file system, LabVIEW uses a special data type called a *path*. A path control or indicator, available from the **String & Path** palette, looks and feels similar to a string control or indicator. You can specify the absolute or relative path name to a file.

Figure 15.1

The little folder icon to the right of the path control (Figure 15.1) is a "Browse Button." Clicking on this button will pull up the file system browser so that you can easily locate a file or directory rather than typing it in.

The difference with a string type is that in a path control, you can only enter a path or filename according to the syntax of your operating system.. A full path name for the file `sample.txt` might look something like the following, for each operating system:

Windows   `C:\TRAVIS\DATA\SAMPLE.TXT`

MacOS   `PowerHD:Travis:Data:Sample Text file`

Unix   `usr/travis/data/sample_text_file`

Whenever you open or create a file, you need to specify the path for that file. If you don't wire a path to the appropriate file function, LabVIEW will pop up a dialog window prompting you to find the file when the **Open File** or **New File** function is called. You can also make LabVIEW prompt the user for a filename with a custom prompt using the **File Dialog** function. We'll look at these functions in a moment.

15.1.2 The Three-Step Process

When you write file I/O functionality into your programs, you should always follow the three-step process: *Open, read* or *write*, and *close*. The old filing cabinet analogy works well here. Whether you are pulling out a folder or putting some papers into a folder, you need to open the file cabinet first. Then you do whatever you need to do with your files. Finally, you could leave your filing cabinet drawer open—but why? Coffee might spill into it, someone else might take some files or disorganize them, and it just looks plain messy! Closing your files on the computer as well will ensure your data integrity.

The diagram in Figure 15.2 illustrates this process. The functions **Open/Create/Replace File, Write File,** and **Close File** are all accessible from the **File I/O** palette. Notice the data dependency from each function to the next. After the **Open/... File** operation is completed, LabVIEW generates a file *refnum* (file reference number) that it passes on to **Write File**. When the string is written to the file, **Write File** in turn passes a refnum to **Close File**. Finally, an **Error Handler** VI at the end is almost always mandatory so that you will know if something went wrong.

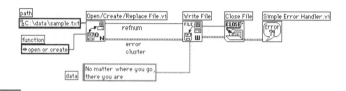

Figure 15.2

LabVIEW uses the refnum to keep track of what file your VI is referring to. You generally don't need to care what these refnums are or how they work—just be sure to wire them between file I/O functions. Wiring the refnum terminals between these VIs has the added advantage of enforcing data dependency, making things happen automatically in the right order.

There are a few file I/O functions that you will probably be using a lot and should become familiar with. These are described next.

The **File Dialog** function (**File I/O >Advanced File Functions** palette) displays a file dialog box for file selection. This dialog is used for the selection of new or existing files or directories. You can specify the **prompt** message that will appear in the dialog box. We will discuss the use of the **datalog type** input later in this chapter.

Figure 15.3
File Dialog.

The **Open File** function (**File I/O>Advanced File Functions** palette) opens an existing file. You can optionally wire a valid path to the **file path** input. This function is

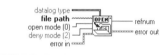

Figure 15.4
Open File.

not capable of creating or replacing files. It opens only *existing* files. The **datalog type** input is used only when opening LabVIEW datalog files.

The **New File** function (**File I/O>Advanced File Functions** palette) creates and opens a new file for reading or writing. You must connect a path to the **file path** input of this function, and it must be a path to a *nonexistent* file. The **datalog type** input is used only when creating new LabVIEW datalog files.

The **Write File** function (**File I/O** palette) writes data to an open file. The behavior of this function varies slightly depending on if you are writing data to a byte stream file or a LabVIEW datalog file. The **header** input is used with binary file types and is ignored for ASCII files.

The **Read File** function (**File I/O** palette) reads data from an open file. When reading byte stream files, you can use the **byte stream type** input to indicate how LabVIEW should interpret data in the file. We'll discuss byte stream files in more detail later in this chapter.

The **Close File** function (**File I/O** palette) closes the file associated with **refnum**.

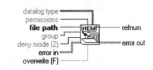

Figure 15.5
New File.

Figure 15.6
Write File.

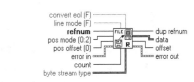

Figure 15.7
Read File.

Figure 15.8
Close File.

Open/Create/Replace File (File I/O palette) is a utility file function that programmatically lets you open a file, create a new one, or replace an old one of the same name. You can optionally specify things such as a dialog prompt string, start path, etc. This VI calls several of the previously mentioned file I/O functions.

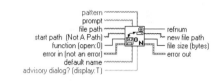

Figure 15.9
Open/Create/Replace File.vi.

15.1.3 Writing and Reading Text Files

If you're going to write text files to save your data, and high-speed disk streaming is not required, save yourself some time by using one of the higher-level file I/O functions shown in Figure 15.10 (these were discussed in Chapter 9). These VIs do all the opening, closing, and error checking for you.

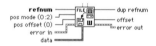

Figure 15.10

However, there may be times when you want to write your own customized file I/O routine. The heart of this process is the **Write File** function (Figure 15.11).

Figure 15.11
Write File. Writes data to the file specified by refnum.

Pos mode:
0: Relative to start of file
1: Relative to end of file
2: Relative to current position

If pos offset is wired, pos mode defaults to 0. If pos offset is unwired, pos mode defaults to 2.

The **data** input to **Write File** are polymorphic: You can wire numeric, string, or cluster data types. The type of data you wire to this function determines what file format you'll be using. Obviously, for text files, you should only wire a string to this input or nothing at all.

The Basics

The actual file management for text files is pretty easy (open, write, close!), but you should be prepared to ask yourself a couple of questions:

1. Where will you get your path and filename from? Will LabVIEW create it automatically? Will the user input it from the front panel or a dialog box?
2. How will your program handle a file error (disk full, invalid path, etc.)? Should the program stop? Should it continue and simply notify the user?

And, of course, you need to know how to convert your data to a string. Fortunately, LabVIEW has all kinds of string conversion and manipulation functions (just browse the **String** palette or see Chapter 9).

Let's suppose you have written a simple VI, called **Acquire Data,** that reads one data point and returns it as a floating-point number. You want to periodically acquire the datapoint and stream it to a file, until a stop button is pressed. The user should be able to choose the filename at runtime. We might start by building a diagram like the one shown in Figure 15.12.

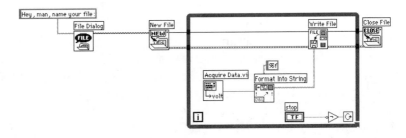

Figure 15.12

The user is prompted to select a filename, a new file is created, and then the While Loop begins. Each time the loop goes around, a data point is converted into string format with a **Format Into String** function and is written to the file. When the stop button is pressed, the loop ends and the file is closed. Simple enough, eh? Well, this VI *will* work as it stands. But, frankly, it's a pretty poor file management VI! It can be improved quite a bit. For starters, notice that nothing is being done to separate each string that represents a data point. The file is going to be just one long jumbled string! Also, what would happen if the user pressed the CANCEL button at the File dialog?

The next diagram (Figure 15.13) makes our program a little more robust. Notice that a carriage return and linefeed are appended to each number (the "\r\n" characters)—this ensures that each number will be on a new line of text. We took the possibility of the user pressing the CANCEL button into account by adding a case statement based on the output of **File Dialog**. We also used the error clusters to check for a file I/O error; if one occurs, it stops the loop.

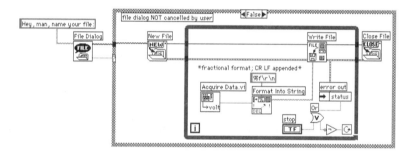

Figure 15.13

There's still another problem with this code that's easy to overlook. It has to do with the **Write File** inside the While Loop. Can you guess what it is? Unless the **Acquire Data** VI has a **Wait** function, this program will write to the file as often as it can push the microprocessor—maybe thousands of times per second! You need to put a **Wait** function inside the loop to limit how often you want to write to your file, say, once a second or something reasonable.

When writing your file I/O routines, you may have runtime problems if you don't take into account "worst case" possibilities. Try to implement graceful escapes if a file I/O error or anomaly occurs.

As you see, a lot of things can go wrong or at least get messy when you're dealing with file I/O. Don't be discouraged, however. Most of the time you can probably use LabVIEW high-level VIs for reading and writing text files. If you do need to do specialized text file I/O, be sure you understand what you're doing—and always consider the "what ifs"!

Reading back a text file is very similar to writing one. You use the **Read File** function in a complementary fashion to **Write File**.

Figure 15.14
Read File.

Again, you don't need to wire anything to the **pos mode** or **pos offset** if you just want to write new data at the end of the file. You can also, of course, use some of the simpler text file functions mentioned in Chapter 9.

Nerd-Level File I/O Stuff

Pos mode and **pos offset** are related to your computer's file system and can be quite confusing, to make an understatement. To keep track of file I/O, an invisible variable called the *file mark* is used by your operating system. The file mark normally points to the current location of where data were last stored inside the file, measured in bytes from some reference point. Whenever you write data to an existing file, LabVIEW writes data at **pos offset** bytes from the reference point determined by **pos mode**.

This reference point can be the beginning of the file, the end of the file, or the current location of the last write. With **Write File**, you can manipulate the new data insertion point and move it around using **pos mode** and **pos offset**. This powerful feature is what allows random access to the data in a file. If you leave both of these unwired, by default new data are appended after the most recent data—which is what you likely will want to do most of the time.

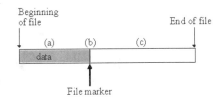

Figure 15.15

To help you understand these concepts, consider Figure 15.15. The file marker points to the place where data were last written. The labels (a), (b), and (c) indicate three of the possible places to which you can write in the file. Here's one possible way you would access each one:

a. Setting **pos mode** to 0 (relative to start of file) and setting **pos offset** to a positive number

b. Leaving **pos mode** and **pos offset** unwired

c. Setting **pos mode** to 2 (which specifies a write relative to the current location of file marker) and setting **pos offset** to a positive number

Many possibilities other than those just mentioned exist, of course. If all this "pos" stuff confuses you, don't worry about it— you can get by fine without it if you don't need rapid random access to files. Leaving both unwired will let you read from the beginning of a file and write to the position following the file mark, which is the way you might expect to normally handle files.

Activity 15-1: Reading and Writing Text Files

Build a VI that has the ability to both read and write ASCII files. The data type input should be a 2D array of numbers. A Boolean switch determines whether the VI will read a file or write a new one. The front panel of the VI is shown in Figure 15.16.

Save this VI as **ASCII Read/Write.vi**. One simple solution is given on the CD with the same name.

*1. To read the file, use an easy file I/O function, such as **Read Characters from File**.*
*2. Use the **Array Subset** function (in the **Array & Cluster** palette) to break up the array contents into units you can convert into strings.*

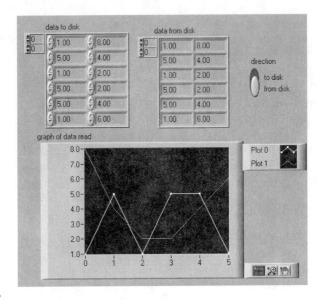

Figure 15.16

15.1.4 Writing and Reading Datalog Files

Datalog files store data from a LabVIEW object (such as a cluster, string, Boolean array, etc.) in a special binary format. Each time you write to a datalog file, LabVIEW normally appends a *record* of the data. A record is sort of like a file within a file: The operating system only sees one file, but from within LabVIEW you can "see" several separate records in a datalog file. The nice thing about this structure is that you can randomly access any record in the file. Another advantage of using records is that you don't have to know how many bytes you need to "skip" to get to a particular set of data, just the record number. Datalog files are especially useful for storing mixed data types, such as a Boolean and an array. There are two ways to create a datalog file:

1. By using front panel datalogging: Use the built-in datalogging features found under the **Operate** menu. These commands allow you to log the whole front panel into a file, without writing any file I/O code. The data are logged either at the completion of the VI or by a user command in the **Operate Menu**.

2. By reading and writing datalog files with the **Read File** and **Write File** functions. In this case, you can specify the stored data to be whatever you want (instead of the whole front panel) and store the data whenever you want.

Front Panel Datalogging

Front panel datalogging is very easy to use and doesn't involve any block diagram programming on your part. When you enable datalogging, Lab-VIEW saves the data in all front panel controls and a timestamp to a datalog file. You can have several separate files, each filled with logged data from different tests. You can later retrieve these data with the same VI you saved data from, or in another VI using the file I/O functions.

To have your VI log front panel data, choose **Data Logging>Log...** from the **Operate** menu. The first time you log data, you will be asked for a log file "binding," which is the file that will contain all your datalog records. This is nothing more than the name of your file, so give it any name you like.

You can also enable your VI to log data automatically every time it completes its execution by selecting **Operate>Log at Completion**. Every time you log data to the same log file, you create a new record in that file.

To view the logged data interactively, select **Operate>Data Logging>Retrieve...** . The Toolbar will change to become a data retrieval toolbar as shown in Figure 15.17. All the front panel controls and indicators will suddenly change their values to display the saved data.

Figure 15.17

The highlighted number tells you which record you're currently viewing. The number range in brackets to the right tells you how many records exist. You can switch records using the arrow buttons. As you switch records, the front panel objects will show the data corresponding to the record. To the right of the record number, a timestamp shows you when this record was logged. You can delete an individual record by choosing that record number and clicking on the trash can icon. Click OK to exit the data retrieval mode.

An easy way to programmatically retrieve information in a datalog file is to use the VI that logged the data as a subVI. Popping up on the subVI, you can choose the **Enable Database Access** option. When you do this, a funny-looking yellow frame appears around your VI. (They actually call this a *halo* in the LabVIEW manuals. It makes us wonder if we could start classifying VIs into angelic and demonic types.) If you run this database-access subVI, it does not execute. Instead, it returns the saved data from its front panel as a cluster, according to the record number wired at the input. This cluster contains all the data on the front panel. Each item in the cluster appears in the same cluster order as the front panel order of objects.

Figure 15.18

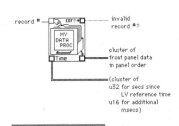

Figure 15.19

Programmatic Datalog File I/O

Besides creating and reading datalog files using the built-in LabVIEW data-logging features, you can do a little more advanced datalog file I/O using the **Read File** and **Write File** VIs.

With *programmatic* datalog files, you don't need to store *all* the front panel data. You can store just some of it. In fact, the data needn't even be on the front panel; they could be generated at the block diagram. The purpose of datalog files is to let you store in one file multiple records of one LabVIEW data type.

Even though you can only store one type of data in a datalog file, you can combine several different variables into one cluster, which is a valid data type. Or the data type can be an array, a string, a numeric, or a Boolean. The importance of datalog files lies in the fact that you can read and write directly into LabVIEW variables without any concern for conversion to text, headers, etc.

Although you use the **Write File** and **Read File** to store datalog files in a similar manner as you do for text files, the meaning of the **pos mode** and **pos offset** inputs changes. Instead of these inputs relating to the file mark, they relate to the record number. Thus, in **Read File**, for example, you can wire the record number to the **pos offset** input to retrieve a specific record. The **count** input specifies how many records you want to read, not how many bytes.

*The data or data type input on the file I/O VIs is called **data, byte stream type,** or **datalog type**, depending on the specific function. In the Help window, this input always appears as a thick brown wire, indicating a polymorphic input. For simplicity's sake, we'll refer to this input as "data type."*

You should always wire the data type input to all the auxiliary file functions (such as **New File** or **Open File**—but not the **Read File** VI) when using datalog files. Doing so is crucial because it tells the read and write VIs that you're dealing with datalog files (as opposed to binary files), which is important because the **pos mode** and **pos offset** input change their behavior. This fact is possibly one of the most confusing aspects we've ever seen about LabVIEW. It's easier to remember what to do with this rule:

> For datalog files, *always* wire the "data type" (**data, datalog type**) on all the relevant file I/O VIs, *except* **Read File**. You must leave the "data type" (**byte stream type**) input unwired on **Read File** to specify a datalog file.
>
> For binary files, wire the "data type" (**byte stream type, data**) input only on the **Read File** or **Write File**. You must leave the "data type" (**datalog type**) unwired on the **New File, File Dialog,** etc., to specify a binary file.

Intuitive, isn't it? We thought so.

The block diagram of an example included on the CD called **Simple Datalogger.vi** is shown in Figure 15.20. This VI stores datalog files that include a time and date stamp with a numeric type representing temperature. This VI incorporates some error handling, which is always a good idea when you're messing with file I/O. Notice again how the datalog type is wired to all the file I/O functions (except **Close File**, where the type is irrelevant).

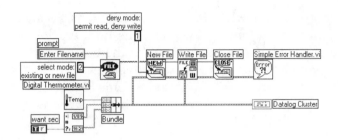

Figure 15.20

The VI for retrieving these same datalog files is **Simple Temp Datalog Reader**, also on the CD. Its block diagram is shown in Figure 15.21.

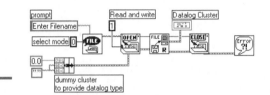

Figure 15.21

Notice how a block diagram constant of the appropriate cluster type was created to provide the **data type** input. Care is required here, because you need to specify the exact data type that was originally stored. For example, if the numeric type were an **I32** type, attempting to read the data with the above VI would not work. To make sure you access the right variable, it is also important to remember the cluster order.

15.1.5 Writing and Reading Binary Files

Binary files, also called *byte stream* files, are to text files what a Porsche might be to an 18-wheeler: They're much smaller and much faster. The disadvantage is that if you want to read a binary file, you'd better know every detail about how the file was written, or else you can forget about it. All the raw data may be there, but it's up to you, the programmer, to provide a way to interpret the data when it's time to read them. Because there's no explicit information in the binary file about the data types or convenient things like headers, you have to know what's in there.

In summary, binary files are not very easy to use, compared to other file types. However, LabVIEW includes some higher-level binary file VIs that let you read and write numeric data. These functions are found in the **Binary File VIs** in the **File I/O** palette (Figure 15.22).

Figure 15.22

The first two VIs let you read and write **I16** numeric data; the second pair lets you read and write **SGL** numeric types. All VIs assume the data are in array format (either one or two dimensions). The following is a description of their functionality:

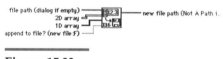

Figure 15.23
Write To I16 File.vi.

Writes a 1D or 2D array of unsigned integers (**I16**) to a binary file specified by **file path**. You can either create a new file or append to an old one with the Boolean input **append to file?**.

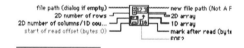

Figure 15.24
Read From I16 file.vi.

Reads a binary file whose data type is an array of **I16** integers. If you know the number of rows and columns that you want in the 2D array output, you can wire these inputs to get the correct array size returned at **2D array**. Otherwise, leave the inputs unwired to read the entire data file into a 1D array returned at **1D array**. You can optionally specify a byte offset at **start of read offset** for random access to the file.

The other two VIs in the palette, **Write To SGL File** and **Read From SGL File**, have identical functionality except for the numeric data type.

As a simple example, look at this **Binary Read/Write.vi**, which acquires a waveform and stores it in a file or allows you to retrieve these data from the file. In either case, the data are passed to a graph.

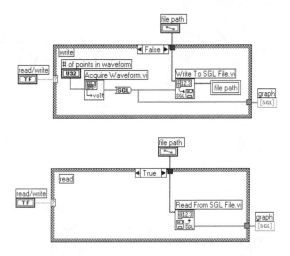

Figure 15.25

Notice the conversion function to a **SGL** type in the write feature. With these binary files, it's very important to wire the correct data type, or else you may not be able to retrieve the data correctly.

For the Curious Nerds

If you'd like to take a peek at how to use binary files with the lower-level file I/O VIs, read on.

The **Write File** function, as used in the text file section examples, skips over any header information that LabVIEW uses to store the string in memory and simply writes the contents of the string data to the file. In fact, the **Write File** function does not distinguish an ASCII string from a binary string when writing the data to the file; it simply places the data in the file. In other words, there's no *functional* distinction between text files and binary text files: Whether the string is meaningful as text or as binary information depends on how you interpret it.

But remember, the **data** terminal of the **Write File** function is *polymorphic*, meaning that it will adapt to any kind of data you wire into it. For example, you can wire a two-dimensional array of numbers into the **data** input, just as you do with datalog files.

Figure 15.26

However, datalog files and binary files are very different in structure and behavior, even though you use the same LabVIEW functions to work with them. Because it's so confusing, we'll repeat the rule for working with binary versus datalog files.

In the block diagram of Figure 15.26, the **Write File** function simply places a byte-for-byte copy of the input data into the file. It does not convert the numbers to ASCII, nor does it place any information about the number of rows or columns in the array into the file. If the **data** array had 3 rows and 2 columns, then the file would be 3 rows $\times$ 2 columns $\times$ 4 = 24 bytes in size (recall that single-precision floating-point numbers use 4 bytes of memory each).

The data saved by the **Write File** function are the same as those obtained by removing the header information from the output of the **Flatten To String** function and then writing the resulting string to the file. Recall that the **Flatten To String** function collects all of the data in its input terminal and places them in a binary string. It precedes the data in the output string with header information necessary to decode the string back into the original data type.

This example exposes a very important aspect of binary files. If you do not store some kind of header information in the file, it will be virtually impossible to successfully interpret the file. In this example, the VI saved 24 bytes of data to a file. Even if you know that the data were originally single-precision floating-point numbers, you cannot successfully reconstruct the two-dimensional array. How's that, you say? You know that you have a total of 6 values in the array (24 / 4 = 6), but how do you know whether the original data were stored in a one-dimensional array of 6 elements, an array with 1 row and 6 columns, or a table with 3 columns and 2 rows?

When you work with binary files, you will find that storing header information is vital to working with the file. Fortunately, the **Write File** function has a very simple way for you to generate headers to include in the file. If you wire a Boolean value of TRUE to the **header** input of the **Write File** function, it will write exactly the same data to the file as if you had created a binary string with the **Flatten To String** function and written *that* string to the file.

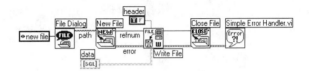

Figure 15.27

Writing binary data to a file with header information using the **header** feature.

You could use the code in Figure 15.28 to read the file. Note that the **byte stream type** input of the **Read File** function has a dummy two-dimensional array of single-precision numbers wired to it. The **Read File** function uses only the *data type* of this input. In essence, when you use the **Read File** function in this manner, it operates on file data in the same way that the **Unflatten From String** function acts on an input string. Recall that the **Unflatten From String** function, given data string and data type inputs, converts the data stored in the string back into the given data type, provided the header information in the binary data is correct.

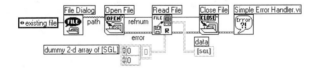

Figure 15.28

Reading the file created in the previous examples.

When you create binary files, designing an appropriate header for the file is often the single most important task. Also, you must clearly document all information necessary to interpret the header so that you can work with the file in the future.

One big plus of using binary files is that you have random access to the file. For example, if you store arrays of numeric data in a file, you may find it necessary to access data at random locations in the file. Random access in

ASCII files is hampered by the presence of negative signs, a varying number of digits in individual data points, and other factors. Such obstacles do not occur with binary files.

In a binary file, the flattened format of a number in LabVIEW is just a binary image of the number itself. Therefore, each number in the array uses a fixed number of bytes of storage on disk. If you know that a file stores single-precision numbers, which use four bytes per number, it is quite simple to read an arbitrary group of elements from the array, as shown in Figure 15.29.

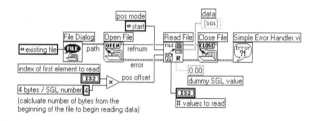

Figure 15.29

Let's look at an example.

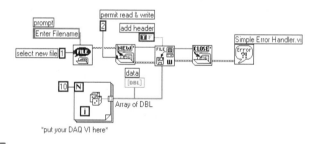

Figure 15.30

The **Write Binary File** VI in Figure 15.30 takes an array of random numeric (**DBL**) data and writes it to a binary file. Notice how the **data type** input is only wired to the **Write File** VI. To retrieve the binary data, we use **Read Binary File** (Figure 15.31).

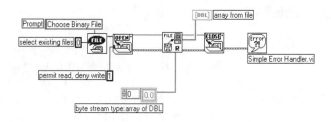

Figure 15.31

Reading the file is a little trickier. To rebuild the numeric array, we have to wire the **count** input to **Read File.** The **EOF** VI is used to count the total number of bytes in the file, which is divided by 8 to figure the number of array elements.

If some or all of this file I/O business is confusing, don't feel bad—you're not alone. Even seasoned LabVIEW programmers have difficulties grasping the subtleties of file markers, headers, etc. The best advice is to use good examples such as those in this book and those included with LabVIEW to do your file management.

Activity 15-2: Using Text, Binary, and Datalog Files

Using some of your own DAQ VIs, write a VI that performs three file I/O functions: one that reads/writes to a binary file, another that reads/writes to an ASCII (text) file, and finally a third that acts as a datalogger (to datalog file types). Then, use the **File/Directory Info** VI (from the **File I/O>Advanced** palette) to measure the file size for each of the three generated files. It should prove to be an interesting comparison. As an optional feature, incorporate a way to measure the time it takes to write or read the file.

Save your VI as **File Type Comparison**.

Figure 15.32
File/Directory Info.

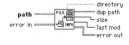

15.1.6 **Working with Waveform Data in Files**

If you mainly need to store and retrieve data in LabVIEW's *waveform* format (see Chapter 8), for example, if you are acquiring waveform data from a DAQ device that you want to log to disk, then you can use the **Waveform File I/O** functions, in the **Waveform** palette (Figure 15.33). These functions are really just working with datalog files; the difference is that they can take care of the low-level formatting for you, so that you don't have to worry about header information, etc.

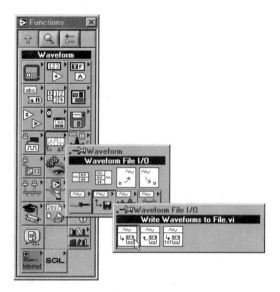

Figure 15.33

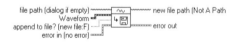

Figure 15.34
Write Waveforms to File.vi.

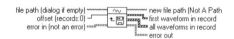

Figure 15.35
Read Waveform from File.vi.

Creates a new datalog file, writes the specified number of records to the file, and then closes it and checks for errors. Each record is a cluster that contains a string and an array of single-precision numbers.

Opens a datalog file created with the **Write Waveforms to File** VI, reads one record at a time until all are read, and then closes the file and checks for errors. The record is a cluster that con-

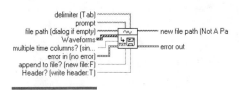

Figure 15.36
Export Waveforms to Spreadsheet File.vi.

tains a string and an array of single-precision numbers.

Converts a 2D or 1D array of single-precision numbers to a text string and writes the string to a new byte stream file or appends the string to an existing file. You can optionally transpose the data. This VI opens or creates the file beforehand and closes it afterwards. You can use this VI to create a text file readable by most spreadsheet applications.

15.2 Putting It in Writing: Printing in LabVIEW

As we've seen before in Chapter 5, you can always print the active window in LabVIEW by choosing **Print Window...** from the **File** menu. This method isn't exactly thrilling if you want your application to automatically generate a large number of printouts. You may want an application that has a "Print" button on the front panel or that prints a plot when a certain pattern in the data occurs. Fortunately, you can print programmatically in LabVIEW.

LabVIEW gives you the option for automatically printing the front panel whenever the VI finishes executing. Choose **Print at Completion** from the **Operate** menu to do this. This is a piece of cake if all you needed was a printout of your front panel each time your VI ran. But what if you only wanted to print part of your front panel or to print something different? Or if you wanted to just print on certain occasions, not necessarily when your VI finishes executing? It's still easy to do this using the **Print Panel VI** function (located in the **Application Control** palette).

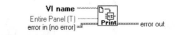

Figure 15.37
Print Panel.

Produces the same printout as programmatic print at completion only if the target VI's front panel is open before it executes. **Print Panel** can be called from other VIs and at times other than at completion. By default, it prints the entire panel, not just what is visible in the window. This VI assumes that the VI is loaded but does not require the window to be open.

Let's do a simple activity to see how to print programmatically

Activity 15-3: Printing Programmatically

In this activity, you will print a front panel programmatically using the **Print Panel** function. You will allow users to press a button to print the panel when they wish.

1. Open the **Temp Limit (Max/Min).vi** you created in Chapter 8. If you don't have it, you can find it on the CD in CH8.LLB.
2. Save a copy of it as "**Temp Limit with Print.vi**."
3. Add a button (with Mechanical Action set to "Latch When Released") that says "Print" on the front panel.
4. Modify the block diagram as shown in Figure 15.38. Be sure and type in the exact name and punctuation of your VI.

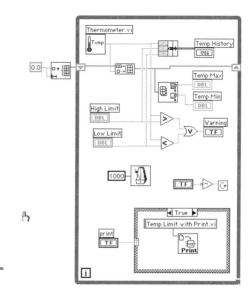

Figure 15.38

5. Test the VI.

You can also control printing functions programmatically using VI Server methods and properties. We talked about the VI Server in Chapter 13. For example, with the VI Class, you have access to methods such as **Print to Printer, Print to HTML,** and **Print to RTF.** With the Application Class, you can set properties like whether it should print a border around the VI, etc. For more information, see the LabVIEW documentation.

15.3 Reports from LabVIEW

While printing front panels of a VI is often useful as a documented result, sometimes you need to create text-based reports that require a precise layout and the use of different font attributes (bold, italic) or data tables. In this case, you can use the **Report** functions on the **Report Generation** palette.

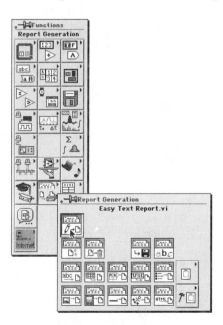

Figure 15.39

The **Report Generation** functions allow you to create text (including RTF) or HTML reports that you can either save as a file or send directly to a printer.

15.4 Wrap It Up!

In this chapter, you took a deeper look at file I/O and at printing. We learned that LabVIEW supports different file types: ASCII, Datalog, Binary.

ASCII, or text, files are the easiest to work with but take up the most disk space.

Binary files are very efficient but require some extra work to format properly in order to read them later on.

Datalog files are a LabVIEW-specific type of file that lets you store records of LabVIEW data types.

Each of these file types is suited for particular purposes, but most of the time you can use the same file I/O functions from the **File I/O** palette to work with them. You examined how you can use the open, read, write, and close functions, as well as how to create new files or prompt the user for a file location.

Finally, we took a brief look at some programmatic printing capability functions in LabVIEW.

OVERVIEW

This chapter focuses on techniques, tips, and suggestions for making better Lab-VIEW applications. We start with the appearance of the front panel of your application, making it look as cool and user-friendly as possible. The whole concept of an intuitive, appealing graphical user interface (GUI) is embedded in LabVIEW. We'll show you a few tips and techniques for taking it a bit further with suggestions on panel arrangement, decorations, customized controls, dynamic help windows, and more. Some common programming problems and their solutions will be presented. We will also cover issues such as performance, memory management, and platform compatibility to help you write better VIs. Finally, we'll examine some general aspects of good programming techniques and style that you will find especially useful in managing large LabVIEW projects.

GOALS

- Become familiar with general guidelines and recommendations for creating an aesthetically pleasing, professional-looking graphical interface
- See how you can import external pictures onto your front panel and into picture rings
- Build custom controls using the control editor
- Learn to dynamically open and close the Help window
- Become familiar with some nifty solutions to common LabVIEW programming challenges (the kind that will make you go, "a-ha!")
- Be able to boost performance, gobble less memory, and make your VIs platform-independent when you need to
- Know some proposed guidelines for writing an awesome application
- Become the coolest LabVIEW programmer in town

KEY TERMS

- Decorations
- Custom controls
- Control Editor
- Importing pictures
- Custom help
- Aesthetics
- Common questions
- Memory usage
- Performance
- Platform dependency
- Good style

The Art of LabVIEW Programming

16

16.1 Why Worry about the Graphical Interface Appearance?

In our culture, image counts for a lot—even in software. People are often more impressed by what a program looks like on the screen than how it actually works. I can think of at least a couple of reasons why you would want to polish and improve your graphical user interface: 1) It impresses and convinces someone else (your supervisor, customers, your spouse) about the quality of your software and its goals. 2) It makes your software more intuitive and easier to use for the end user. 3) It's easier than doing actual work. And therein lies the beauty of LabVIEW—even if you're a novice user you can put together an impressive graphical interface on the front panel (never mind that it doesn't do anything yet) before a guru C programmer can blink.

LabVIEW's front panel objects are already pretty cool looking: knobs, sliders, LEDs, etc. But this chapter will teach you a few more tricks for organizing and putting your objects on the front panel in such a way to create an even better interface. You will also see how to add online help so that the end user can figure out what each control does without a manual. Making a good GUI is not just about aesthetics—it's also about saving the user time and effort.

Look at the VI shown in Figure 16.1, a Temperature System Monitor. It doesn't look too bad. But almost anybody would agree that this same VI with the interface depicted in Figure 16.2 would be easier to work with and certainly have greater visual appeal.

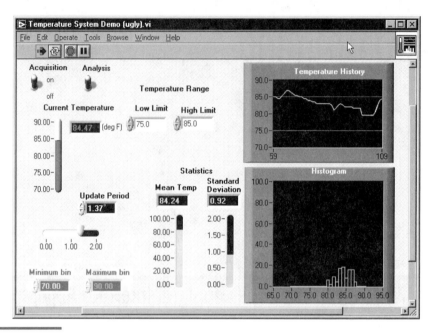

Figure 16.1

16.2 Arranging, Decorating, Grouping, and Locking

One way to improve the appearance of your front panel is to organize the objects by deciding where to physically place them on the panel. You can do this by aligning and distributing your objects evenly and by grouping sets of objects that are logically related onto or inside a decoration.

LabVIEW provides you with the ability to *align, distribute, group,* and *lock* objects, much like a drawing program, as we saw in Chapter 4. You have a variety of choices for how to align your objects. Distribution of objects refers to the spacing arrangement among objects. Grouping objects allows you to

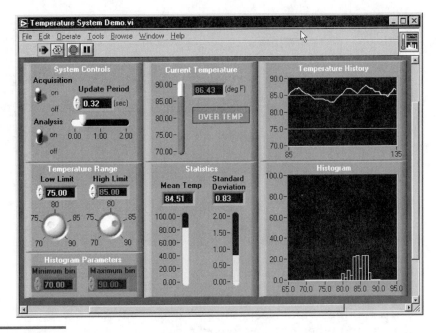

Figure 16.2

easily move a set around without disturbing their relative arrangement. And locking objects helps keep them from being accidentally selected or resized. You can also change the z-order, or depth placement (what's in front and what's behind), of any object. All of this graphical control is done with the toolbar set you'll find on the VI window:

Alignment ——————— 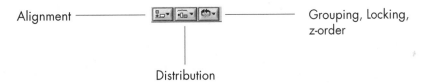 ——————— Grouping, Locking, z-order

Distribution

Remember that the label of an object can be placed anywhere, but it remains attached to that object and can be used as a reference point for alignment and distribution.

A palette in the **Controls** palette you may have noticed by now is the **Decorations** set (Figure 16.3).

Figure 16.3

These front panels are just that: They don't do anything, and they have no corresponding block diagram terminal. You can use these boxes, lines, frames, and other features to encapsulate and group your front panel controls and indicators into logical units. You might also use them to approximate the front panel of an instrument more closely.

Often you'll have to make use of the previously mentioned z-order functions, such as **Move to Back**, to place the decorations "behind" your controls. The reason for these commands is obvious once you create a decoration to enclose a set of controls: Decorations aren't normally transparent, so if you place a decoration on the front panel *after* you created controls, the decoration will obscure them since it is at the "front."

Clicking on a decoration is not the same as clicking on a blank part of the front panel. When using decorations behind controls, it's easy to unintentionally select the decoration with the cursors in the edit mode.

16.3 Vive l'Art: Importing Pictures

You can paste pictures right into LabVIEW and include them on your front panel. For example, you might want to make a block on your instrument that has your company logo. Or you might want to be a little more elaborate, adding piping and valve pictures to represent some process control loop (you can actually make your valve pictures Boolean controls as explained in Section 16.4).

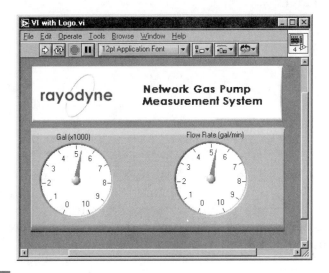

Figure 16.4

To import a picture into LabVIEW under Windows or MacOS, simply copy the picture from the application in which it's open, and paste it onto the front panel. LabVIEW then treats the picture in the same way as a decoration: You can resize it, move it behind or in front of other objects, etc. However, you can't edit it from LabVIEW.

On a Mac system, LabVIEW converts your graphic to the PICT format, which preserves a fairly high amount of resolution.

On Windows systems, LabVIEW normally converts the graphic in the clipboard to a bitmap image. The disadvantage of a bitmap image over another popular Windows graphic file format, a metafile, is that you lose some resolution if you resize the graphic. However, LabVIEW does support enhanced metafile graphics, a newer graphic format for Windows. An enhanced metafile can be stretched without "distortion" and can also have transparent sections, so you don't need to worry about matching the background color of your front panel. If you import an enhanced metafile graphic into LabVIEW, LabVIEW will accept the graphic without converting it to bitmap format.

Another place where you can put pictures is in a **Picture Ring** or a **Picture and Text Ring,** available from the **List & Ring** palette. Two examples are shown in Figure 16.5. Using the picture rings allows you to get very creative in presenting the user with a customized *graphical* set of options that you can index and track.

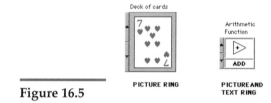

Figure 16.5

To add a picture to a picture ring, first copy your picture into the clipboard. Then choose **Import Picture** from the pop-up menu on the ring control. To add more pictures, choose either **Import Picture After** or **Import Picture Before** from the pop-up menu. Choosing **Import Picture** overwrites the current picture.

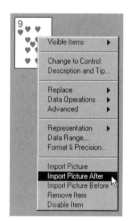

Figure 16.6

16.4 Custom Controls and Indicators

Q: What do all the objects in Figure 16.7 have in common?

A: They are all LabVIEW controls!

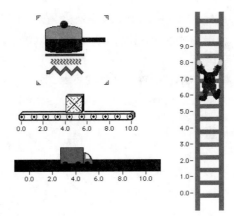

Figure 16.7

You can customize controls and indicators to make them better suited for your application while displaying a more impressive graphical interface. In the previous examples, you might want to display the position of a box on a conveyor belt to represent the production stage of some product. Or you might want to turn on and off a furnace by clicking on the "fire" in the stove picture control shown in Figure 16.7.

You can save a custom control or indicator ("control" from here on for simplicity) in a directory, or VI library, just as you do with VIs and globals. The standard convention is to name custom control files with a .ctl extension. You can use the saved control in other VIs as well as in the one in which it was created. It's possible to even create a master copy of the control if you need to use it in many places in the same VI by saving the control as a *type definition*. When you make changes to a type definition, LabVIEW automatically updates all the VIs that use it.

You may also want to place a frequently used custom control in the **Controls** palette. Do this by clicking on the **Options** button on the palette.

OK, so how do you create a custom control?

Activity 16-1: Custom Controls

Usually, you will want to import pictures for your custom control. So first have your picture files available, maybe along with a graphics program. LabVIEW doesn't have any sort of picture editor. When you create a custom control, you always base it on the form of an existing control, such as a Boolean LED or a Numeric Slide control. In this activity, we'll create a custom Boolean that looks like a valve that is open or closed.

1. Place a Boolean LED control on a Front Panel, and select it.

2. Launch the Control Editor by selecting **Customize Control...** from the **Edit** menu.

3. The Control Editor window will show the Boolean.

4. In Edit Mode, the Control window works just like the Front Panel. You can pop up on the control to manipulate its settings (things like scale, precision, etc.).

Figure 16.8

5. In the Customize mode, which you access by clicking on the tool button, you can resize, color, and replace the various picture components of the control.

Figure 16.9

6. Copy the picture file of an open valve or use the one provided on the CD (named `closed.bmp`). In the Customize mode, pop up on the Boolean and select **Import Picture**.

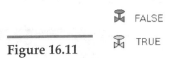

Figure 16.10

7. Repeat step 6 for the TRUE case of the Boolean, using a different valve picture. The TRUE, or open valve, picture is also on the CD (`open.bmp`).

FALSE

Figure 16.11   TRUE

8. Save the custom control by selecting **Save** from the **File** menu. By convention, controls are named with the `.ctl` extension

9. The front panel shown in Figure 16.12 is found in the examples of the full version of LabVIEW (`examples\apps\demos.llb\Control Mixer Process.vi`). This front panel uses the same Boolean controls you just created, as well as some others.

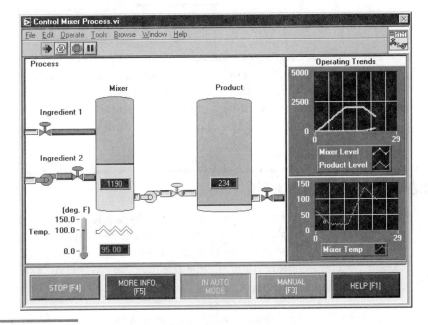

Figure 16.12

An important concept to understand is that you cannot change the behavior of a control; you can only modify its appearance. That also means that generally you can't change the way a control displays its data (a custom control based on a slide will always have something that slides in one dimension, for example). Despite these limitations, you can produce some fancy graphical interfaces, especially if you're artistically inclined and willing to experiment.

If you want to create a master copy of your control so that all the VIs that contain it are automatically updated when you make a change to the master copy, select **Type Def.** from the ring at the Control Editor's toolbar.

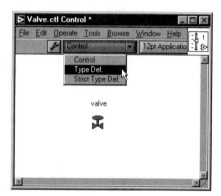

Figure 16.13

A type definition forces the control data type to be the same everywhere it is used, but different copies of the type definition can have their own name, color, size, etc. This is useful because only the master copy can be modified in behavior, thus preventing accidental changes. A *strict type definition* forces almost everything about the control to be identical everywhere it is used.

16.5 Adding Online Help

LabVIEW's Help window is a godsend when you need to learn something in a hurry; most people come to depend on it for wiring their diagram, and this is a good habit to develop. You can make your own application as easy to learn by adding your own entries for the Help window as well as links to a hypertext Help document. Three levels of customized help are available:

1. Tooltip Help (also called "hesitation help"): the text in a yellow box that shows up when you pause the cursor over a control

2. Window Help: the comments that appear in the Help window that describe controls and indicators as you move the cursor over them

3. Online help to a Help (hypertext) document: text where you can programmatically create a link to bring up an external help file

Providing customized help in the Tooltip format and Help window is fairly easy. We covered this briefly in Chapter 5. To do this, choose **Description and Tip...** from the pop-up menu of a control or indicator. Enter the tip and description in the dialog box, and click **OK** to save it. LabVIEW displays this description whenever you have the Help window open and move the cursor over the front panel object.

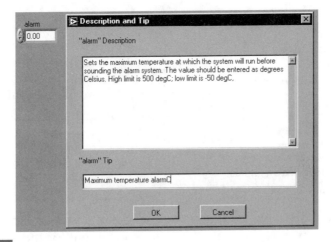

Figure 16.14

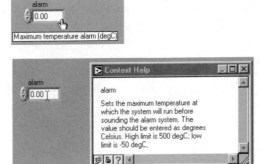

Figure 16.15

If you document all your front panel controls and indicators by including their descriptions, the end user can open the Help window and quickly peruse the meaning of your front panel objects.

To provide a description for VIs, enter information in the text box that appears when you choose **Documentation** from the **VI Properties...** menu. These comments will show up in the Help window, along with the wiring diagram, whenever the cursor is positioned over the VI's icon in the block diagram of another VI.

You can also bring up, position, and close the Help window programmatically. Use the **Control Help Window** and the **Get Help Window Status** functions, available from the **Help** subpalette of the **Application Control** palette.

Figure 16.16
Control Help Window.

The Boolean **Show** input closes or opens the Help window; the cluster input consists of two numerical indicators that determine the top and left pixel position of the window.

Figure 16.17
Get Help Window Status.

Returns the state and position of the Help window.

The more advanced help method, calling a link to an external help file, is more elaborate. To create the source document for the external help file, you will need a help compiler for your specific platform. Windows help compilers are available from Microsoft as well as from a number of third-party companies, which provide compilers such as *RoboHelp* (Blue Sky Software) and *Doc-to-Help* (WexTech Systems). For Macintosh, you can use *QuickHelp* from Altura Software. In UNIX, you can use *HyperHelp* from Bristol Technologies. All the help compilers include tools to create help documents. To call these help files programmatically, you can use the remaining function in the **Help** subpalette.

Figure 16.18
Control Online Help.

Manipulates an external help file. You can display the contents or the index or jump to a specific part of the help file.

16.6 Pointers and Recommendations for a "Wow!" Graphical Interface

It's the attention to detail that often makes a graphical interface so startling that people go, "Wow!" To achieve this, we've collected a "bag o'tricks" over the years from experience and from observing other people's fancy VIs. Many VIs start out as a scratchpad with no concern for order or aesthetics and never get cleaned up later. Believe me, it is worth the effort; even if you're not out to impress anybody, you and others will have an easier time working with your own VI. What if you have to modify that graphical spaghetti you threw together last year?

The following list is just a set of recommendations for a cool GUI; they don't necessarily always apply to your particular situation—perhaps you'll find better ones—but it should help you get started.

Panel Layout

- If possible, make your top-level front panel and window size such that it fills up the whole screen, as long as you know the monitor size and its resolution.
- Try to use neatly aligned square decorations as "modules" on the front panel, neatly aligned to group objects.

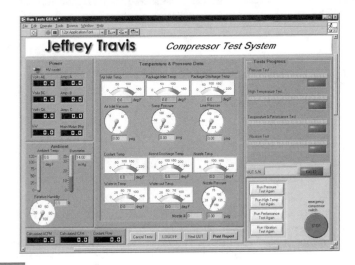

Figure 16.19

- If you have empty space, consider filling it up with a decoration "module" as something you could use later; or fill it up with your company's logo or some cool picture.

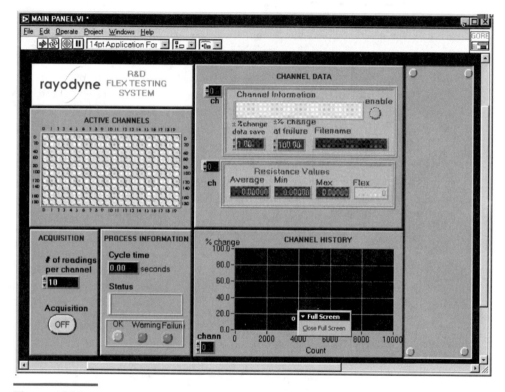

Figure 16.20

- Label the decoration "modules."

Text, Fonts, and Color

- Use color, but don't overdo it, and be *consistent* with a coloring scheme. If you use red for a "hi" alarm in one panel, use the same red for the "hi" alarms in other panels (and please, no puke green or other weird colors...).

- Choose a background color different from your decoration box colors. Often a darker gray or black background seems to fit nicely to add a nice border effect. Alternatively, if you don't use the decoration modules, you may find that white or a soft, light color works best for the background.

- Select the "transparent" option for object label boxes from the **Options...** command under the Tools menu. Labels generally look much better without their gray box.

Figure 16.21

- For many labels in general, and especially for numeric indicators and controls, coloring the background of the text black and making the text a vivid green or yellow makes them stand out. Choosing bold for the text usually helps as well.
- Use bright, highlighted colors only for items that are very important, such as alarms or error notifications.
- LabVIEW has three standard fonts (*Application, System,* and *Dialog*). In general, stick to these if you are writing a cross-platform application. A custom font style or size you selected on a Windows machine may look very different (usually too big or too small) on a Mac.

Graphs and Imported Pictures

- Make your panel look truly customized by importing your company's logo picture somewhere onto the front panel.
- Be aware that pictures and graph indicators can significantly slow down the update of your VI, especially if you have placed any controls or indicators on top of a picture or graph indicator.
- In general, you will want to select the **Use Smooth Updates** option from the **Options...** command to avoid the nasty "flickering" that would otherwise appear on graphs and other image-intensive operations.

Miscellaneous

- Include units on numerical controls if appropriate. Remember LabVIEW has built-in units, if you choose to use these. You can show the unit label for numeric control or indicator by popping up on it and selecting **Visible Items>> Unit Label.** See Chapter 13 for a discussion of units.

- Include custom controls when you can for neat visual effects.
- If you want to get really creative, for multimedia-type effects, use sounds and property nodes such as the position property.
- Use property nodes to hide or "gray out" controls when you want to indicate that these controls shouldn't be used.
- The **VI Properties** options, discussed in Chapter 13, are very useful for making your front panel fill the screen, center itself, etc.

- Never forget to spend enough time on the icon editor—make cool icons!
- When you see a good front panel design, learn and imitate!

Some Final Words on the GUI

Most of you at some time or another have probably had to give a presentation to your manager or some other audience about how your software works. The success of this "dog and pony show," in my experience, usually depends far more on the razzle and dazzle of the GUI than on whether the software is efficient or even works properly. Perhaps it shouldn't be this way, but that's often the way projects get accepted or rejected. The story goes that at one large semiconductor manufacturer, some engineers had recently decided to revise a piece of software written in C that had taken two years to develop. Another engineer put together a LabVIEW front panel that demonstrated what the new software should look like. One of the top executives in the company was so impressed by the GUI that he immediately approved the engineer's proposal for all their worldwide plants—despite the fact that the front panel was just a demo and didn't do anything yet!

The moral of the story is: Spend some time making the best, coolest graphical interface you can. You never know how far it might carry you.

16.7 How Do You Do That in LabVIEW?

This section is more about programming solutions than about LabVIEW features. Common problems arise in developing many applications and we show you some of them along with their solution in the following pages. Some of the solutions are very simple, and some are quite ingenious. By no means do all or even most programming challenges appear here; rather, these examples have been collected from a variety of sources to give you a starting point on some applications or at least provide some food for thought.

You might take some of these problems as a challenge: Try to come up with a programming solution on your own first before looking at the answer!

How Do You Acquire Data from Different Channels at Different Sampling Rates?

Things work fine when you're sampling all your data channels at the same rate. It's more tricky when one channel has to be sampled at 10 kHz and another channel only requires sampling at 50 Hz. There's not really any magical solution to the multiple-sampling rate requirements, but you have a couple of options:

1. If you have more than one DAQ board, try assigning the channels so that each board can handle the channels with the same sampling rate.

2. Sample everything at the highest rate, and then throw away the extra points on the lower-rate channels. One efficient way to do this is to use the **Decimate Array** function (**Array** palette).

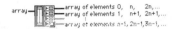

Figure 16.22
Decimate 1D Array. Divides input elements into output arrays, much like the way a dealer distributes cards. Input must be one dimensional. The function is resizable.

How Can you Create a Set of Boolean Controls so That Only One Can Be True at Any Time?

This is the common "Radio Buttons" or "Menu Bar" problem. You'll find this solution extremely useful in many applications. The basic problem setup goes somewhat like this: You want to give the user a list of options (Boolean controls) that will each go to a different subVI or subroutine. For example, you might want a top-level screen for a test program that will allow the user to run a test, edit the test parameters, view a test result, or exit. The user should only be able to select one of these options at a time. What is an efficient way to do this? You certainly don't want four parallel While Loops or some complicated Boolean logic diagram. Fortunately, one solution to this problem comes with the LabVIEW examples. It's found in the `examples\general\controls\booleans.llb` library and is called **Simulating Radio Buttons.vi**.

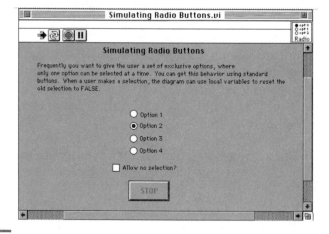

Figure 16.23

The heart of the **Simulating Radio Buttons** VI is the subVI **Manage Radio Buttons**, which is located in the same library in the full version of LabVIEW. The block diagram of this VI is shown in Figure 16.25 to give you an idea of how it works; however, we recommend just using this VI directly instead of reinventing the wheel.

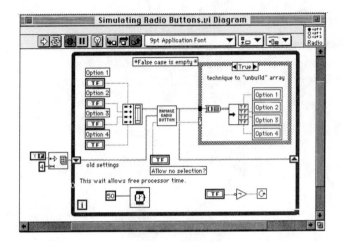

Figure 16.24

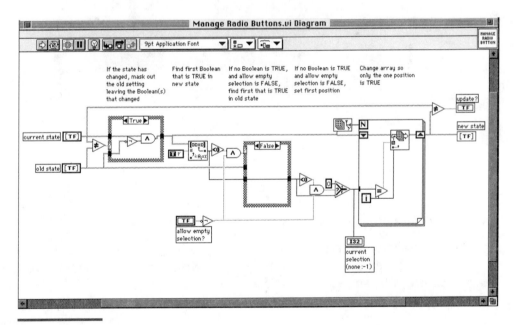

Figure 16.25

How Do I Clear a Chart Programmatically?

Wire an empty array to the History Data attribute of a chart. The data type of this empty array is the same as the data type wired to the chart. There is an excellent example that illustrates this technique in `Examples\General\Graphs\Charts.llb\How to Clear Charts and Graphs.vi`.

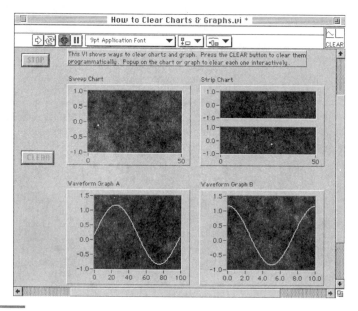

Figure 16.26

You can also pop up on the chart in the front panel and select **Clear Chart**.

How Can I Display a Waterfall Plot?

The basic idea to creating a waterfall plot, or a succession of plots stacked over each other in time, is to build a multiplot graph, where you create the z axis effect by successively adding a constant offset to each plot. Look for examples at http://zone.ni.com.

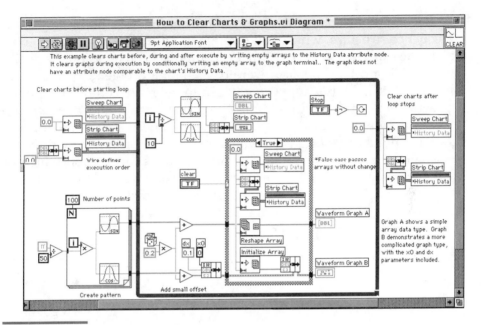

Figure 16.27

Is There a Way to Put More Than One *Y* Axis on a Graph for Two or More Plots?

Yes. On the graph, pop up and select **Visible Items>>Scale Legend.** Then resize the Scale Legend to add more *Y* axes.

My Graphs Flicker Every Time They Are Updated. It's Getting Really Annoying. Is There Any Way to Fix This?

Fortunately, yes. From the **Tools** menu, choose **Options....** Select the **Front Panel** item. Check the box that says "Use smooth updates during drawing." This will turn off the flicker; the trade-off is that more memory will be used.

Is There Any Way to Jump to Any Given Frame in a Sequence Structure (Like a State Machine)?

This is a very interesting problem whose solution, which can be a challenge to the novice user, nevertheless presents us with a powerful programming structure—the state machine.

The problem statement goes somewhat like this: Suppose you have a se-quence structure, which you arrange to normally perform some operations in the expected sequence. However, suppose you occasionally need to skip a frame if a certain Boolean were true. "Not a problem," you say; just add a case statement. But suppose it began to get more complicated; perhaps you actually need to go from frame 4 to frame 2 again under certain conditions. This is why it's called a state machine: You need a structure that can be in a finite number of *states*, determined by your algorithm. Any time you have a chain of events where one operation is dependent upon which operation previously occurred, a state machine is a good way to do the job.

Although LabVIEW doesn't have a built-in state machine structure, you can create one easily enough using a Case structure inside a While Loop and adding a few shift registers.

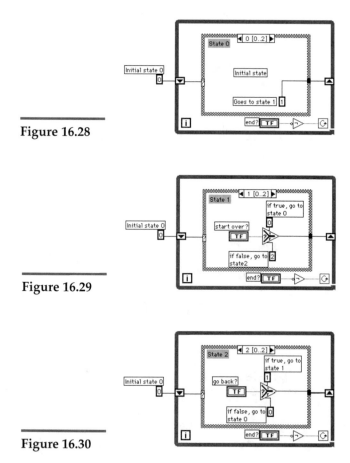

Figure 16.28

Figure 16.29

Figure 16.30

Each frame of the state machine's Case structure can transfer control to any other frame that is allowed according to the algorithm on the next iteration, or it can cause the While Loop to terminate. By placing each "state" in a separate case, you can jump around to different states in any order you wish.

How Can You Create a Toolbar-Type Feature That Will Pop Up Different Sets of Controls and Indicators on the Same Window?

This is a pretty neat trick. The idea is to create clusters of controls or indicators, one for each toolbar button. The clusters should be the same size, and then you literally stack them on top of each other. By using property nodes, you make all clusters invisible except the one that is selected. Toolbar buttons should be set up to work as radio buttons.

This is an excellent alternative when your front panel requires many controls or indicators but it would be to confusing to display them all at once. The toolbar buttons serve as "menus" that appear to bring up a logically grouped set of objects.

Can I Access the Parallel Port in LabVIEW?

Yes. Use the serial port VIs (never mind that parallel is the opposite of serial). Just as you use port 0 for COM1, port 1 for COM2, etc., in LabVIEW for Windows, port 10 is LPT1, port 11 is LPT2, and so on. You can even send data to a printer connected to a parallel port, by using **Serial Port Write**, although you may need to know the printer's special control characters. Another good use for accessing the parallel port is to do some digital I/O, without a plug-in board: You get eight digital lines for free! (but a hardware buffer is recommended to protect your computer).

Can I Do Object-Oriented Programming in LabVIEW?

Yes, to an extent. If you are familiar with object-oriented concepts, check out the GOOP (G Object Oriented Programming) VIs at http://ni.com/goop.

Can I Turn My VIs into Standalone Executables So That People Can Run Them without LabVIEW?

Yes. You need to have either the Professional version of LabVIEW, or you should purchase the Application Builder Toolkit.

How Do I Monitor or Control My VIs over the Web?

If you just need to monitor your VIs, use the built-in Web server. To control your VIs, use the free LabVNC utility on the CD. See Chapter 14 for more information.

Can LabVIEW Cook Me Dinner?

Probably not yet, although LabVIEW can do just about anything else you might think of. Don't assume "LabVIEW can't ..." until you check out the documentation; you'll be surprised at how flexible and powerful LabVIEW is.

16.8 Memory, Performance, and All That

Becoming a better LabVIEW programmer means you know how to make applications that are mean and lean. Sure, if you have oodles of RAM and a dual terahertz processor, you may not need to worry about these sort of things very much. But on average computers, critical or real-time applications are going to work better if you follow some simple guidelines for increasing performance and reducing memory consumption. Even if you don't foresee your application needing to conserve memory and processor speed, programmers who never take these sort of issues into account are—well—just plain sloppy. Nobody likes debugging sloppy programs.

16.8.1 Curing Amnesia and Slothfulness

Let's face it—LabVIEW does tend to make applications gobble memory, but you can make the best of it by knowing some tips. Memory management is generally an advanced topic, but it quickly becomes a concern when you have large arrays and/or critical timing problems. Read anything you can find about improving LabVIEW performance and saving memory. Here are some tips:

- Are you using the proper data types? Extended precision (EXT) floats are fine where the highest accuracy is needed, but they waste memory if a smaller type will do the job. This is especially important where large arrays are involved.

- Globals use a significant amount of memory. Minimize not only the creation of globals but the amount of times you read or write to them.

- Don't use complicated, hierarchical data types (such as an array of clusters of arrays) if you need more memory efficiency and speed.

- Avoid unnecessary coercion (the gray dots on terminals). Coercion indicates that the expected data type is different than the data type wired to the terminal. Although LabVIEW does an astounding job of accepting the data anyway in most cases (polymorphism), the result is a loss of speed and increased memory usage because copies of the data must be made. This is especially true of large arrays.

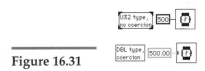

Figure 16.31

- How are you handling arrays and strings? Are you using loops where you don't have to? Sometimes there is a built-in function that can do the job, or several loops can be combined into one. Avoid putting unnecessary elements into loops, as shown in Figure 16.32.

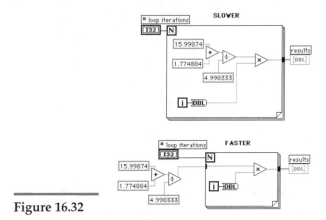

Figure 16.32

- Where possible, avoid using **Build Array** inside loops, thus avoiding repetitive calls to the Memory Manager. Every time you call the **Build Array** function, a new space in memory for the whole "new" array is allocated. Use auto-indexing or **Replace Array Element** with a pre-sized array instead. Similar problems occur with **Concatenate Strings**.

Obviously, this is not an option if you need to display the array data in real time as it is being built. In this case, initialize the array to the maximum size first, and use the **Replace Array Subset** function instead of **Build Array**.

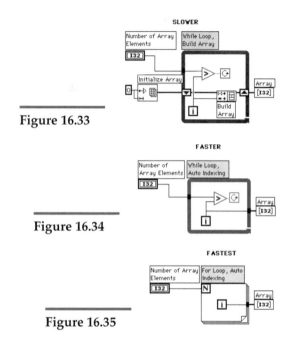

Figure 16.33

Figure 16.34

Figure 16.35

- Consider indicator overhead. Minimize the use of graphics (especially graphs and charts) when speed is extremely important.
- Update controls and indicators outside of loops whenever possible, that is, whenever it is not critical to view the object's value until the loop is finished executing.

16.8.2 The Declaration of Independence

LabVIEW does just an amazing job of porting VIs among different platforms. For the most part, you can take a VI created on a Sun and use it on a MacOS or Win95 system (as long as they use the same LabVIEW version or a later one). However, not all parts of a block diagram are portable. If you want to design your VIs to be platform-independent, there are a few things you should know about portability.

- Be aware that the LabVIEW *application* and everything included with it (such as all the VIs in the `vi.lib` directory) are not portable. What the LabVIEW for each OS does is to *convert* VIs from other platforms, using its internal code, to VIs for its own platform.

- Some of the VIs in the **Advanced** palette (such as the **System Exec** VI in LabVIEW for Windows), as well as VIs in the **Communication** palette (like AppleEvents or ActiveX), are system-specific and thus are not portable.

- VIs that contain CINs are not immediately portable. But if you write your source code to be platform-independent and recompile it on the new operating system, the CIN should work fine.

- Keep in mind such things as filenames that have their own special rules for each OS, and don't use characters such as brackets, colons, and slashes, which are path delimiters in different operating systems.

- The end-of-line (EOL) character is different on each platform (`\r` for MacOS, `\r\n` for Windows, and `\n` for Sun). The easiest solution is to use LabVIEW's **End of Line** constant (🔲), from the **String** palette.

- Fonts can be a real mess when porting among systems. If you can, stick to the three standard LabVIEW font schemes (Application, Dialog, System), because custom fonts that look fine on one platform may look huge, tiny, or distorted on another.

- Screen resolution can also be a nuisance. Some people recommend sticking to using a 800×600 screen resolution, which will make VI windows fit fine on most monitors.

If you really want to peer into the memory consumption of your VIs, you can use the **Profile VIs** tool (from the **Tools>>Advanced** menu).

16.9 Programming with Style

Programming really is an art, and it can be especially fun in LabVIEW! This section is a collection of final reminders and guidelines for writing a *good* LabVIEW application. You can find some of these concepts and many more in *The LabVIEW Style Guide: A Guide to Better LabVIEW Applications*, written by Gary W. Johnson and Meg F. Kay, also available at the National Instruments Web page (`http://www.ni.com`).

16.9.1 Modularize and Test Your VIs

Although it's theoretically possible to be *too* modular in designing your program, this rarely happens. Make all but the most simple and trivial functions and procedures subVIs. This gives you a chance to test each of your individual pieces of code before working with the big piece. It also lets you easily reuse code, keep yourself organized, and make the block diagram size manageable. Don't forget to test each subVI as a top-level VI—and be thorough: Test all sorts of strange input combinations. If you know all your subVIs are working, it should be very easy to debug any problems with your top-level VI.

 Often LabVIEW programmers won't test certain VIs because they require DAQ hardware or other externally generated inputs. Don't wait until you have the hardware. Write a simple "dummy data" VI to pass data to these VIs in the meantime so that you can at least test part of their functionality.

16.9.2 Document as You Go Along

Please document your work! Many programmers shun documentation, thinking it will be done later or is not needed... until a user (possibly yourself) two years later is trying to figure out how this VI works. Take advantage of LabVIEW's built-in documentation capabilities:

1. **VI Documentation**. At the very least, write a short description for each VI you create. This is extremely valuable when someone is looking at the subVI icons in the block diagram and needs an idea of what they do.

2. **Descriptions**. Ideally, write a help statement for each control and indicator using the pop-up **Description...** command. These invaluable comments will appear on the Help window if a user points to the control or indicator in question.

3. **VI History**. This option, available from the **Windows** menu, is a more sophisticated tool for larger projects. It allows you to enter comments about the changes you've made to a VI along the way. The

History Window can be quite helpful when more than one person works on a project, since it keeps track of the user, time, and date.

4. **Front Panel Text**. For important indications, just write some text (perhaps with a bold or large font) on the front panel itself. Users can't miss that one!

16.9.3 One More Time: Dataflow!

As you get more comfortable with LabVIEW, you will begin to take more advantage of the way dataflow programming works. Some tips to remember are:

• Dataflow means that data are carried across wires. When data reach a terminal of a subVI or a function, that subVI or function will only start executing after *all* of its input terminals have received data.

• Two or more objects or groups of objects on the block diagram that are not connected have no specific execution sequence. Many people new to LabVIEW have a feeling that execution should take place left-to-right or top-to-bottom. That is not true! There is no way to predict in what order two or more pieces of block diagram will occur unless it's specified by dataflow.

• When you need to "force" an execution sequence, you have the option of using a Sequence structure, which can be awkward for large diagrams, or an artificial data dependency structure, as shown in Figure 16.36.

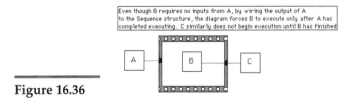

Figure 16.36

• You've noticed that many LabVIEW functions have common "threads": refnums, taskID, error clusters, etc. These threads are designed to string several common VIs together in a natural dataflow sequence. You can also use threads, such as error clusters, for your own VIs. Using threads will reduce the need for Sequence structures with awkward Sequence Locals.

16.10 Wrap It Up!

This chapter gave you some instructions on the *art* of graphical programming. We looked first at the front panel end: suggestions, guidelines, and reasons for making an exciting graphical user interface (GUI). Second, we focused on the block diagram: programming solutions, performance, memory, and style.

Creating a good GUI is important for "selling" your program to your customer or your boss, as well as making it much easier to use. LabVIEW's "art" enhancements include decoration modules, align and distribution commands, and layered objects.

Custom controls and indicators can add value to your GUI by providing graphical simulation and animation tools. The Control Editor lets you modify standard LabVIEW controls and indicators and import picture files to represent the new objects.

The Help window is not just for you, the programmer. The end user can use the Help window to examine front panel object descriptions. You can open or close the Help window programmatically.

Many common questions and problems arise when you are programming in LabVIEW. Some of the solutions were discussed in this chapter. You can find more ideas for solving your specific problem by consulting some of the resources listed in Appendix A.

When you need to improve the speed and/or memory performance of LabVIEW, follow the guidelines discussed in this chapter. Watch out always for unnecessary operations inside loops, especially with arrays.

Although for the most part LabVIEW VIs are platform independent, you do need to be aware of a few obstacles that can creep up—such as CINs or system-specific functions.

Finally, to be a good LabVIEW programmer, you need to be systematic in making modular VIs that are tested thoroughly. Good documentation is essential to making a quality, maintainable piece of software. Last but not least, learn to use LabVIEW's distinctive hallmark unknown to other languages—dataflow—to your advantage.

16.11 Concluding Remarks

This is the end of the book! (Well, there is an appendix, and if you're really bored, you can read the index.) By now, you've gained a solid understanding of how LabVIEW works. You've also begun to see how it can work for you, whether your application is teaching an electrical engineering class or building a process control system for a large plant.

Where do you go from here? More than anything else, hands-on experience is the best teacher. Experiment with a VI. Build a prototype. Look at examples. Be creative. Above all, have fun. If you decided to buy the full version of LabVIEW, don't be afraid to go through the manuals and built-in examples. They can be an invaluable reference for details on your application. Finally, please check out my website at http://jeffreytravis.com. I will be posting the errata and updates for this book there.

Good luck!

Appendix:
Resources for LabVIEW

You have many options for getting help and finding additional information about LabVIEW and virtual instrumentation. The following is a list of websites, organizations, publications, and other resources for LabVIEW users.

LabVIEW Documentation and Online Help

Sometimes it's easy to overlook the obvious places to get help. The Lab-VIEW online and printed manuals provide excellent tutorials to get you up to speed with LabVIEW. They also contain plenty of reference information. You can find the answers to almost all of your questions in these manuals. In addition, LabVIEW has extensive online help to assist you as you build your application.

Jeffrey Travis Studios

When all else fails, you can always hire "the guy who wrote the book" to help. At Jeffrey Travis Studios, I offer consulting services that include software system design, strategy assessment, on-site training and mentoring, customized

LabVIEW coding standards and reviews, and contract LabVIEW software development. If you need to ensure the success of your project with the best technical expertise available, e-mail me at <u>consulting@jeffreytravis.com</u>.
You can also reach me at

Jeffrey Travis Studios
713 E. 43rd St
Austin, Texas 78751
USA
Voice/Fax (208) 293-9072
Web: <u>http://jeffreytravis.com</u>

You can also find updates on this book and my other books and courses at <u>http://jeffreytravis.com/books</u>.

National Instruments

National Instruments is the company that creates and sells LabVIEW. They offer extensive technical support by phone, e-mail, and their website. You can contact them at

National Instruments
6504 Bridge Point Parkway
Austin, Texas 78730-5039
USA
Tel. (512) 794-0100
Fax (512) 794-8411
Website: <u>http://ni.com</u>
Technical resource site: <u>http://zone.ni.com</u>

Info-LabVIEW Mailing List

Info-LabVIEW is a user-sponsored Internet mailing list that you can use to communicate with other LabVIEW users. It is probably one of the best unbiased resources you can find on LabVIEW. You can post messages containing questions, answers, and discussions about LabVIEW on this mailing list. These messages will be sent to LabVIEW users worldwide.

To subscribe to Info-LabVIEW, send an e-mail message to

info-labview-request@pica.army.mil.

requesting that your e-mail address be added to the subscription list. All messages posted to Info-LabVIEW will then be forwarded to your e-mail address. You can cancel your subscription by sending a message to the above address requesting that your e-mail address be removed from the list.

You can get more information about Info-LabVIEW at

http://labview.pica.army.mil.

You can also search the Info-LabVIEW archives at

http://infolv.brianrenken.com.

LabVIEW Technical Resource (LTR)

LabVIEW Technical Resource, or *LTR* as it is often called, is a paid-subscription quarterly newsletter that provides technical information for LabVIEW systems developers. It offers solutions to common problems, programming tips, tools, and techniques, and each issue comes with a CD containing LabVIEW programs. For more information, see the LTR website,

http://www.ltrpub.com.

Find All VIs (FAVI) Project

The FAVI Project aims to be an exhaustive list of all "off-the-shelf" available VIs for LabVIEW, from free example code to complete commercial toolkits. The catalog even includes, for example, the VIs on this book's CD! You can access the FAVIs page at

http://www.mooregoodideas.com/FAVIs/.

Add-on Toolkits for LabVIEW

You can purchase special add-on toolkits to increase LabVIEW's functionality. In addition, new toolkits are created frequently, so if you have a particular goal, it's worthwhile to check and see if a toolkit already exists to accomplish it. Some toolkits are sold by National Instruments; others are created by third-party companies (often referred to as "Alliance members").

The following toolkits are available from National Instruments (http://ni.com):

- LabVIEW Report Generation Toolkit for Microsoft Office
- LabVIEW Database Connectivity Toolset
- LabVIEW Enterprise Connectivity Toolset
- LabVIEW PID Control Toolset
- LabVIEW Order Analysis Toolset
- LabVIEW System Simulation and Design Toolset
- LabVIEW Signal Processing Toolset
- LabVIEW Sound and Vibration Toolset
- LabVIEW Vision Development Module
- LabVIEW Motion Control Software
- LabVIEW Application Builder
- LabVIEW IVI Driver Toolset
- Instrument Driver Network
- LabVIEW Datalogging and Supervisory Control Module

LabVIEW Open Source Tools (LOST) Project

This is an initiative to encourage development of open source tools. You can contribute to the discussions, download free software, or even participate in writing open source software yourself. The LOST Web page and discussion group are at

http://groups.yahoo.com/group/openG.

Other Books

There are a number of good books that explore specific LabVIEW topics or application areas in much further detail than *LabVIEW for Everyone*; below are some that are recommended.

Internet Applications in LabVIEW by Jeffrey Travis, 2000, Prentice-Hall. ISBN: 0130141445.

LabVIEW Graphical Programming by Gary Johnson and Richard Jennings, 2001, McGraw-Hill. ISBN: 0071370013.

LabVIEW for Data Acquisition by Bruce Mihura, 2001, Prentice-Hall. ISBN: 0130153621.

Virtual Bio-Instrumentation: Biomedical, Clinical, and HealthCare Applications in LabVIEW by Jon Olansen and Eric Rosow, 2001, Prentice-Hall. ISBN: 0130652164.

LabVIEW Advanced Programming Techniques by Rick Bitter, Taqi Mohiuddin, and Matthew Nawrocki, 2000, CRC Press. ISBN: 0849320496.

Glossary

Symbols

∞	Infinity.
π	Pi.
Δ	Delta; difference. Δx denotes the value by which x changes from one index to the next.

A

Absolute Path

File or directory path that describes the location relative to the top level of the file system.

Active Window

Window that is currently set to accept user input, usually the frontmost window. The title bar of an active window is highlighted. You make a window active by clicking on it, or by selecting it from the Windows menu.

A/D	Analog-to-digital conversion. Refers to the operation electronic circuitry does to take a real-world analog signal and convert it to a digital form (as a series of bits) that the computer can understand.
ADC	See *A/D*.
ANSI	American National Standards Institute.
Array	Ordered, indexed set of data elements of the same type.
Array Shell	Front panel object that houses an array. It consists of an index display, a data object window, and an optional label. It can accept various data types.
Artificial Data Dependency	Condition in a dataflow programming language in which the arrival of data, rather than its value, triggers execution of a node.
ASCII	American Standard Code for Information Interchange. Refers to a seven-bit encoding scheme for alphanumeric characters.
Asynchronous Execution	Mode in which multiple processes share processor time. For example, one process executes while others wait for interrupts during device I/O or while waiting for a clock tick.
Auto-Indexing	Capability of loop structures to disassemble and assemble arrays at their borders. As an array enters a loop with auto-indexing enabled, the loop automatically disassembles it with scalars extracted from one-dimensional arrays, one-dimensional arrays extracted from two-dimensional arrays, and so on. Loops assemble data into arrays as they exit the loop according to the reverse of the same procedure.
Autoscaling	Ability of scales to adjust to the range of plotted values. On graph scales, this feature determines maximum and minimum scale values as well.

Autosizing   Automatic resizing of labels to accommodate text that you enter.

B

Block Diagram   Pictorial description or representation of a program or algorithm. In LabVIEW, the block diagram, which consists of executable icons called nodes and wires that carry data between the nodes, is the source code for the VI. The block diagram resides in the block diagram window of the VI.

Boolean Controls   Front panel objects used to manipulate and display or input and output Boolean (TRUE or FALSE) data. Several styles are available, such as switches, buttons, and LEDs.

Breakpoint   A pause in execution. You set a breakpoint by clicking on a VI, node, or wire with the Breakpoint tool from the Tools palette.

Breakpoint Tool   Tool used to set a breakpoint on a VI, node, or wire.

Broken VI   VI that cannot be compiled or run; signified by a broken arrow in the run button.

Bundle Node   Function that creates clusters from various types of elements.

Byte Stream File   File that stores data as a sequence of ASCII characters or bytes.

C

Case   One subdiagram of a Case Structure.

Case Structure   Conditional branching control structure, which executes one and only one of its subdiagrams based on its input. It is the combination of the IF, THEN, ELSE, and CASE statements in control flow languages.

Channel	Pin or wire lead to which an analog signal is read from or applied.
Chart	See scope chart, strip chart, and sweep chart.
CIN	See *Code Interface Node*.
Cloning	To make a copy of a control or some other LabVIEW object by clicking the mouse button while pressing the <ctrl> (Windows); <option> (Macintosh); <meta> (Sun); or <alt> (Linux) key and dragging the copy to its new location.
	(Sun and Linux) You can also clone an object by clicking on the object with the middle mouse button and then dragging the copy to its new location.
Cluster	A set of ordered, unindexed data elements of any data type including numeric, Boolean, string, array, or cluster. The elements must be all controls or all indicators.
Cluster Shell	Front panel object that contains the elements of a cluster.
Code Interface Node (CIN)	Special block diagram node through which you can link conventional, text-based code to a VI.
Coercion	The automatic conversion LabVIEW performs to change the numeric representation of a data element.
Coercion Dot	Glyph on a node or terminal indicating that the numeric representation of the data element changes at that point.
Color Tool	Tool you use to set foreground and background colors.
Color Copy Tool	Copies colors for pasting with the Color tool.
Compile	Process that converts high-level code to machine-executable code. LabVIEW automatically compiles VIs before they run for the first time after creation or alteration.

Conditional Terminal	The terminal of a While Loop containing a Boolean value that determines whether the VI performs another iteration.
Connector	Part of the VI or function node that contains its input and output terminals, through which data passes to and from the node.
Connector Pane	Region in the upper right corner of a front panel window that displays the VI terminal pattern. It underlies the icon pane.
Constant	See *universal constant* and *user-defined constant*.
Continuous Run	Execution mode in which a VI is run repeatedly until the operator stops it. You enable it by clicking on the continuous run button.
Control	Front panel object for entering data to a VI interactively or to a subVI programmatically.
Control Flow	Programming system in which the sequential order of instructions determines execution order. Most conventional text-based programming languages, such as C, Pascal, and BASIC, are control flow languages.
Controls Palette	Palette containing front panel controls and indicators.
Conversion	Changing the type of a data element.
Count Terminal	The terminal of a For Loop whose value determines the number of times the For Loop executes its subdiagram.
CPU	Central processing unit.
Current VI	VI for which the front panel, block diagram, or Icon Editor is the active window.
Custom PICT Controls	Controls and indicators for which parts can be replaced by graphics and indicators you supply.

D

D/A	Digital-to-analog conversion. The opposite operation of an A/D.
Data Acquisition (DAQ)	Process of acquiring data, usually by performing an analog-to-digital (A/D) conversion. Its meaning is sometimes expanded to include data generation (D/A).
Data Dependency	Condition in a dataflow programming language in which a node cannot execute until it receives data from another node. See also *artificial data dependency*.
Data Logging	Generally, to acquire data and simultaneously store it in a disk file. LabVIEW file I/O functions can log data.
Data Storage Formats	The arrangement and representation of data stored in memory.
Data Type Descriptor	Code that identifies data types; used in data storage and representation.
Dataflow	Programming system consisting of executable nodes in which nodes execute only when they have received all required input data and produce output automatically when they have executed. LabVIEW is a dataflow system.
Datalog File	File that stores data as a sequence of records of a single, arbitrary data type that you specify when you create the file. Although all the records in a datalog file must be of a single type, that type can be complex; for instance, you can specify that each record is a cluster containing a string, a number, and an array.
DataSocket	A communications protocol supported by LabVIEW for sharing live data across a network.
DC	Direct current. The opposite of AC (alternating current). Refers to a very low frequency signal, such as one that varies less than once a second.

Device	A plug-in DAQ board.
Device Number	Number assigned to a device (DAQ board) in the NI-DAQ configuration utility.
Description Box	Online documentation for a LabVIEW object.
Destination Terminal	See *sink terminal*.
Dialog Box	An interactive screen with prompts in which you specify additional information needed to complete a command.
Differential Measurement	Way to configure a device to read signals in which the inputs need not be connected to a reference ground. The measurement is made between two input channels.
Dimension	Size and structure attribute of an array.
DMA	Direct memory access. A method by which you can transfer data to computer memory from a device or memory on the bus (or from computer memory to a device) while the processor does something else. DMA is the fastest method of transferring data to or from computer memory.
Drag	The act of moving the mouse cursor on the screen to select, move, copy, or delete objects.

E

Empty Array	Array that has zero elements but has a defined data type. For example, an array that has a numeric control in its data display window but has no defined values for any element is an empty numeric array.
EOF	End of file. Character offset of the end of file relative to the beginning of the file (that is, the EOF is the size of the file).
Execution Highlighting	Feature that animates VI execution to illustrate the dataflow in the VI.

F

FFT	Fast Fourier transform.
File Refnum	An identifier that LabVIEW associates with a file when you open it. You use the file refnum to specify that you want a function or VI to perform an operation on the open file.
Flattened Data	Data of any type that have been converted to a string, usually for writing it to a file.
For Loop	Iterative loop structure that executes its subdiagram a set number of times. Equivalent to conventional code:

```
For I = 0 to n - 1, do . . .
```

Formula Node	Node that executes formulas that you enter as text. Especially useful for lengthy formulas that would be cumbersome to build in block diagram form.
Frame	Subdiagram of a Sequence Structure.
Free Label	Label on the front panel or block diagram that does not belong to any other object.
Front Panel	The interactive user interface of a VI. Modeled from the front panel of physical instruments, it is composed of switches, slides, meters, graphs, charts, gauges, LEDs, and other controls and indicators.
Function	Built-in execution element, comparable to an operator, function, or statement in a conventional language.
Functions palette	Palette containing block diagram structures, constants, communication features, and VIs.

G

G	The LabVIEW graphical programming language.
Global Variable	Nonreentrant subVI with local memory that uses an uninitialized shift register to store data from one execution to the next. The memory of copies of these sub-

	VIs is shared and thus can be used to pass global data among them.
Glyph	A small picture or icon.
GPIB	General purpose interface bus. Also known as HP-IB (Hewlett-Packard Interface Bus) and IEEE 488.2 bus (Institute of Electrical and Electronic Engineers standard 488.2), it has become the world standard for almost any instrument to communicate with a computer. Originally developed by Hewlett-Packard in the 1960s to allow their instruments to be programmed in BASIC with a PC. Now IEEE has helped define this bus with strict hardware protocols that ensure uniformity across instrument.
Graph Control	Front panel object that displays data in a Cartesian plane.
Ground	The common reference point in a system; i.e., ground is at 0 volts.

H

Help Window	Special window that displays the names and locations of the terminals for a function or subVI, the description of controls and indicators, the values of universal constants, and the descriptions and data types of control attributes. The window also accesses LabVIEW's Online Reference.
Hertz, Hz	Cycles per second.
Hex	Hexadecimal. A base-16 number system.
Hierarchical Palette	Menu that contains palettes and subpalettes.
Hierarchy Window	Window that graphically displays the hierarchy of VIs and subVIs.
Housing	Nonmoving part of front panel controls and indicators that contains sliders and scales.

I

Icon	Graphical representation of a node on a block diagram.
Icon Editor	Interface similar to that of a paint program for creating VI icons.
Icon Pane	Region in the upper-right corner of the front panel and block diagram that displays the VI icon.
IEEE	Institute for Electrical and Electronic Engineers.
Indicator	Front panel object that displays output.
Inf	Digital display value for a floating-point representation of infinity.
Instrument Driver	VI that controls a programmable instrument.
I/O	Input/output. The transfer of data to or from a computer system involving communications channels, operator input devices, and/or data acquisition and control interfaces.
Iteration Terminal	The terminal of a For Loop or While Loop that contains the current number of completed iterations.

L

Label	Text object used to name or describe other objects or regions on the front panel or block diagram.
Labeling Tool	Tool used to create labels and enter text into text windows.
LabVIEW	Laboratory Virtual Instrument Engineering Workbench.
LED	Light-emitting diode.
Legend	Object owned by a chart or graph that displays the names and plot styles of plots on that chart or graph.

Line	The equivalent of an analog channel—a path where a single digital signal is set or retrieved.

M

Marquee	A moving, dashed border that surrounds selected objects.
Matrix	Two-dimensional array.
Menu Bar	Horizontal bar that contains names of main menus.
Modular Programming	Programming that uses interchangeable computer routines.

N

NaN	Digital display value for a floating-point representation of not a number, typically the result of an undefined operation, such as log(–1).
NI-DAQ	Driver software for National Instruments DAQ boards and SCXI modules. This software acts as an interface between LabVIEW and the devices.
NI-MAX	National Instruments Measurement & Automation Explorer. A configuration utility that interacts with NI-DAQ, allowing you to configure your hardware, set up virtual channels, and test your I/O from the desktop.
Nodes	Execution elements of a block diagram consisting of functions, structures, and subVIs.
Nondisplayable Characters	ASCII characters that cannot be displayed, such as new line, tab, and so on.
Not-a-Path	A predefined value for the path control that means the path is invalid.
Not-a-Refnum	A predefined value that means the refnum is invalid.

Numeric Controls and Indicators	Front panel objects used to manipulate and display or input and output numeric data.
NRSE	Nonreferenced single-ended.
NRSE Measurement	All measurements are made with respect to a common reference. This reference voltage can vary with respect to ground.
Nyquist Frequency	One-half the sampling frequency. If the signal contains any frequencies above the Nyquist frequency, the resulting sampled signal will be aliased or distorted.

0

Object	Generic term for any item on the front panel or block diagram, including controls, nodes, wires, and imported pictures.
Object Pop-up Menu Tool	Tool used to access an object's pop-up menu.
Octal	A base-eight numbering system.
Operating Tool	Tool used to enter data into controls as well as operate them. Resembles a pointing finger.

P

Palette	Menu of pictures that represent possible options.
Platform	Computer and operating system.
Plot	A graphical representation of an array of data shown either on a graph or a chart.
Polymorphism	Ability of a node to automatically adjust to data of different representation, type, or structure.
Pop Up	To call up a special menu by clicking (usually on an object) with the right mouse button (Windows, Sun, Linux) or while holding down the command key (MacOS).

Pop-up Menus	Menus accessed by popping up, usually on an object. Menu options pertain to that object specifically.
Port	A collection of digital lines that are configured in the same direction and can be used at the same time.
Positioning Tool	Tool used to move, select, and resize objects.
Probe	Debugging feature for checking intermediate values in a VI.
Probe tool	Tool used to create probes on wires.
Programmatic Printing	Automatic printing of a VI front panel after execution.
Pseudocode	Simplified language-independent representation of programming code.
Pull-down Menus	Menus accessed from a menu bar. Pull-down menu options are usually general in nature.

R

Reentrant Execution	Mode in which calls to multiple instances of a subVI can execute in parallel with distinct and separate data storage.
Representation	Subtype of the numeric data type, of which there are signed and unsigned byte, word, and long integers, as well as single-, double-, and extended-precision floating-point numbers, both real and complex.
Resizing Handles	Angled handles on the corner of objects that indicate resizing points.
Ring Control	Special numeric control that associates 32-bit integers, starting at 0 and increasing sequentially, with a series of text labels or graphics.
RS-232	Recommended Standard #232. A standard proposed by the Instrument Society of America for serial communications. It's used interchangeably with the term "serial communication," although serial communications more generally refers to communicating one bit at a

time. A few other standards you might see are RS-485, RS-422, and RS-423.

RSE   Referenced single-ended.

RSE Measurement   All measurements are made with respect to a common ground; also known as a grounded measurement.

S

Sample   A single analog input or output data point.

Scalar   Number capable of being represented by a point on a scale. A single value as opposed to an array. Scalar Booleans and clusters are explicitly singular instances of their respective data types.

Scale   Part of mechanical action, chart, and graph controls and indicators that contains a series of marks or points at known intervals to denote units of measure.

Scope Mode   Mode of a waveform chart modeled on the operation of an oscilloscope.

Scroll Tool   Tool used to scroll windows.

SCXI   Signal conditioning extensions for instrumentation. A high-performance signal conditioning system devised by National Instruments, using an external chassis that contains I/O modules for signal conditioning, multiplexing, etc. The chassis is wired into a DAQ board in the PC.

Sequence Local   Terminal that passes data between the frames of a Sequence Structure.

Sequence Structure   Program control structure that executes its subdiagrams in numeric order. Commonly used to force nodes that are not data-dependent to execute in a desired order.

Shift Register   Optional mechanism in loop structures used to pass the value of a variable from one iteration of a loop to a subsequent iteration.

Sink Terminal	Terminal that absorbs data. Also called a destination terminal.
Slider	Moveable part of slide controls and indicators.
Source Terminal	Terminal that emits data.
State Machine	A method of execution in which individual tasks are separate cases in a Case Structure that is embedded in a While Loop. Sequences are specified as arrays of case strings.
String Controls and Indicators	Front panel objects used to manipulate and display or input and output text.
Strip Mode	Mode of a waveform chart modeled after a paper strip chart recorder, which scrolls as it plots data.
Structure	Program control element, such as a Sequence, Case, For Loop, or While Loop.
Subdiagram	Block diagram within the border of a structure.
SubVI	VI used in the block diagram of another VI; comparable to a subroutine.
Sweep Mode	Similar to scope mode—except a line sweeps across the display to separate old data from new data.

T

Terminal	Object or region on a node through which data pass.
Tool	Special LabVIEW cursor you can use to perform specific operations.
Toolbar	Bar containing command buttons that you can use to run and debug VIs.
Tools Palette	Palette containing tools you can use to edit and debug front panel and block diagram objects.
Top-level VI	VI at the top of the VI hierarchy. This term distinguishes the VI from its subVIs.

Trigger	A condition for starting or stopping a DAQ operation.
Tunnel	Data entry or exit terminal on a structure.
Typecast	To change the type descriptor of a data element without altering the memory image of the data.
Type Descriptor	See *data type descriptor*.

U

| Universal Constant | Uneditable block diagram object that emits a particular ASCII character or standard numeric constant, for example, pi. |
| User-defined Constant | Block diagram object that emits a value you set. |

V

VI	See *virtual instrument*.
VI Library	Special file that contains a collection of related VIs for a specific use.
VI Server	A feature in LabVIEW that allows you to programmatically and remotely control the behavior of VIs and controls.
Virtual Instrument	LabVIEW program; so called because it models the appearance and function of a physical instrument.

W

Waveform	A data type in LabVIEW that usually represents an analog signal; it bundles the Y-axis data with the timing information (Xo and delta-X).
While Loop	Loop structure that repeats a section of code until a condition is met. Comparable to a Do Loop or a Repeat-Until Loop in conventional programming languages.
Wire	Data path between nodes.

| Wiring Tool | Tool used to define data paths between source and sink terminals. |

Index

About the Author

Jeffrey Travis has extensive experience with software development, Web applications, Internet technologies, virtual instrumentation, and LabVIEW. He provides consulting to companies who need to make decisions in the areas of Web applications, e-commerce, and remote instrumentation systems. He has been a guest speaker at symposiums and conferences on instrumentation and Internet technologies and has published award-winning articles in technical journals. Jeffrey Travis is also the author of *Internet Applications in LabVIEW* (2000, Prentice-Hall) and the "LabVIEW Internet Applications" course.

He first became acquainted with LabVIEW while a graduate student at the University of Texas at Austin, where he received his B.S. in Electrical Engineering (1993) and a Master's degree in Biomedical Engineering in 1995. Since then, he has been involved with developing LabVIEW-based systems in different capacities. He continues to provide consulting, training, and development for clients through his company, Jeffrey Travis Studios.

In addition to software programming and writing books, Jeffrey has spent time directing and producing independent films. He co-created "*busy signals," his first short, which was completed and released in the summer of 2001. Jeffrey enjoys mountain biking, racquetball, skiing, reading literature classics, and playing classical guitar (he has a classical guitar degree from the Schnabel Conservatory of Music in City Bell, Argentina). Having grown up in Mexico and Argentina, Jeffrey is a native Spanish speaker and is a fan of Latin American magical realism literature. His genealogical claim to fame is his relative, William B. Travis, who defended the Alamo.

Jeffrey lives with his wife and three children in Austin, Texas. You can reach him at author@jeffreytravis.com, or visit the Web page
http://JeffreyTravis.com.

Timely, Technical Information for LabVIEW™ Users!

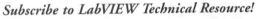

LABVIEW™ TECHNICAL RESOURCE

THE ONLY LABVIEW SUBSCRIPTION WITH VI SOFTWARE INCLUDED

ORDER FORM

WHAT IS LTR?

LabVIEW Technical Resource (LTR) is a quarterly journal for LabVIEW users and developers available by subscription from LTR Publishing, Inc. Each LTR issue presents powerful LabVIEW tips and techniques and includes a Resource CD packed with VI source code, utilities, and documentation. Technical articles on LabVIEW programming methodology, in-depth tutorials, and time-saving tips and techniques address everyday programming issues in LabVIEW.

In its eighth year of publication, LTR has subscribers in over 50 countries and is well-known as a leading independent source of LabVIEW-specific information.

Purchase the LabVIEW Technical Resource CD Library of Back Issues, Version 3.0 and browse this searchable CD-ROM for easy access to over 250 articles and VIs from LTR Volumes 1-8.

To subscribe to the LabVIEW Technical Resource or to order the CD-ROM Library of Back Issues, fax this form to LTR Publishing at **(214) 706-0506.**

Visit the LTR web page at **www.ltrpub.com** to download a free sample issue.

TEL: 214-706-0587   FAX: 214-706-0506

CONTACT INFORMATION

Name _____ Company _____

Address _____

City _____ State _____

Country _____ Zip/Post Code _____

Tel (required) _____ FAX _____ E-mail _____

ORDER INFORMATION

QTY	MAC/PC	PRODUCT	U.S.	INTL.	EXTENDED PRICE
		1 year subscription (4 issues / 4 Resource CDs)	$95	$120	
		2 year subscription (8 issues / 8 Resource CDs)	$175	$215	
		CD-ROM library of back issues (28 issues / over 250 VIs)	$350	$375	
		Back issues – [Article Index available at **www.ltrpub.com**]	$25	$30	
		Server Version CD-ROM library of back issues Version 3.0 (5 user license)*	$495	$530	
		10 user license Add-On pack (for Server Version)*	$295	$325	

*contact LTR for additional licensing information.

PAYMENT INFORMATION

✔	PAYMENT METHOD		
	Check enclosed (U.S. BANK ONLY* – Make check payable to LTR Publishing) (Texas residents please add 8.25% sales tax)		
	Bill company(U.S. Only) / (fax of P.O. required) ▶ PO#		
	Visa / MC / AMEX Card Number ▶		Exp.
	Signature ▶		
	* Wire information available for international orders		

Fill out the form above and Fax it to: 214-706-0506 with your credit card information and signature,
OR fill out the form above and send order form with U.S. check to:

LTR Publishing, Inc., 860 Avenue F, Suite 100 Plano, Texas 75074.
Tel: 214.706.0587 • Fax: 214.706.0506 • email: ltr@ltrpub.com

You may also include your own Federal Express or Airborne #. If you are ordering
a product for delivery within Texas, please include Texas Sales Tax at 8.25%

WWW.LTRPUB.COM

informIT

YOUR GUIDE TO IT REFERENCE

Articles

Keep your edge with thousands of free articles, in-depth features, interviews, and IT reference recommendations – all written by experts you know and trust.

Online Books

Answers in an instant from **InformIT Online Book's** 600+ fully searchable on line books. Sign up now and get your first 14 days **free**.

POWERED BY

Safari

Catalog

Review online sample chapters, author biographies and customer rankings and choose exactly the right book from a selection of over 5,000 titles.

LICENSE AGREEMENT AND LIMITED WARRANTY

READ THE FOLLOWING TERMS AND CONDITIONS CAREFULLY BEFORE OPENING THIS SOFTWARE PACKAGE. THIS LEGAL DOCUMENT IS AN AGREEMENT BETWEEN YOU AND PRENTICE-HALL, INC. (THE "COMPANY"). BY OPENING THIS SEALED SOFTWARE PACKAGE, YOU ARE AGREEING TO BE BOUND BY THESE TERMS AND CONDITIONS. IF YOU DO NOT AGREE WITH THESE TERMS AND CONDITIONS, DO NOT OPEN THE SOFTWARE PACKAGE. PROMPTLY RETURN THE UNOPENED SOFTWARE PACKAGE AND ALL ACCOMPANYING ITEMS TO THE PLACE YOU OBTAINED THEM FOR A FULL REFUND OF ANY SUMS YOU HAVE PAID.

1.   **GRANT OF LICENSE:** In consideration of your payment of the license fee, which is part of the price you paid for this product, and your agreement to abide by the terms and conditions of this Agreement, the Company grants to you a nonexclusive right to use and display the copy of the enclosed software program (hereinafter the "software") on a single computer (i.e., with a single CPU) at a single location so long as you comply with the terms of this Agreement. The Company reserves all rights not expressly granted to you under this Agreement.

2.   **OWNERSHIP OF SOFTWARE:** You own only the magnetic or physical media (the enclosed software) on which the software is recorded or fixed, but the Company retains all the rights, title, and ownership to the software recorded on the original software copy(ies) and all subsequent copies of the software, regardless of the form or media on which the original or other copies may exist. This license is not a sale of the original software or any copy to you.

3.   **COPY RESTRICTIONS:** This software and the accompanying printed materials and user manual (the "Documentation") are the subject of copyright. You may <u>not</u> copy the Documentation or the software, except that you may make a single copy of the software for backup or archival purposes only. You may be held legally responsible for any copying or copyright infringement which is caused or encouraged by your failure to abide by the terms of this restriction.

4.   **USE RESTRICTIONS:** You may <u>not</u> network the software or otherwise use it on more than one computer or computer terminal at the same time. You may physically transfer the software from one computer to another provided that the software is used on only one computer at a time. You may <u>not</u> distribute copies of the software or Documentation to others. You may <u>not</u> reverse engineer, disassemble, decompile, modify, adapt, translate, or create derivative works based on the software or the Documentation without the prior written consent of the Company.

5.   **TRANSFER RESTRICTIONS:** The enclosed software is licensed only to you and may <u>not</u> be transferred to any one else without the prior written consent of the Company. Any unauthorized transfer of the software shall result in the immediate termination of this Agreement.

6.   **TERMINATION:** This license is effective until terminated. This license will terminate automatically without notice from the Company and become null and void if you fail to comply with any provisions or limitations of this license. Upon termination, you shall destroy the Documentation and all copies of the software. All provisions of this Agreement as to warranties, limitation of liability, remedies or damages, and our ownership rights shall survive termination.

7.   **MISCELLANEOUS:** This Agreement shall be construed in accordance with the laws of the United States of America and the State of New York and shall benefit the Company, its affiliates, and assignees.

8.   **LIMITED WARRANTY AND DISCLAIMER OF WARRANTY:** The Company warrants that the software, when properly used in accordance with the Documentation, will operate in substantial conformity with the description of the software set forth in the Documentation. The Company does not warrant that the software will meet your requirements or that the operation of the software will be uninterrupted or error-free. The Company warrants that the media on which the software is delivered shall be free from defects in materials and workmanship under normal use

for a period of thirty (30) days from the date of your purchase. Your only remedy and the Company's only obligation under these limited warranties is, at the Company's option, return of the warranted item for a refund of any amounts paid by you or replacement of the item. Any replacement of software or media under the warranties shall not extend the original warranty period. The limited warranty set forth above shall not apply to any software which the Company determines in good faith has been subject to misuse, neglect, improper installation, repair, alteration, or damage by you. EXCEPT FOR THE EXPRESSED WARRANTIES SET FORTH ABOVE, THE COMPANY DISCLAIMS ALL WARRANTIES, EXPRESS OR IMPLIED, INCLUDING WITHOUT LIMITATION, THE IMPLIED WARRANTIES OF MERCHANTABILITY AND FITNESS FOR A PARTICULAR PURPOSE. EXCEPT FOR THE EXPRESS WARRANTY SET FORTH ABOVE, THE COMPANY DOES NOT WARRANT, GUARANTEE, OR MAKE ANY REPRESENTATION REGARDING THE USE OR THE RESULTS OF THE USE OF THE SOFTWARE IN TERMS OF ITS CORRECTNESS, ACCURACY, RELIABILITY, CURRENTNESS, OR OTHERWISE.

IN NO EVENT, SHALL THE COMPANY OR ITS EMPLOYEES, AGENTS, SUPPLIERS, OR CONTRACTORS BE LIABLE FOR ANY INCIDENTAL, INDIRECT, SPECIAL, OR CONSEQUENTIAL DAMAGES ARISING OUT OF OR IN CONNECTION WITH THE LICENSE GRANTED UNDER THIS AGREEMENT, OR FOR LOSS OF USE, LOSS OF DATA, LOSS OF INCOME OR PROFIT, OR OTHER LOSSES, SUSTAINED AS A RESULT OF INJURY TO ANY PERSON, OR LOSS OF OR DAMAGE TO PROPERTY, OR CLAIMS OF THIRD PARTIES, EVEN IF THE COMPANY OR AN AUTHORIZED REPRESENTATIVE OF THE COMPANY HAS BEEN ADVISED OF THE POSSIBILITY OF SUCH DAMAGES. IN NO EVENT SHALL LIABILITY OF THE COMPANY FOR DAMAGES WITH RESPECT TO THE SOFTWARE EXCEED THE AMOUNTS ACTUALLY PAID BY YOU, IF ANY, FOR THE SOFTWARE.

SOME JURISDICTIONS DO NOT ALLOW THE LIMITATION OF IMPLIED WARRANTIES OR LIABILITY FOR INCIDENTAL, INDIRECT, SPECIAL, OR CONSEQUENTIAL DAMAGES, SO THE ABOVE LIMITATIONS MAY NOT ALWAYS APPLY. THE WARRANTIES IN THIS AGREEMENT GIVE YOU SPECIFIC LEGAL RIGHTS AND YOU MAY ALSO HAVE OTHER RIGHTS WHICH VARY IN ACCORDANCE WITH LOCAL LAW.

ACKNOWLEDGMENT

YOU ACKNOWLEDGE THAT YOU HAVE READ THIS AGREEMENT, UNDERSTAND IT, AND AGREE TO BE BOUND BY ITS TERMS AND CONDITIONS. YOU ALSO AGREE THAT THIS AGREEMENT IS THE COMPLETE AND EXCLUSIVE STATEMENT OF THE AGREEMENT BETWEEN YOU AND THE COMPANY AND SUPERSEDES ALL PROPOSALS OR PRIOR AGREEMENTS, ORAL, OR WRITTEN, AND ANY OTHER COMMUNICATIONS BETWEEN YOU AND THE COMPANY OR ANY REPRESENTATIVE OF THE COMPANY RELATING TO THE SUBJECT MATTER OF THIS AGREEMENT.

Should you have any questions concerning this Agreement or if you wish to contact the Company for any reason, please contact in writing at the address below.

Robin Short
Prentice Hall PTR
One Lake Street
Upper Saddle River, New Jersey 07458

About the CD-ROM

The CD-ROM included with *LabVIEW For Everyone, 2nd edition*, includes the "Everyone" directory containing the activities (examples and exercises) from the book. Each chapter's activities are in the corresponding LabVIEW library (.llb) file; for example, you will find the activities for Chapter 9 in CH9.LLB.

This disk also provides an evaluation version of LabVIEW 6.0 that you can use for most of the activities if you do not own a copy of LabVIEW. Please see its accompanying documentation for information on its limitations.

In addition to the LabVIEW Evaluation version, you will find some other software goodies that are part of the LabVIEW Open Source Tools (LOST) initiative.

The CD-ROM can be used on Microsoft Windows® 95/98/NT®/2000 and MacOS.

Note: There are no CH1.LLB, CH2.LLB, or CH10.LLB files, since there are no corresponding software files for these chapters.

License Agreement

Use of the software accompanying *LabVIEW For Everyone, 2nd edition*, is subject to the terms of the License Agreement and Limited Warranty, found on the previous two pages.

Technical Support

Prentice Hall does not offer technical support for any of the programs on the CD-ROM. However, if the CD-ROM is damaged, you may obtain a replacement copy by sending an e-mail describing the problem to: disc_exchange@prenhall.com.